Oxford Graduate Texts in Mathematics

OXFORD GRADUATE TEXTS IN MATHEMATICS

1. Keith Hannabuss: *An introduction to quantum theory*

An Introduction to Quantum Theory

Keith Hannabuss
Balliol College, Oxford

OXFORD
UNIVERSITY PRESS
Great Clarendon Street, Oxford OX2 6DP
Oxford University Press is a department of the University of Oxford.
It furthers the University's objective of excellence in research, scholarship, and education by publishing worldwide in
Oxford New York
Athens Auckland Bangkok Bogotá Buenos Aires Calcutta
Cape Town Chennai Dar es Salaam Delhi Florence Hong Kong Istanbul
Karachi Kuala Lumpur Madrid Melbourne Mexico City Mumbai
Nairobi Paris São Paulo Singapore Taipei Tokyo Toronto Warsaw
with associated companies in Berlin Ibadan

Published in the United States
by Oxford University Press Inc., New York

Reprinted 1999

First published 1997

A catalogue record for this book is available from the British Library

Library of Congress Cataloging in Publication Data
(Data available)
ISBN 0 19 853794 8

Typeset by the author using LaTeX

Printed in Great Britain by
Biddles Ltd, Guildford and King's Lynn

To my parents

Preface

One morning about the 10 July 1925 I suddenly saw light: Heisenberg's symbolic manipulation was nothing but the matrix calculus well-known to me since my student days.

MAX BORN, in *My life*, 1978

These notes are based on lectures given to second- and third-year mathematics undergraduates at Oxford and aim to take advantage of the algebraic background which British mathematics students have acquired by the time they have the opportunity to study quantum mechanics. As such this book differs from most of the introductory texts, which are aimed primarily at physicists and have to develop the necessary mathematics as they go along. They presume an acquaintance with the basic ideas of vector spaces and inner product spaces, which are to be found in most elementary textbooks on linear algebra. (Though the first appendix contains a brief summary of the main definitions and results which are needed.) On the other side, I thought it wiser not to demand too great a familiarity with special functions, since mathematicians are no longer trained to recognize the hypergeometric function lurking behind a long calculation and camouflaged by unexpected changes of variable. Although not strictly necessary, it would be useful for the reader to have encountered the basic ideas of Hamiltonian mechanics, as found, for example, in Chapter 4 of N.M.J. Woodhouse's *Introduction to analytical dynamics*, Oxford, 1987.

There are, of course, dangers with this approach. Quantum mechanics (in contrast to quantum field theory) can be subjected to a perfectly rigorous mathematical treatment, but this requires a more extensive knowledge of mathematics than the average undergraduate is likely to have acquired by this stage, and would probably hinder rather than help understanding. However, the basic structure of the theory can be set out in terms of elementary linear algebra even if some of the detail is missing. I have therefore tried to give the flavour of the correct argument, noting where there are genuine difficulties, but trying to avoid lies. It is hoped that, besides placing quantum theory firmly in the context of the mathematics course, it will also give students a new perspective on algebra by providing them with a different range of examples.

On the other hand, in a first course in quantum theory I feel that it is best to build up confidence with simple examples before launching into the mathematical structure. For that reason the selection of topics is fairly standard, covering most of the elementary properties, including relativistic quantum mechanics, but little scattering theory and avoiding quantum field theory altogether. I have, however, indulged my own interests by including a brief introduction to the algebraic quantum theory in Chapter 11. Chapters 9 and 19 on symmetry in quantum mechanics provide an opportunity to exploit the knowledge of group theory gleaned from algebra courses. For that reason the exposition concentrates on the groups, rather than the Lie algebras which appear in most physics texts. The material in these and other chapters and sections which are marked with an asterisk is not required for any of the other unmarked sections, and can be omitted on a first reading. I have included two approaches to perturbation theory, supplementing the usual Rayleigh–Schrödinger theory with an iterative approach which is more closely linked to other approximation techniques. I have also included a brief and, no doubt, tendentious account of some of the paradoxes of quantum mechanics, since there is no doubt that most students are keen to link the technical theory to popular accounts which they may have read.

The development of quantum theory was anything but straightforward. It took the best part of thirty years to obtain a reasonable theory of systems with a finite number of degrees of freedom, and the search for an equally consistent quantum theory of fields continues to this day. During the early years the founders of the theory were often discouraged; witness Einstein's 1912 remark that 'the more successes the quantum theory has the sillier it looks'. Even as the new theory became clearer in the mid 1920s the physicists who were creating it chased down many blind alleys and poured scorn on ideas which later proved to be correct. To try to give a bit more feel for the historical perspective I have included at the head of each section a quotation from one of those involved in this revolution in our understanding of the laws of physics. Most of these date from the time when the new laws were being uncovered and their discoverers were either still confused or still elated at their progress, and I hope that they will help to give an impression of what went on. (I reluctantly omitted their pithy criticisms of each other, replete with earthy expletives and cries of 'trash' and 'rubbish'.)

The end of a proof is signalled by the symbol □. Exercises marked with a ° are based on questions asked in the Final Honour School of Mathematics at Oxford.

The first part of this book appeared as Mathematical Institute Lecture Notes in 1988. I am grateful to several generations of students who read earlier versions of these notes, struggled with badly worded exercises and made helpful suggestions. I have included hints or partial solutions for those exercises which seemed to cause the most difficulties.

Oxford K.C.H.
May 1996

Contents

1 Introduction

Today I have made a discovery which is just as important as the discovery of Newton.

MAX PLANCK to his son Erwin, Autumn 1900

He [de Broglie] has lifted one corner of the great veil.

ALBERT EINSTEIN, letter to Paul Langevin, 1924

In the first quarter of this century physicists were slowly and somewhat reluctantly forced to realize that the view of the physical world, so painstakingly built up over the preceding four centuries, was in serious need of revision. In particular, the laws that governed the motion of particles on the atomic scale were not those of classical mechanics. Physics found itself catapulted into the strange new world of quantum theory in which the very notion of position or velocity of a particle became blurred and statistical laws dominated everything.

With the benefit of another sixty years we can see that the gains far outweigh the extra difficulties of visualization that quantum theory introduced, for it made possible a real unification of physical theories that had previously been quite distinct. The same laws that governed the motion of a projectile or a planet also determined the shape of molecules, the energy of chemical reactions and the optical and electronic properties of matter. The invention of the laser and the transistor were both inspired by quantum theory, and superconductors are also quantum mechanical devices.

To appreciate why quantum theory was needed, it is helpful to trace its development from the closing years of the last century. At that time there was a growing confidence that the basic goals of physics were all but achieved. Newton's laws of gravity and motion were now well established. The experimental discovery of radio waves had confirmed the predictions of Maxwell's theory, which united electricity, magnetism and light. The next task was to apply statistical mechanics to explain how radiation interacted with matter and then the fundamental laws would all be known. Experiments aimed at improving the newly invented electric light bulb showed clearly that at any given temperature a hot body radiated a characteristic mix of light of different wavelengths, giving it a particular hue. By the spring of 1900 the experimental results began to diverge from the theoretical prediction made by Willy Wien: classical physics was unable to explain the electric light.

In the closing weeks of the last century Max Planck realized that this predicament could be avoided if radiation were emitted only in packets or *quanta*. For, if the energy E of such a packet was proportional to its frequency ω,

$$E = \hbar\omega, \tag{1.1}$$

then there would be insufficient energy to make high frequency packets, and the colour of the radiation would be modified. (This is now known as *Planck's law.*) By taking the constant of proportionality $\hbar = 1.0546 \times 10^{-34}\,\mathrm{J\,s}$ it was possible to obtain excellent agreement with experiment. (The quantity $h = 2\pi\hbar$ is known as Planck's constant. We have quoted a modern measurement rather than the value known to Planck.) The microscopic value of $\hbar$ meant that the bunching into packets would not normally be perceptible; the quanta for sodium yellow street lights, for example, carry an energy of $3 \times 10^{-19}\,\mathrm{J}$, and 100 watts would be enough power to eject 3×10^{20} such quanta every second.

Planck seems to have thought that these packets of energy simply represented the mode by which atoms were able to release energy. The process seemed analogous to the way in which a plucked string of a particular length emits a note of a certain pitch, so that it might give a clue to the structure of atoms. It was Albert Einstein who realized in 1905 that the quanta were not a feature of atoms but of light itself, and he showed how this would explain another puzzling phenomenon, the photoelectric effect.

Solar panels have familiarized us all with the way in which light striking certain materials can eject electrons and cause a current to flow. The puzzle was that increasing the frequency of the light increased the energy of the electrons but not their number (below a certain frequency none were emitted at all), whilst increasing the intensity of the light affected the numbers but not the energy. Einstein pointed out that if one thought of the light as a stream of quanta or *photons* whose energy was proportional to their frequency then at low frequency no photon would have enough energy to eject an electron at all, but as the frequency and energy of each photon increased, so it could transfer more energy to an electron. Since each photon would eject a single electron one could see why the numbers of electrons would not increase. Changing the intensity of the light, on the other hand, increased the number of photons available without affecting the energy of each individual one, and so ejected more electrons of the same energy.

Einstein's suggestion was revolutionary, for it flew in the face of almost a century of experimentation that seemed to show conclusively that light consists of waves and not particles. Now it seemed that one had to consider it as both. For that reason the physical reality of photons was not widely accepted until new experimental data on the scattering of photons by electrons became available in the 1920s. Nonetheless, further confirmation of the link between matter and radiation came in 1913 when Niels Bohr combined the idea of quanta with Rutherford's nuclear model of the atom to provide an explanation of atomic spectra.

Retrospectively it may seem strange that Einstein did not take the next step of suggesting that if waves could mimic particles then perhaps particles could also behave like waves. This was proposed by Louis de Broglie in 1923–4, and he showed that Einstein's theory of relativity provided a unique consistent procedure for associating a wave to a particle. De Broglie's idea was soon developed into a far wider and more powerful theory by Schrödinger, Heisenberg, Born, Dirac and others.

The reality of de Broglie's particle waves was soon demonstrated by Davisson and Germer and by G.P. Thomson who showed that electrons passing through thin foil produce the same kind of interference and diffraction patterns as light. (In fact physicists had been observing such interference patterns for years without realizing their significance.) There are now many more sophisticated demonstrations of the same phenomenon using such devices as the neutron interferometer (a carefully crafted crystal that splits a neutron beam into two and then recombines them to produce interference effects).

Richard Feynman once remarked that all quantum effects are ultimately a consequence of the wave nature of matter. This is therefore an appropriate point at which to begin the mathematical discussion.

Exercises

1.1 Calculate the energy of a single photon of the following kinds of electromagnetic radiation:

(i) a radio wave of frequency 200 kHz ($= 2 \times 10^5$ Hz),

(ii) red light of frequency 4.95×10^{14} Hz,

(iii) X-rays of frequency 10^{20} Hz,

(iv) gamma radiation of frequency 10^{23} Hz.

[Note that in Planck's formula ω is an angular frequency so 1 Hz corresponds to $\omega = 2\pi$ radians per second.]

1.2 A radio station broadcasts on a frequency of 200 kHz. Estimate the number of photons striking a 1 square metre aerial each second at a distance of 1000 km from a 200 kW transmitter.

How would your answer change if the aerial were on a space probe at a distance of 3000 million km from the earth?

2 Wave mechanics

At the moment I am struggling with a new atomic theory. If only I knew more mathematics!

ERWIN SCHRÖDINGER, letter to Willy Wien, 27 December 1925

2.1. Schrödinger's equation

According to Planck and Einstein the energy and frequency of light are related by $E = \hbar\omega$. De Broglie observed that to be consistent with the theory of relativity the momentum $\mathbf{p}$ of a wave should be $\hbar\mathbf{k}$ where $\mathbf{k}$ is the wave vector of magnitude 2π/wavelength, perpendicular to the wave front. (This is because the special theory of relativity combines E and $\mathbf{p}$ and also ω and $\mathbf{k}$ into four-dimensional vectors, and linear relations between particular components force the same relations for all other components as well. This is explained in more detail in Section 17.1, which is independent of the rest of the book.)

In the case of light the fixed propagation speed c leads to the relations $\omega = c|\mathbf{k}|$ and $E = c|\mathbf{p}|$, so that $|\mathbf{p}| = \hbar|\mathbf{k}|$ follows directly from Planck's law. Thus for photons de Broglie's relationship adds nothing new, and it had indeed already been introduced by Johannes Stark in 1909. However, de Broglie's other crucial idea was that these relationships might apply to other elementary particles such as the electron as well.

De Broglie was able to use his new relationship together with geometrical arguments to solve some simple problems, but further advances had to wait until Erwin Schrödinger had discovered the wave equation satisfied by de Broglie's 'matter wave'. Consider first a plane wave $\psi(t,x) = \exp[-i(\omega t - \mathbf{k}\cdot\mathbf{x})]$. By direct differentiation we obtain

$$i\hbar\frac{\partial\psi}{\partial t} = \hbar\omega\psi = E\psi \tag{2.1}$$

and

$$\frac{\hbar}{i}\nabla\psi = \hbar\mathbf{k}\psi = \mathbf{p}\psi. \tag{2.2}$$

The normal relations between energy and momentum then provide a differential equation for ψ. For example, a particle of mass m in a potential $V(\mathbf{x})$ has energy $E = |\mathbf{p}|^2/2m + V(\mathbf{x})$, so that

$$i\hbar\frac{\partial\psi}{\partial t} = E\psi = \left(\frac{|\mathbf{p}|^2}{2m} + V(\mathbf{x})\right)\psi = -\frac{\hbar^2}{2m}\nabla^2\psi + V\psi. \tag{2.3}$$

Unfortunately this is not consistent, because the solutions of this equation are not usually plane waves. Schrödinger therefore suggested that one should regard the partial differential equation

$$i\hbar\frac{\partial\psi}{\partial t} = -\frac{\hbar^2}{2m}\nabla^2\psi + V\psi \tag{2.4}$$

as more basic and simply use its solutions, $\psi(t, \mathbf{x})$, to describe the waves.

Definition 2.1.1. The equation

$$i\hbar\frac{\partial\psi}{\partial t} = -\frac{\hbar^2}{2m}\nabla^2\psi + V\psi$$

is known as *Schrödinger's equation.*

The equation can easily be modified for particles moving in one or two dimensions by substituting the appropriate form of ∇^2.

One natural approach to finding solutions of this equation is to look first for separable solutions that are products of functions $T(t)$ and $\Psi(\mathbf{x})$. Substituting $\psi = T\Psi$ into Schrödinger's equation and dividing by $T\Psi$ to separate the variables then gives

$$i\hbar\frac{dT}{dt}/T = \left(-\frac{\hbar^2}{2m}\nabla^2\Psi + V\Psi\right)/\Psi. \tag{2.5}$$

Since the two sides of this equation depend on different sets of variables, each must be a constant, and it will turn out to be consistent with previous notation to call this constant E. This gives us two equations

$$i\hbar\frac{dT}{dt} = ET \tag{2.6}$$

and

$$-\frac{\hbar^2}{2m}\nabla^2\Psi + V\Psi = E\Psi. \tag{2.7}$$

The first equation for T can immediately be integrated to give

$$T(t) = e^{-iEt/\hbar}T(0) \tag{2.8}$$

so that

$$\psi(t, \mathbf{x}) = e^{-iEt/\hbar}\psi(0, \mathbf{x}). \tag{2.9}$$

Multiplying the first equation by Ψ and the second by T casts these into another useful form

$$i\hbar\frac{\partial\psi}{\partial t} = E\psi = -\frac{\hbar^2}{2m}\nabla^2\psi + V\psi. \tag{2.10}$$

We have thus recovered the relation between the energy and the wave function postulated by Schrödinger, and so retrospectively justified the use of the notation E for the constant of separation.

Definition 2.1.2. The equation

$$-\frac{\hbar^2}{2m}\nabla^2\psi + V\psi = E\psi$$

that determines the spatial form of the wave function is also known as Schrödinger's equation or, more precisely, as *Schrödinger's time-independent equation.*

Remark 2.1.1. Both of Schrödinger's equations are linear; that is, when ψ_1 and ψ_2 are solutions so is $c_1\psi_1 + c_2\psi_2$, for any complex numbers c_1 and c_2. In more algebraic language this says that the solutions form a complex vector space, a fact that will often be useful in the examples that follow.

2.2. The square well

As a first example let us consider a particle moving within the interval $[0, a]$ on the x-axis under the influence of the potential $V(x) = 0$. The zero potential allows it to move freely within the interval.

Schrödinger's time-independent equation for a particle moving in one dimension with energy E is

$$\frac{-\hbar^2}{2m}\frac{d^2\psi}{dx^2} + V\psi = E\psi, \tag{2.11}$$

and to avoid problems at the endpoints we make ψ vanish there. This leaves us with the differential equation

$$-\frac{\hbar^2}{2m}\frac{d^2\psi}{dx^2} = E\psi \tag{2.12}$$

for x in the interval $(0, a)$, and the boundary conditions

$$\psi(0) = 0 = \psi(a). \tag{2.13}$$

This equation is the same as that which arises when we look for the normal modes of oscillation of a string fixed at the points 0 and a, and we can solve it in the same way. The general solution of the differential equation takes the form

$$\psi(x) = \begin{cases} A\cosh(\sqrt{2m|E|}x/\hbar) + B\sinh(\sqrt{2m|E|}x/\hbar), & \text{if } E < 0; \\ A + Bx, & \text{if } E = 0; \\ A\cos(\sqrt{2mE}x/\hbar) + B\sin(\sqrt{2mE}x/\hbar), & \text{if } E > 0; \end{cases} \tag{2.14}$$

where A and B are constants.

In each of the three cases the condition that $\psi(0) = 0$ forces A to vanish. When $E \leq 0$ the other boundary condition, $\psi(a) = 0$, also forces B to vanish, and leaves only the trivial solution $\psi \equiv 0$. However, for positive E the boundary condition at a can also be satisfied if $\sqrt{2mE}/\hbar = n\pi/a$ for some integer n, giving

$$\psi(x) = \psi_n(x) = B\sin(n\pi x/a). \tag{2.15}$$

For $n = 0$ we again get the trivial solution $\psi \equiv 0$ and the solutions for n and $-n$ are essentially the same (since a sign can always be absorbed into the constant B), so we may as well take n to be a positive integer.

The corresponding energy is given by

$$E = E_n = \frac{n^2\pi^2\hbar^2}{2ma^2}. \tag{2.16}$$

This reveals one of the remarkable features of quantum mechanics: the energy of the system can take only certain discrete values.

Definition 2.2.1. When the possible energies of a quantum system are discrete and bounded below then the lowest possible energy is called the *ground state energy*. The higher energies are known in increasing order as *the first excited state energy, the second excited state energy*, and so on.

A wave function corresponding to the ground state is called a *ground state wave function*, and a wave function corresponding to the k-th excited state energy is called a *k-th excited state wave function.*

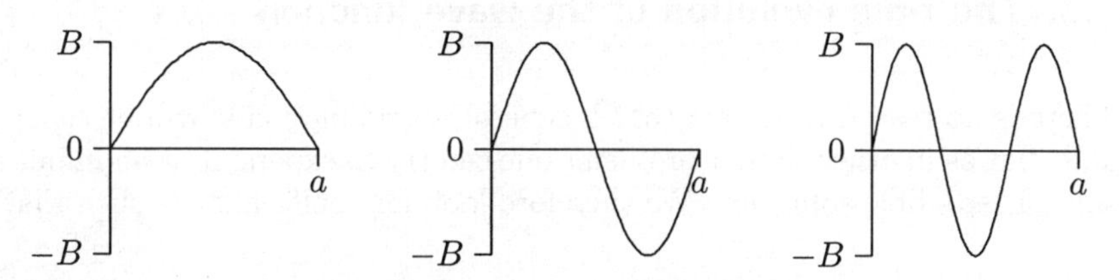

FIGURE 2.1. The three lowest energy wave functions for a particle in a finite box ($n = 1$, 2, and 3).

Example 2.2.1. The ground state energy in the infinite square well potential is E_1, and the k-th excited state energy is E_{k+1}. □

The constant B that appears in ψ is arbitrary, but it is customary, for reasons that will become apparent later, to choose it so that

$$\int |\psi(x)|^2 dx = 1. \tag{2.17}$$

The wave function is then said to be *normalized.* In this case the normalization condition amounts to

$$1 = |B|^2 \int_0^a \sin^2 \frac{n\pi x}{a} dx = \frac{1}{2} a|B|^2, \tag{2.18}$$

so that

$$\psi_n(x) = \sqrt{\frac{2}{a}} \sin \frac{n\pi x}{a} \tag{2.19}$$

is an appropriate wave function (see the graphs in Figure 2.1).

Combining this spatial form with the time development for a separable solution we arrive at

$$\begin{aligned} \psi_n(t, x) &= \sqrt{\frac{2}{a}} e^{-iE_n t/\hbar} \sin \frac{n\pi x}{a} \\ &= \sqrt{\frac{2}{a}} e^{-in^2\pi^2\hbar t/2ma^2} \sin \frac{n\pi x}{a}. \end{aligned} \tag{2.20}$$

Although this example may seem rather unrealistic, it is now possible to make quantum dots which are two-dimensional analogues of square wells, and to check the predictions directly. D. Eigler and collaborators working in an IBM laboratory have constructed a circular palisade of iron atoms, 14 nanometres in diameter, on a copper surface and checked the form of standing waves within the circle using a scanning tunnelling electron microscope. A photograph of the waves appears on the cover of the November 1993 issue of the magazine *Physics Today.*

2.3. The time evolution of the wave function

There is no reason to expect that a general wave function ψ will be separable, but as in other similar problems one can try to expand it as an infinite sum of separable solutions. We therefore look for coefficients c_n such that

$$\psi(t,x) = \sum_{n=1}^{\infty} c_n \psi_n(t,x) = \sqrt{\frac{2}{a}} \sum_{n=1}^{\infty} c_n e^{-in^2\pi^2\hbar t/2ma^2} \sin\frac{n\pi x}{a}. \tag{2.21}$$

If we know the initial wave function, $\psi(0,x) = f(x)$, then we require that

$$f(x) = \sqrt{\frac{2}{a}} \sum_{n=1}^{\infty} c_n \sin\frac{n\pi x}{a}. \tag{2.22}$$

We know from the theory of Fourier series that such an expansion is possible for suitable functions f (in fact it is sufficient for f to be normalized), and that the coefficients are given by

$$\sqrt{\frac{2}{a}} c_n = \frac{2}{a} \int_0^a f(x) \sin\frac{n\pi x}{a} dx, \tag{2.23}$$

that is

$$c_n = \sqrt{\frac{2}{a}} \int_0^a f(x) \sin\frac{n\pi x}{a} dx = \int_0^a f(x)\psi_n(x) dx. \tag{2.24}$$

(We shall see later that this last simplification is no accident.) It is easy to check that when the Fourier series of f'' obtained by differentiating twice is uniformly convergent, so too is the series for $\partial^2\psi(t,x)/\partial x^2$, and the resulting function $\psi(t,x)$ does satisfy the general form of Schrödinger's equation.

2.4. The interpretation of the wave function

The usefulness of the wave equation in determining the quantized energy levels of the system makes it natural to ask what the wave function ψ means physically. Max Born, modifying suggestions of de Broglie and Schrödinger, came up with what is now the accepted interpretation:

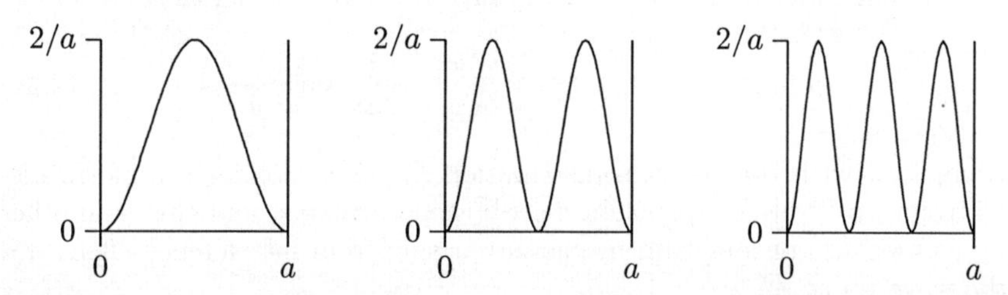

FIGURE 2.2. The probability densities for the three lowest energy wave functions in a square well. For large values of n the graph resembles a typical interference pattern.

Assumption 2.4.1. Quantum mechanical particles are governed by statistical laws and for normalized wave functions, ψ,

$$\rho(x) = |\psi(x)|^2$$

gives the probability density for the position of the particle.

Thus the probability that the particle described by $\psi(x)$ is in the subset S of $\mathbf{R}$ is

$$\int_S |\psi(x)|^2 \, dx, \tag{2.25}$$

with similar expressions for two- and three-dimensional systems. Since $|\psi(x)|^2$ is non-negative and we have agreed to normalize it so that

$$\int_{\mathbf{R}} |\psi(x)|^2 \, dx = 1, \tag{2.26}$$

this assumption is consistent with the elementary requirements of probability theory.

This interpretation makes good sense for the square well. In particular, since $\psi(x)$ vanishes outside the well, the particle is almost sure to be found inside the well. Within the well the probability density is

$$f_n(x) = \frac{2}{a}\sin^2\frac{n\pi x}{a} = \frac{1}{a}\left(1 - \cos\frac{2n\pi x}{a}\right). \tag{2.27}$$

Figure 2.2 shows the graphs of f_n for $n = 1, 2$ and 3.

The distribution function can be derived by integrating the density

$$F_n(x) = \int_0^x f_n(x)\,dx = \left(\frac{x}{a} - \frac{1}{2n\pi}\sin\frac{2n\pi x}{a}\right). \tag{2.28}$$

Compared with the classical distribution $F_n(x) = x/a$ for a uniform distribution in $[0, a]$ the quantum distribution oscillates, and this is another typical feature of quantum mechanics arising from interference effects for the wave function.

Example 2.4.1. To see how all this works in practice let us calculate the probability that the quantum mechanical particle confined in the box $[0, a]$ is actually within $\frac{1}{4}a$ of the centre of the box; in other words the probability that it is within $[\frac{1}{4}a, \frac{3}{4}a]$.

Solution. When the wave function is ψ_n the probability is

$$\begin{aligned}\int_{\frac{1}{4}a}^{\frac{3}{4}a} |\psi_n(x)|^2\,dx &= F_n(3a/4) - F_n(a/4)\\ &= \left[\frac{3}{4} - \frac{1}{2n\pi}\left(\sin\frac{3n\pi}{2}\right)\right] - \left[\frac{1}{4} - \frac{1}{2n\pi}\left(\sin\frac{n\pi}{2}\right)\right].\end{aligned}$$

Since $\sin 3n\pi/2 = -\sin n\pi/2$, this gives for the probability

$$\frac{1}{2} + \frac{1}{n\pi}\sin\frac{n\pi}{2} = \begin{cases}\frac{1}{2} & \text{if } n \text{ is even}\\ \frac{1}{2} + (-1)^{\frac{1}{2}(n-1)}/n\pi & \text{if } n \text{ is odd.}\end{cases}$$

It is also easy to work out the mean and variance for the particle's position. The mean is

$$\begin{aligned}\int_0^a x f_n(x)\,dx &= \Big[xF_n(x)\Big]_0^a - \int_0^a F_n(x)\,dx\\ &= a - \int_0^a \left(\frac{x}{a} - \frac{1}{2n\pi}\sin\frac{2n\pi x}{a}\right) dx\\ &= a - \left[\frac{x^2}{2a} + \frac{a}{4n^2\pi^2}\cos\frac{2n\pi x}{a}\right]_0^a\\ &= \frac{1}{2}a.\end{aligned}$$

This result is also easily visible from the symmetry of the distribution about its midpoint.

The second moment is

$$\begin{aligned}
\int_0^a x^2 f_n(x)\,dx &= \Big[x^2 F_n(x)\Big]_0^a - 2\int_0^a x F_n(x)\,dx \\
&= a^2 - 2\int_0^a \left(\frac{x^2}{a} - \frac{x}{n\pi}\sin\frac{2n\pi x}{a}\right) dx \\
&= a^2 - \frac{2}{3}a^2 \\
&\quad + \frac{1}{n\pi}\left\{\left[-\frac{xa}{2n\pi}\cos\frac{2n\pi x}{a}\right]_0^a + \int_0^a \frac{a}{2n\pi}\cos\frac{2n\pi x}{a}\,dx\right\} \\
&= \frac{1}{3}a^2 - \frac{a^2}{2n^2\pi^2}.
\end{aligned}$$

From this we may deduce that the variance of the distribution is

$$\frac{1}{3}a^2 - \frac{a^2}{2n^2\pi^2} - \left(\frac{1}{2}a\right)^2 = \left(\frac{1}{12} - \frac{1}{2n^2\pi^2}\right)a^2. \qquad \square$$

Remark 2.4.1. It will be noted that for large values of n the quantum mechanical variance approximates ever more closely to the variance $\frac{1}{12}a^2$ of the classical uniform distribution. The tendency of quantum mechanical formulae to approach those of classical theory when the quantum number is large has been called by Bohr the *correspondence principle*.

2.5. Currents and probability conservation

So far our discussion has glossed over one important point that could jeopardize the whole statistical interpretation of the wave function: is probability conserved? That is, if we arrange for

$$\int_{\mathbf{R}^3} |\psi(0,\mathbf{x})|^2\, d^3\mathbf{x} = 1 \tag{2.29}$$

and then let ψ evolve according to Schrödinger's equation, will

$$\int_{\mathbf{R}^3} |\psi(t,\mathbf{x})|^2\, d^3\mathbf{x} = 1 \tag{2.30}$$

for all values of t, or are particles merely evanescent with a fluctuating probability of existence?

For stationary states it is clear that the probability must be conserved, as the above example illustrates, since

$$|\psi_n(t,\mathbf{x})|^2 = |e^{-iE_nt/\hbar}\psi_n(0,\mathbf{x})|^2 = |\psi_n(0,\mathbf{x})|^2. \tag{2.31}$$

In general, for any wave function ψ satisfying Schrödinger's equation in a real potential V,

$$\begin{aligned}
\frac{\partial}{\partial t}|\psi(t,\mathbf{x})|^2 &= \overline{\frac{\partial}{\partial t}(\psi(t,\mathbf{x}))}\psi(t,\mathbf{x}) + \overline{\psi(t,\mathbf{x})}\frac{\partial}{\partial t}\psi(t,\mathbf{x}) \\
&= \overline{\left[-\frac{i}{\hbar}\left(-\frac{\hbar^2}{2m}\nabla^2\psi + V\psi\right)\right]}\psi + \overline{\psi}\left[-\frac{i}{\hbar}\left(-\frac{\hbar^2}{2m}\nabla^2\psi + V\psi\right)\right] \\
&= \frac{i}{\hbar}\left(-\frac{\hbar^2}{2m}\nabla^2\overline{\psi} + V\overline{\psi}\right)\psi - \overline{\psi}\left[\frac{i}{\hbar}\left(-\frac{\hbar^2}{2m}\nabla^2\psi + V\psi\right)\right] \\
&= \frac{i\hbar}{2m}\left(\overline{\psi}\nabla^2\psi - \psi\nabla^2\overline{\psi}\right) \\
&= \frac{i\hbar}{2m}\operatorname{div}\left(\overline{\psi}\operatorname{grad}\psi - \psi\operatorname{grad}\overline{\psi}\right).
\end{aligned} \tag{2.32}$$

Definition 2.5.1. The *probability current*, $\mathbf{j}$, is defined by

$$\mathbf{j}(t,\mathbf{x}) = \frac{\hbar}{2mi}\left(\overline{\psi}\operatorname{grad}\psi - \psi\operatorname{grad}\overline{\psi}\right).$$

Our previous calculation can now be summarized as a theorem.

Proposition 2.5.1. The probability density and probability current satisfy the continuity equation

$$\frac{\partial\rho}{\partial t} + \operatorname{div}\mathbf{j} = 0.$$

Like its analogues in other areas of applied mathematics this equation leads directly to a conservation law for probability.

Proposition 2.5.2. Suppose that for all t the probability current $j(t, \mathbf{x})$ tends to 0 faster than $|\mathbf{x}|^{-2}$ as $|\mathbf{x}| \to \infty$. Then

$$\int_{\mathbf{R}^3} \rho(t, \mathbf{x})\, d^3\mathbf{x}$$

is independent of t.

Proof. If D is the volume enclosed by a surface S then for suitably well-behaved functions ψ

$$\frac{\partial}{\partial t} \int_D \rho\, d^3\mathbf{x} = \int_D \frac{\partial \rho}{\partial t}\, d^3\mathbf{x} = \int_D (-\mathrm{div}\mathbf{j})\, d^3\mathbf{x} = -\int_S \mathbf{j}.d\mathbf{S}. \tag{2.33}$$

By considering a sphere S of radius R we see that if $\mathbf{j}$ tends to 0 faster than $1/R^2$ for large R then this last surface integral tends to 0 as $R \to \infty$, and so

$$\frac{\partial}{\partial t} \int_{\mathbf{R}^3} \rho\, d^3\mathbf{x} = 0. \tag{2.34}$$

Thence,

$$\int_{\mathbf{R}^3} \rho(t, \mathbf{x})\, d^3\mathbf{x} \tag{2.35}$$

is independent of t. (In particular, if it is 1 when $t = 0$, then it is 1 for all values of t.) □

When considering the meaning of the probability current it is worth noting that

$$\mathbf{j}(t, \mathbf{x}) = \frac{\hbar}{2mi}(\overline{\psi}\mathrm{grad}\psi - \psi\mathrm{grad}\overline{\psi}) = \frac{1}{m}\mathrm{Re}\left(\overline{\psi}\frac{\hbar}{i}\mathrm{grad}\psi\right). \tag{2.36}$$

According to Schrödinger the vector operator $\mathbf{P} = -i\hbar\nabla$ applied to a wave function ψ gives its momentum, so that $\mathbf{P}/m$ can be thought of as its velocity. Thus $\mathbf{j}$ looks like the density times the velocity of the wave, which is similar to the way in which the electric current is defined as the charge density times the velocity.

This becomes particularly clear if ψ is a plane wave $\exp(i\mathbf{k}.\mathbf{x})$. Although it is impossible to normalize the wave function since $\rho(\mathbf{x}) = |\psi(\mathbf{x})|^2 = 1$, we nonetheless have

$$\mathbf{j}(\mathbf{x}) = \frac{1}{m}\mathrm{Re}\left(e^{-i\mathbf{k}.\mathbf{x}}\hbar\mathbf{k}e^{i\mathbf{k}.\mathbf{x}}\right) = \frac{\hbar}{m}\mathbf{k}. \tag{2.37}$$

In other words, the probability current is constant and equal to the momentum of the wave divided by its mass.

Generally only normalized wave functions are ascribed any direct physical significance, but plane waves are a useful approximation when one has a beam of particles whose energy and momentum are known very accurately. We shall consider examples of that in Chapter 5.

Definition 2.5.2. To distinguish their different physical roles we call the normalizable wave functions *bound states* of the system, and the wave functions that cannot be normalized, *scattering states.*

2.6. The statistical distribution of energy

It is not only the position of the particle which is subject to statistical laws. In Section 2.3 we saw that a general wave function for the square well can be expanded in terms of separable wave functions:

$$\psi(t,x) = \sum c_n \psi_n(t,x) = \sqrt{\frac{2}{a}} \sum c_n \sin \frac{n\pi x}{a}. \tag{2.38}$$

In this situation $\partial\psi/\partial t$ is no longer of the form $E\psi$ so that the energy is no longer unambiguously defined. Heisenberg and Born therefore proposed the following:

Assumption 2.6.1. The probability of measuring the value E_n is $|c_n|^2$.

Remark 2.6.1. Parseval's theorem for Fourier series gives

$$\frac{2}{a}\int_0^a |\psi(t,x)|^2\, dx = \sum_{n=1}^{\infty} \left| \sqrt{\frac{2}{a}} c_n \right|^2, \tag{2.39}$$

so, since ψ is normalized, we have

$$\sum |c_n|^2 = 1. \tag{2.40}$$

This shows both that it is a reasonable definition, and that there is a zero probability of finding an energy other than one of the E_n.

Example 2.6.1. As an example, suppose that the wave function ψ inside the square well is just the constant $1/\sqrt{a}$ (the value being chosen to ensure that ψ is normalized). Then

$$\begin{aligned} c_n &= \sqrt{\frac{2}{a}} \int_0^a \sqrt{\frac{1}{a}} \sin \frac{n\pi x}{a}\, dx \\ &= \frac{\sqrt{2}}{a} \left[-\frac{a}{n\pi} \cos \frac{n\pi x}{a} \right]_0^a \\ &= \frac{\sqrt{2}}{n\pi} [1 - (-1)^n]. \end{aligned}$$

The probability of measuring the energy E_n is therefore zero if n is even, and is given by

$$E_n = \left(\frac{2\sqrt{2}}{n\pi} \right)^2 = \frac{8}{n^2\pi^2}$$

if n is odd. □

2.7. Historical notes

Schrödinger's attention was apparently drawn to de Broglie's thesis by Einstein and he gave a colloquium on it in Zürich towards the end of 1925. According to Felix Bloch at the end of Schrödinger's lecture his colleague Peter Debye remarked that the proper way to deal with waves was by using wave equations. Schrödinger soon found an equation, but was discouraged when it predicted the wrong spectral lines for the hydrogen atom. Over Christmas he went up to the mountain resort of Arosa with a girl friend. Whilst there he returned to the equation and noticed that if one used the classical relation between the energy and momentum (as we did in Section 2.1), rather than its relativistic analogue, then the hydrogen atom spectrum came out correctly, as did many other problems. (This was a lucky accident. When in 1928 Paul Dirac discovered the correct relativistic equation it turned out that the relativistic effects that Schrödinger had suppressed are exactly cancelled by the spin angular momentum of the electron that he had also neglected.)

The previous June, whilst recuperating from a severe attack of hay fever on the island of Helgoland, Werner Heisenberg had already come up with the crucial idea that he, Max Born, Pascual Jordan, and Dirac had developed into an algebraic quantum theory. By the beginning of November Wolfgang Pauli, overcoming his initial distaste for abstract algebra, had already shown that this approach gave the correct spectral lines for the hydrogen atom. None of these was initially impressed by Schrödinger's work. ('I quite agree with your criticism of Schrödinger's ... wave theory of matter', Heisenberg wrote to Dirac on 26 May 1926, 'this theory must be inconsistent'.) However, before long Schrödinger and then Pauli had shown that the two apparently quite different theories using wave equations and algebra are actually equivalent.

Exercises

2.1 A sodium atom with a mass of 3.82×10^{-26} kg emits a photon with a wavelength of 5.89×10^{-7} m. Find the recoil velocity of the atom. [Recall that $\hbar = 1.0546 \times 10^{-34}$ J s. Otto Frisch measured this recoil in 1933.]

2.2 A particle of mass m moves in the interval $[0, a]$ under the influence of the constant potential $V(x) = V_0$. Show that the possible energies of the system are

$$E = V_0 + \frac{n^2\pi^2\hbar^2}{2ma^2}.$$

2.3° A particle of mass m moves in the rectangle $[0, a] \times [0, b]$ in the xy-plane under the influence of a zero potential. The wave function vanishes at the boundaries of the region. By separating the variables x and y in Schrödinger's equation show that the permitted energies of the system are

$$E_{j,k} = \frac{\pi^2\hbar^2}{2m}\left(\frac{j^2}{a^2} + \frac{k^2}{b^2}\right)$$

where j and k are positive integers.

In the case when $a = b$ find two normalized wave functions corresponding to the energy $5\pi^2\hbar^2/2ma^2$. Find the probability in each case that the particle lies in the region $\{(x, y) \in \mathbf{R}^2 : x \leq y\}$.

2.4 A particle of mass m moves within a ball of radius a in $\mathbf{R}^3$ under the influence of the potential $V(r) = 0$. Show that there are continuous wave functions of the form $\psi(r)$ (independent of angles) that satisfy Schrödinger's equation with energy

$$E_n = \frac{n^2\pi^2\hbar^2}{2ma^2}$$

for $n = 1, 2, \ldots$. What is the probability of finding the particle within a distance $\frac{1}{2}a$ of the centre?

[You may assume that the wave function vanishes on the boundary.]

2.5 Show that the mean position of the particle in the previous question is at the origin. Find the variance of its height above the centre.

2.6° A particle of mass m moves freely within the interval $[0, a]$ on the x-axis. Initially the wave function is

$$\frac{1}{\sqrt{a}}\sin\left(\frac{\pi x}{a}\right)\left[1 + 2\cos\left(\frac{\pi x}{a}\right)\right].$$

Show that at a later time t the wave function is

$$\frac{1}{\sqrt{a}}e^{-i\left(\pi^2\hbar t/2ma^2\right)}\sin\left(\frac{\pi x}{a}\right)\left[1 + 2e^{-3i\left(\pi^2\hbar t/2ma^2\right)}\cos\left(\frac{\pi x}{a}\right)\right].$$

Hence, or otherwise, find the probability that at time t the particle lies within the interval $[0, \frac{1}{2}a]$.

2.7° A particle, moving freely between impenetrable barriers at $x = 0$ and $x = a$, is in its lowest energy stationary state when the barrier at $x = a$ is suddenly displaced to $x = 2a$. By expanding the original wave function in terms of separable solutions for the motion within $[0, 2a]$, find the wave function at a subsequent time t, and show that it is a superposition of states of energies

$$E_n = \frac{n^2\pi^2\hbar^2}{8ma^2},$$

for $n = 2$ and $n = 1, 3, 5, \ldots$. Show that the probability of finding the particle energy unchanged is $\frac{1}{2}$.

2.8 Let $\rho(t, x) = |\psi(t, x)|^2$ be the probability density for a particle moving on the x-axis, and

$$j(t, x) = \frac{\hbar}{2mi}\left(\overline{\psi}\frac{\partial\psi}{\partial x} - \psi\frac{\partial\overline{\psi}}{\partial x}\right)$$

the probability current. Show that

$$\frac{\partial\rho}{\partial t} + \frac{\partial j}{\partial x} = 0.$$

Show further that j vanishes identically if and only if there exists a function $\lambda(t)$, such that $\lambda(t)\psi(t, x)$ takes only real values.

2.9 Show that the probability current for a wave function of the form

$$\psi(t, \mathbf{x}) = \sum_s A_s e^{i\mathbf{k}_s.\mathbf{x}},$$

with $\mathbf{k}_s \in \mathbf{R}^3$, can be expressed as

$$\mathbf{j}(t, \mathbf{x}) = \frac{\hbar}{2m} \sum_{r,s} \overline{A_r} A_s (\mathbf{k}_r + \mathbf{k}_s) e^{i(\mathbf{k}_s - \mathbf{k}_r).\mathbf{x}}.$$

Deduce that the probability current associated to the wave function $\psi(t, \mathbf{x}) = A \exp(i\mathbf{k}{\cdot}\mathbf{x}) + B \exp(-i\mathbf{k}{\cdot}\mathbf{x})$ is

$$\mathbf{j}(t, \mathbf{x}) = \frac{\hbar}{m} \left(|A|^2 - |B|^2\right) \mathbf{k}.$$

3 Quadratic and linear potentials

If anything like mechanics were true then one would never understand the existence of atoms. Evidently there exists another 'quantum mechanics'.

WERNER HEISENBERG, letter to Wolfgang Pauli, 21 June 1925

3.1. The harmonic oscillator

Although the classical examples of harmonic oscillators such as springs or the simple pendulum are of limited use in the submicroscopic world of quantum mechanics, there are plenty of other applications that make the oscillator worth studying. Atoms in a crystal, for instance, oscillate about their mean positions and it is their vibrations that carry sound and heat through the material. Indeed, we know from classical theory that by using normal coordinates we may regard any system performing small oscillations about a point of stable equilibrium as a collection of harmonic oscillators. This extends even to the normal modes of vibration of a string or of the electromagnetic field. The harmonic oscillator thus provides the key to understanding how wave phenomena such as light can be quantized, and lies at the heart of Einstein's idea of the photon and of Planck's quantum.

Proposition 3.1.1. The permissible energy levels of the harmonic oscillator with potential energy $V(x) = \frac{1}{2}m\omega^2x^2$ and Schrödinger equation

$$\frac{-\hbar^2}{2m}\frac{d^2\psi}{dx^2} + \frac{1}{2}m\omega^2x^2\psi = E\psi$$

form the sequence

$$E_N = \left(N + \frac{1}{2}\right)\hbar\omega$$

for $N = 0, 1, \ldots$. The corresponding wave functions take the form

$$\psi_N(x) = \left(\frac{m\omega}{\pi\hbar}\right)^{\frac{1}{4}} H_N\left(\sqrt{\frac{m\omega}{2\hbar}}x\right) e^{-m\omega x^2/2\hbar},$$

where H_N is a polynomial of degree N.

Proof. This is not the sort of equation whose solutions immediately spring to mind, so we shall adopt the following strategy. We start by considering the behaviour when $|x|$ is large, and looking for an approximate solution ϕ. This guides us to a substitution $\psi = f \cdot \phi$, and a differential equation for f. Since ϕ already takes care of the long range behaviour of the wave function, we are interested in the short range behaviour of f, which suggests the use of a series solution.

At large distances E is small by comparison with the potential energy, so

$$\frac{\hbar^2}{2m}\frac{d^2\psi}{dx^2} = \left(\frac{1}{2}m\omega^2x^2 - E\right)\psi \sim \frac{1}{2}m\omega^2x^2\psi. \tag{3.1}$$

The formula

$$\psi'' \sim \left(\frac{m\omega x}{\hbar}\right)^2 \psi \tag{3.2}$$

suggests that we might start by considering the first-order differential equations,

$$\phi' = \pm\left(\frac{m\omega x}{\hbar}\right)\phi, \tag{3.3}$$

whose solutions are the functions $\phi_\pm = \exp(\pm m\omega x^2/2\hbar)$. Since the oscillator potential represents a strong attractive force, we would expect only a low probability of finding the particle far from the origin, so it is the function $\phi = \phi_-$ which is of real interest. (In any case ϕ_+ is not normalizable.) We can now check by direct differentiation that

$$\frac{-\hbar^2}{2m}\frac{d^2\phi}{dx^2} + \frac{1}{2}m\omega^2x^2\phi = \frac{1}{2}\hbar\omega\phi, \tag{3.4}$$

that is Schrödinger's equation with $E = \frac{1}{2}\hbar\omega$. For large x, where the potential energy dominates E, this is almost the same as the equation for ψ, so we expect that for any energy $\psi \sim \phi$.

We shall therefore try a solution of the form $\psi = f \cdot \phi$. Substituting this into Schrödinger's equation, we obtain

$$-\frac{\hbar^2}{2m}\left(f''\phi + 2f'\phi' + f\phi''\right) + \frac{1}{2}m\omega^2x^2f\phi = Ef\phi. \tag{3.5}$$

Recalling that $\phi' = -(m\omega x/\hbar)\phi$ and $\phi'' = [(m\omega x/\hbar)^2 - (m\omega/\hbar)]\phi$, this reduces to

$$f'' - 2\left(\frac{m\omega x}{\hbar}\right)f' - \frac{m\omega}{\hbar}f = -\frac{2mE}{\hbar^2}f, \tag{3.6}$$

or even more simply to

$$f'' - 2\left(\frac{m\omega x}{\hbar}\right)f' + \frac{2m\omega}{\hbar}\left(\frac{E}{\hbar\omega} - \frac{1}{2}\right)f = 0. \tag{3.7}$$

We can simplify the notation by setting $N = E/\hbar\omega - \frac{1}{2}$, and changing the variable to $\xi = x\sqrt{m\omega/\hbar}$. With these changes, the equation reduces to

$$\frac{d^2 f}{d\xi^2} - 2\xi\frac{df}{d\xi} + 2Nf = 0. \tag{3.8}$$

We now try a series solution of the form

$$f(\xi) = \sum_{n=0}^{\infty} a_n \xi^{n+c}, \tag{3.9}$$

with $a_0 \neq 0$. Substitution gives

$$\sum_{n=0}^{\infty}(n+c)(n+c-1)a_n\xi^{n+c-2} - 2\sum_{n=0}^{\infty}(n+c)a_n\xi^{n+c} + 2N\sum_{n=0}^{\infty}a_n\xi^{n+c} = 0. \tag{3.10}$$

The indicial equation coming from the coefficient of ξ^{c-2} is

$$c(c-1) = 0, \tag{3.11}$$

so that $c = 0$ or 1. The coefficient of ξ^{c-1} gives

$$(c+1)ca_1 = 0. \tag{3.12}$$

If $c = 1$ this forces a_1 to vanish, but if $c = 0$ there is no constraint on a_1. However, by subtracting a suitable multiple of the $c = 1$ series, we can ensure that the coefficient of ξ vanishes anyway, and so without loss of generality we take $a_1 = 0$.

Comparing coefficients of ξ^{n+c-2} we arrive at the recurrence relation

$$(n+c)(n+c-1)a_n = 2(n+c-2-N)a_{n-2} \tag{3.13}$$

for $n \geq 2$. By induction we see that $a_n = 0$ for all odd n, whilst

$$a_n = \frac{2(n+c-2-N)}{(n+c)(n+c-1)}a_{n-2} \tag{3.14}$$

determines the even coefficient in terms of a_0.

Unless the coefficients vanish for $n > N + 2 - c$ they all have the same sign, and we may as well assume that they are all positive. We need only consider the case of even n, and for definiteness we take $c = 0$ (the case of $c = 1$ being similar). Then we can exploit some cancellations to obtain

$$[n/2]!\, a_n = \left(1 - \frac{N+1}{n-1}\right)[(n-2)/2]!\, a_{n-2}. \tag{3.15}$$

For $n > 3N + 4$ the factor $1 - (N+1)/(n-1)$ lies in the interval $(\frac{2}{3}, 1)$, and so, for $n > 2(j+1) > 3N + 4$, we have iteratively

$$\left[\frac{n}{2}\right]! \, a_n > \frac{2}{3}\left[\frac{n-2}{2}\right]! \, a_{n-2} > \ldots > \left(\frac{2}{3}\right)^{(n-2j)/2} j! \, a_{2j}. \tag{3.16}$$

Introducing $K = j! \, (3/2)^j a_{2j}$, for fixed j, we arrive at the inequality

$$a_n > \frac{K}{(\frac{1}{2}n)!}\left(\frac{2}{3}\right)^{n/2}, \tag{3.17}$$

and similar inequalities can be found when $c = 1$. Summing this over even $n = 2m$, and introducing the polynomial

$$P(\xi) = \sum_{m=0}^{j-1}\left[a_{2m} - \frac{K}{m!}\left(\frac{2}{3}\right)^m\right]\xi^{2m}, \tag{3.18}$$

we obtain

$$f(\xi) = \sum_{m=0}^{\infty} a_{2m}\xi^{2m} > \sum_{m=0}^{\infty}\frac{K}{m!}\left(\frac{2}{3}\right)^m \xi^{2m} + P(\xi) = Ke^{\frac{2}{3}\xi^2} + P(\xi). \tag{3.19}$$

Thus $f(\xi)\exp(-\frac{1}{2}\xi^2)$ grows faster than $\exp(\xi^2/6)$ for large ξ, and we are back to precisely the sort of exponentially growing wave function that we were trying to avoid. We can only escape from this dilemma if the coefficients vanish for large n, so that the series actually terminates and f is a polynomial.

This happens if for some even $n \geq 2$ we have

$$N = n + c - 2, \tag{3.20}$$

since then a_n and all successive coefficients vanish. Bearing in mind the fact that c can be either 0 or 1 the above condition just means that N must be a non-negative integer. Returning to the definition of N we are now ready to deduce that the energy levels have the form stated in the proposition. The fact that these are the only possible values for the energy follows from the fact that N must be a non-negative integer. Conversely, for each of these energies the series for f terminates and provides a solution ψ_N for the wave function.

Explicitly, if $N = 0$, so that $E_0 = \frac{1}{2}\hbar\omega$ and $c = 0$, we already have $a_2 = 0$, so that f is just a constant, C, and

$$\psi_0 = Ce^{-m\omega x^2/2\hbar}. \tag{3.21}$$

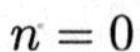

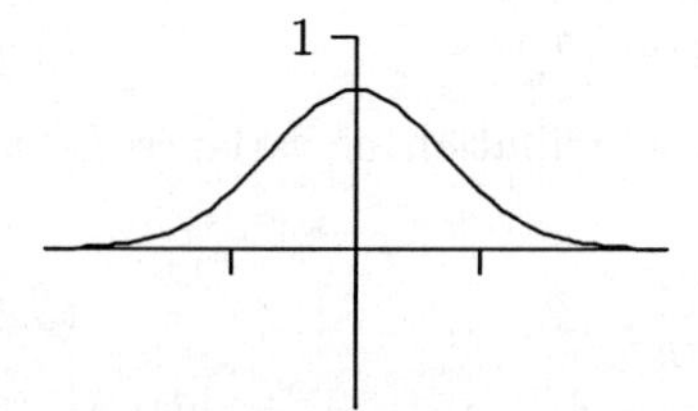

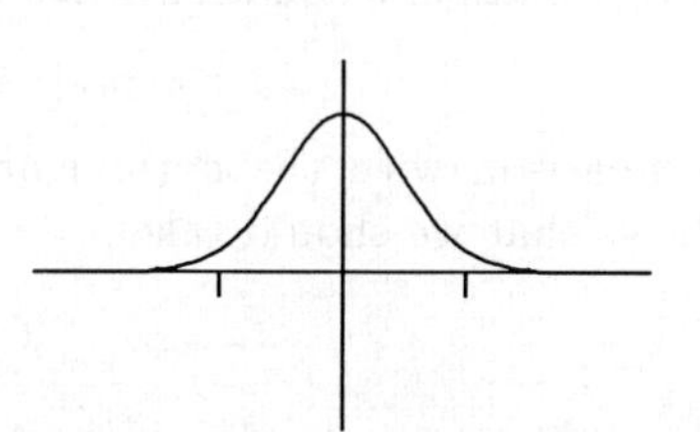

$n = 1$

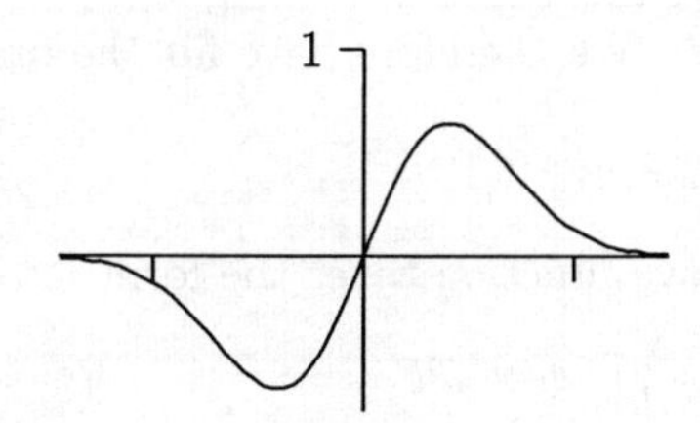

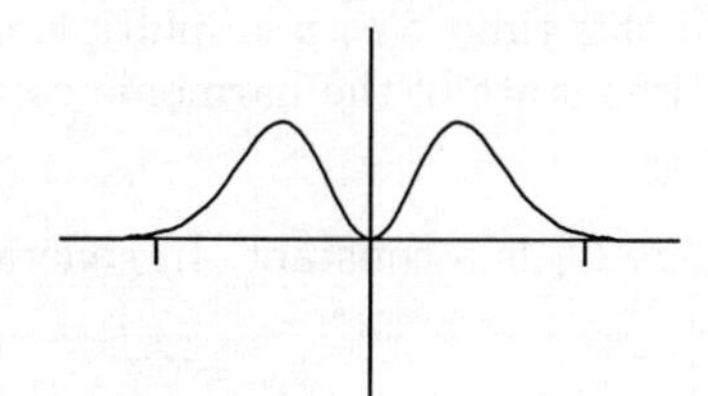

$n = 2$

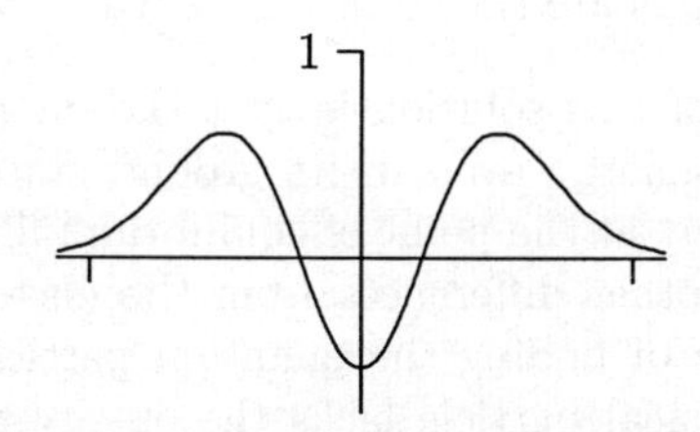

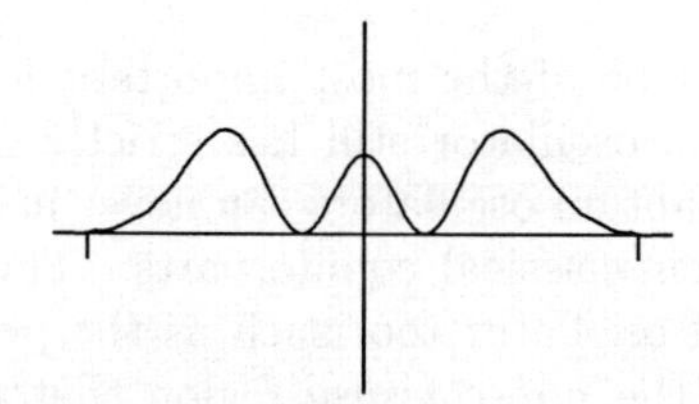

FIGURE 3.1. The wave functions (left) and probability densities (right) for the ground state and first two excited states of the harmonic oscillator. The marks on the horizontal scale show the extreme limits of the motion of a classical harmonic oscillator of the same energy.

(This is, of course, essentially ϕ, which we already knew to be a solution of Schrödinger's equation when $E = \frac{1}{2}\hbar\omega$.) Since

$$|\psi_0|^2 = |C|^2 e^{-m\omega x^2/\hbar}, \tag{3.22}$$

comparison with the normal probability distribution of variance $\hbar/2m\omega$ shows that we should take

$$|C|^2 = \sqrt{\frac{m\omega}{\pi\hbar}}. \tag{3.23}$$

From this the normalized ground state wave function is immediately seen to be

$$\psi_0 = \left(\frac{m\omega}{\pi\hbar}\right)^{\frac{1}{4}} e^{-m\omega x^2/2\hbar}. \tag{3.24}$$

When $N = 1$ we have $E_1 = \frac{3}{2}\hbar\omega$ and we must take $c = 1$. Again $a_2 = 0$, and this time $f(\xi)$ is a multiple of $\xi^c = \xi$. We therefore have for the first excited state of the harmonic oscillator

$$\psi_1 = C_1 x e^{-m\omega x^2/2\hbar}, \tag{3.25}$$

where C_1 is a constant. In general the wave function takes the form

$$\psi_N = C_N H_N\left(\sqrt{\frac{m\omega}{2\hbar}}x\right) e^{-m\omega x^2/2\hbar}. \tag{3.26}$$

where H_N is a polynomial and C_N is another constant. With appropriate normalization H_N is known as the N-th Hermite polynomial. The properties of these polynomials will be discussed in more detail in Section 7.7. The graphs of the first three wave functions are shown in Figure 3.1. □

One of the most important features of this solution is that the quantum oscillator still has strictly positive energy even in its ground state: quantum oscillators can never just sit inert at the point of equilibrium like their classical counterparts. There are other differences from the classical oscillator too, such as the possibility of finding the quantum particle in the *non-classical* region that the classical particle lacks the energy to reach. For example, the non-classical region for the ground state is where $V > E = \frac{1}{2}\hbar\omega$, that is $\frac{1}{2}m\omega^2x^2 > \frac{1}{2}\hbar\omega$, or $|\xi| > 1$. The probability of finding the particle there is

$$\frac{1}{\sqrt{\pi}}\int_{|\xi|>1} e^{-\xi^2}\,d\xi = 0.157 \tag{3.27}$$

Alternatively, one can argue that, since the probability density is that of a normal distribution with variance $\hbar/2m\omega$, the probability of finding the particle in the non-classical region is the probability of being further than $\sqrt{2}$ standard deviations from the mean in a normal distribution, which is 0.157. The probability of being in the non-classical region decreases, albeit slowly, for larger values of n.

3.2. Harmonic oscillators in higher dimensions

Higher-dimensional oscillators can easily be handled by an appropriate separation of variables. For instance, consider a two-dimensional oscillator whose potential energy is $V = \frac{1}{2}m\left(\omega_1^2 x^2 + \omega_2^2 y^2\right)$.

Proposition 3.2.1. The energy levels of the two-dimensional harmonic oscillator whose Schrödinger equation is

$$-\frac{\hbar^2}{2m}\left(\frac{\partial^2\psi}{\partial x^2} + \frac{\partial^2\psi}{\partial y^2}\right) + \frac{1}{2}m\left(\omega_1^2 x^2 + \omega_2^2 y^2\right)\psi = E\psi$$

have the form

$$E = \left(N_1 + \frac{1}{2}\right)\hbar\omega_1 + \left(N_2 + \frac{1}{2}\right)\hbar\omega_2,$$

for N_1, $N_2 = 0, 1, \ldots$. The corresponding wave functions may be written as

$$\psi(x,y) = \left(\frac{m}{\pi\hbar}\right)^{\frac{1}{2}}(\omega_1\omega_2)^{\frac{1}{4}} H_{N_1}\left(\sqrt{\frac{m\omega_1}{\hbar}}x\right) H_{N_2}\left(\sqrt{\frac{m\omega_2}{\hbar}}y\right)$$
$$\times \exp\left(-\frac{m}{2\hbar}\left(\omega_1 x^2 + \omega_2 y^2\right)\right).$$

Proof. Substituting a separable solution $\psi(x,y) = X(x)Y(y)$, we obtain

$$-\frac{\hbar^2}{2m}\left(\frac{X''}{X} + \frac{Y''}{Y}\right) + \frac{1}{2}m\left(\omega_1^2 x^2 + \omega_2^2 y^2\right) = E, \tag{3.28}$$

so that

$$-\frac{\hbar^2}{2m}\frac{X''}{X} + \frac{1}{2}m\omega_1^2 x^2 = E + \frac{\hbar^2}{2m}\frac{Y''}{Y} - \frac{1}{2}m\omega_1^2 y^2. \tag{3.29}$$

Since the two variables have now been separated, each side of the equation must be a constant, which we shall call E_1. This gives

$$-\frac{\hbar^2}{2m}X'' + \frac{1}{2}m\omega_1^2 x^2 X = E_1 X, \tag{3.30}$$

and also, setting $E_2 = E - E_1$,

$$-\frac{\hbar^2}{2m}Y'' + \frac{1}{2}m\omega_2^2 y^2 Y = E_2 Y. \tag{3.31}$$

Thus X and Y both satisfy a one-dimensional oscillator equation, and their respective energy levels are therefore

$$E_j = \left(N_j + \tfrac{1}{2}\right)\hbar\omega_j \tag{3.32}$$

where j can be 1 or 2 and both N_1 and N_2 are negative integers. The total energy is $E = E_1 + E_2$ giving

$$E = \left(N_1 + \tfrac{1}{2}\right)\hbar\omega_1 + \left(N_2 + \tfrac{1}{2}\right)\hbar\omega_2. \tag{3.33}$$

Similarly, the wave functions take the form

$$\begin{aligned}\psi(x,y) &= X(x)Y(y)\\ &= \left(\frac{m}{\pi\hbar}\right)^{\frac{1}{2}}(\omega_1\omega_2)^{\frac{1}{4}} H_{N_1}\left(\sqrt{\frac{m\omega_1}{\hbar}}x\right) H_{N_2}\left(\sqrt{\frac{m\omega_2}{\hbar}}y\right)\\ &\quad \times \exp\left(-\frac{m}{2\hbar}\left(\omega_1 x^2 + \omega_2 y^2\right)\right).\end{aligned}$$

□

Although this provides a complete solution, the problem with which we started was not really typical because the potential energy contained only terms in x^2 and y^2 but no cross terms xy. In fact, we know that even in classical mechanics one must change to normal coordinates before the differential equations for the motion simplify, and the same is true in quantum mechanics. For the sort of problem we have just been considering the kinetic energy is already in a reasonable form, but one must diagonalize the potential V *before* attempting to separate the variables. Provided that the potential energy is diagonalized by an orthogonal transformation the kinetic energy will remain unaffected.

Example 3.2.1. A particle of mass m moves in a plane under the influence of a potential

$$V = m\omega^2(x^2 + xy + y^2).$$

Find the energy levels.

Solution The potential can be written in the matrix form

$$V = \frac{1}{2}m\omega^2 \begin{pmatrix} x & y \end{pmatrix} \begin{pmatrix} 2 & 1 \\ 1 & 2 \end{pmatrix} \begin{pmatrix} x \\ y \end{pmatrix}.$$

The eigenvalues of the matrix

$$\begin{pmatrix} 2 & 1 \\ 1 & 2 \end{pmatrix}$$

are 3 and 1, so there exist coordinates u and v such that

$$\begin{aligned} V &= \frac{1}{2} m\omega^2 \begin{pmatrix} u & v \end{pmatrix} \begin{pmatrix} 3 & 0 \\ 0 & 1 \end{pmatrix} \begin{pmatrix} u \\ v \end{pmatrix} \\ &= \frac{1}{2} m \left(3\omega^2 u^2 + \omega^2 v^2\right). \end{aligned}$$

(Since the matrix associated to V is symmetric we know that it can be diagonalized by an orthogonal change of variables. In fact, by looking at the eigenvalues of the matrix we see that this could be achieved by taking $u = (x+y)/\sqrt{2}$, and $v = (x-y)/\sqrt{2}$.) The possible energies are therefore

$$\left(N_1 + \tfrac{1}{2}\right) \hbar\sqrt{3}\omega + \left(N_2 + \tfrac{1}{2}\right) \hbar\omega,$$

with N_1 and N_2 non-negative integers. (It is worth noting that $\omega_1 = \sqrt{3}\omega$ and $\omega_2 = \omega$ are precisely the normal frequencies of the classical system, so, once we know those, we can immediately deduce the quantum energy levels.) □

In three dimensions and more one can proceed in a similar way. After transferring to normal coordinates the variables are separated (one at a time) to reduce the problem to a number of independent one-dimensional oscillators. The energy is then the sum of their energies and the wave function is the product of their wave functions. (See the exercises at the end of this chapter.)

3.3. Degeneracy

An interesting special case of the original two-dimensional oscillator arises when the two frequencies ω_1 and ω_2 coincide. Then the energy levels take the form

$$E = (N_1 + N_2 + 1)\,\hbar\omega, \tag{3.34}$$

where ω denotes the common value of ω_1 and ω_2. Setting $N = N_1 + N_2$, this may be written as

$$E = (N+1)\hbar\omega. \tag{3.35}$$

The ground state energy is now $\hbar\omega$ and occurs just when $N_1 = N_2 = 0$. In general, however, there is more than one wave function giving the same energy. For example, the first excited state has energy $E = (1+1)\hbar\omega = 2\hbar\omega$, and this occurs both for $N_1 = 1$, $N_2 = 0$, and for $N_1 = 0$, $N_2 = 1$. The wave function corresponding to the first of these possibilities is

$$\begin{aligned}\psi_{10}(x,y) &= \left(\frac{m}{\pi\hbar}\right)^{\frac{1}{2}} (\omega_1\omega_2)^{\frac{1}{4}} H_1\left(\sqrt{\frac{m\omega_1}{\hbar}}x\right) H_0\left(\sqrt{\frac{m\omega_2}{\hbar}}y\right) \\ &\quad \times \exp\left(-\frac{m}{2\hbar}\left(\omega_1 x^2 + \omega_2 y^2\right)\right) \\ &= Ax\exp\left(-\frac{m}{2\hbar}\left(\omega_1 x^2 + \omega_2 y^2\right)\right), \end{aligned} \tag{3.36}$$

where A is a constant. Similarly, for another constant B, we have

$$\psi_{01}(x,y) = By\exp\left(-\frac{m}{2\hbar}\left(\omega_1 x^2 + \omega_2 y^2\right)\right), \tag{3.37}$$

which is clearly not a constant multiple of ψ_{10}. Since any solution with energy $2\hbar\omega$ is a linear combination of the two functions ψ_{10} and ψ_{01}, the solution space is two-dimensional.

Definition 3.3.1. If the space of solutions of Schrödinger's time-independent equation with energy E has dimension $k > 1$ then we say that the energy level is *k-fold degenerate*; if it is one dimensional we say that E is a *non-degenerate* energy level. (One usually says *doubly degenerate* rather than two-fold degenerate.)

Example 3.3.1. Our calculation has thus shown that the first excited state of the two-dimensional oscillator with equal frequencies is doubly degenerate. If we take the general energy level $E = (N+1)\hbar\omega$ then the possible choices for (N_1, N_2) are: $(N, 0)$, $(N-1, 1)$, ..., $(0, N)$. This gives a total of $N+1$ possibilities so that the N-th excited state of energy $E = (N+1)\hbar\omega$ is $(N+1)$-fold degenerate. For the one-dimensional harmonic oscillator and square well all the energy levels were non-degenerate, because the wave functions were uniquely determined up to a multiple.

Definition 3.3.2. If all the normal frequencies of a multi-dimensional harmonic oscillator coincide then the oscillator is said to be *isotropic*.

Remark 3.3.1. Our calculations have shown that all the excited states of a two-dimensional isotropic oscillator are degenerate, and the same holds in higher dimensions. However, it is not only isotropic oscillators which have degenerate energy levels: whenever two of the normal frequencies are rational multiples of each other some energy levels degenerate. (See Exercise 3.2.)

Example 3.3.2. If the normal frequencies of the classical oscillator are 2ω and 3ω then the possible energies of the quantum oscillator are

$$\begin{aligned} E &= \left[2\left(N_1 + \tfrac{1}{2}\right) + 3\left(N_2 + \tfrac{1}{2}\right)\right]\hbar\omega \\ &= \left(2N_1 + 3N_2 + \tfrac{5}{2}\right)\hbar\omega. \end{aligned}$$

The energy level $E = 17\hbar\omega/2$ is degenerate because it can be obtained either with $N_1 = 3$ and $N_2 = 0$ or with $N_1 = 0$ and $N_2 = 2$.

3.4. Momentum space

For some quantum mechanical systems it is useful to use the Fourier transform, which we shall write in the form

$$(\mathcal{F}\psi)(p) = \frac{1}{\sqrt{2\pi\hbar}} \int_{-\infty}^{\infty} e^{-ipx/\hbar}\psi(x)\,dx. \tag{3.38}$$

The appearance of $\hbar$ in the exponential is to ease the physical interpretation later on, and the constant factor outside the integral is chosen so that the inverse transform is just

$$\psi(x) = \frac{1}{\sqrt{2\pi\hbar}} \int_{-\infty}^{\infty} e^{ipx/\hbar}(\mathcal{F}\psi)(p)\,dp. \tag{3.39}$$

If we multiply this formula by $\overline{\phi(x)}$, integrate it from $x = -\infty$ to ∞, and interchange the order of integration on the right, then we obtain a useful result, detailed proofs of which can be found in most analysis books.

Plancherel's theorem 3.4.1.

$$\int_{-\infty}^{\infty} \overline{\phi(x)}\psi(x)\,dx = \int_{-\infty}^{\infty} \overline{\mathcal{F}\phi(p)}\mathcal{F}\psi(p)\,dp.$$

Putting $\phi = \psi$ we see that normalized wave functions have normalized Fourier transforms.

The utility of the Fourier transform in quantum mechanics, as in classical differential equation theory, results from its effect on differential operators, which is summarized in the following result.

Proposition 3.4.2. Let $\mathcal{F}$ denote the Fourier transform; then for all differentiable ψ in $\mathcal{H}$, and all p in $\mathbf{R}$,

$$\frac{\hbar}{i}(\mathcal{F}\psi')(p) = p(\mathcal{F}\psi)(p).$$

Proof. By definition we have

$$\begin{aligned}(\mathcal{F}P\psi)(p) &= \frac{1}{\sqrt{2\pi\hbar}}\int_{-\infty}^{\infty} e^{-ipx/\hbar}(P\psi)(x)\,dx \\ &= \sqrt{\frac{\hbar}{2\pi}}\frac{1}{i}\int_{-\infty}^{\infty} e^{-ipx/\hbar}\frac{d\psi}{dx}\,dx \\ &= \sqrt{\frac{\hbar}{2\pi}}\frac{1}{i}\left\{\left[e^{-ipx/\hbar}\psi(x)\right]_{-\infty}^{\infty}\right. \\ &\qquad \left. + \frac{ip}{\hbar}\int_{-\infty}^{\infty} e^{-ipx/\hbar}\psi(x)\,dx\right\} \\ &= \sqrt{\frac{\hbar}{2\pi}}\frac{1}{i}\left[e^{-ipx/\hbar}\psi(x)\right]_{-\infty}^{\infty} + p(\mathcal{F}\psi)(p).\end{aligned}$$

Any normalizable wave function, ψ, can be approximated by one which tends to zero when $|x|$ is large, so that the first term vanishes leaving the desired answer. □

According to the de Broglie relation $-i\hbar d/dx$ gives the momentum of a wave function. On the Fourier transform space this is just multiplication by p, so we shall often refer to the Fourier-transformed wave functions as *momentum space wave functions.* Of course, one cannot expect to have everything at once, and the Fourier transforms make position calculations harder. In fact,

$$\begin{aligned}(\mathcal{F}X\psi)(p) &= \frac{1}{\sqrt{2\pi\hbar}}\int_{\mathbf{R}} e^{-\frac{ipx}{\hbar}}x\psi(x)\,dx \\ &= i\hbar\frac{\partial}{\partial p}\left(\frac{1}{\sqrt{2\pi\hbar}}\int_{\mathbf{R}} e^{-\frac{ipx}{\hbar}}\psi(x)\,dx\right) \\ &= i\hbar\frac{\partial}{\partial p}(\mathcal{F}\psi)(p),\end{aligned}$$

so that now it is positions which involve differentiation.

For calculations involving momentum in three dimensions there is also an appropriate Fourier transform obtained by transforming each of the three components:

$$(\mathcal{F}\psi)(\mathbf{p}) = (2\pi\hbar)^{-\frac{3}{2}} \int_{\mathbf{R}^3} e^{-i\mathbf{p}.\mathbf{x}/\hbar}\psi(\mathbf{x})\, d^3\mathbf{x}. \tag{3.40}$$

3.5. Motion in a uniform electric field

The potential $V(x) = eFx$ describes the effect on a charge e of a uniform electric field F along the x-axis. Schrödinger's equation is therefore

$$-\frac{\hbar^2}{2m}\frac{\partial^2\psi}{\partial x^2} + eFx\psi = E\psi. \tag{3.41}$$

By Fourier transforming this equation (and writing $\widehat{\psi} = \mathcal{F}\psi$ for brevity), we obtain

$$\frac{1}{2m}p^2\widehat{\psi}(p) + i\hbar eF\frac{\partial\widehat{\psi}}{\partial p} = E\widehat{\psi}(p). \tag{3.42}$$

On introducing an integrating factor this simplifies to

$$i\hbar eF\frac{\partial}{\partial p}\left(e^{p^3/6im\hbar eF}\widehat{\psi}\right) = E\left(e^{p^3/6im\hbar eF}\widehat{\psi}\right), \tag{3.43}$$

and has the solution

$$\left(e^{p^3/6im\hbar eF}\widehat{\psi}\right) = Ne^{-iEp/eF\hbar}. \tag{3.44}$$

Taking moduli we see that $|\widehat{\psi}|^2 = |K|^2$ is constant, so that the wave function is not normalizable. This means that this problem has no bound states. This is physically reasonable, since the particle can always gain energy from the field.

Let us now consider how the wave functions evolve. By Fourier transforming Schrödinger's equation

$$i\hbar\frac{\partial\psi}{\partial t} = -\frac{\hbar^2}{2m}\frac{\partial^2\psi}{\partial x^2} + eFx\psi, \tag{3.45}$$

and introducing the same integrating factor as before, we obtain

$$\frac{\partial}{\partial t}\left(e^{p^3/6im\hbar eF}\widehat{\psi}\right) = eF\frac{\partial}{\partial p}\left(e^{p^3/6im\hbar eF}\widehat{\psi}\right). \tag{3.46}$$

In terms of new variables $u = p - eFt$ and $v = p + eFt$, this becomes

$$\frac{\partial}{\partial u}\left(e^{p^3/6im\hbar eF}\widehat{\psi}\right) = 0, \tag{3.47}$$

so that the general solution is of the form

$$e^{p^3/6im\hbar eF}\widehat{\psi}(t,p) = \Psi(v) = \Psi(p + eFt), \tag{3.48}$$

where the arbitrary function Ψ can be determined from the initial condition

$$e^{p^3/6im\hbar eF}\widehat{\psi}(0,p) = \Psi(p). \tag{3.49}$$

We therefore have the solution

$$\widehat{\psi}(t,p) = e^{-p^3/6im\hbar}e^{(p+eFt)^3/6im\hbar eF}\widehat{\psi}(0,p+eFt), \tag{3.50}$$

which can be rearranged to give

$$\widehat{\psi}(t,p) = \widehat{\psi}(0,p+eFt)\exp\left[-\frac{it}{2m\hbar}\left(p^2 + eFtp + \frac{1}{3}(eFt)^2\right)\right]. \tag{3.51}$$

The wave function can now be found as a function of x by inverting the Fourier transform or by using the convolution theorem.

Exercises

3.1 Find normalized wave functions for the first two excited states of the one-dimensional harmonic oscillator with potential $V = \frac{1}{2}m\omega^2x^2$.

3.2 A particle moves in two dimensions under the influence of the potential

$$V(x,y) = \frac{1}{2}m\omega^2(10x^2 + 12xy + 10y^2).$$

Find the energy levels and calculate the associated degeneracy of each level.

3.3 A particle of mass m moves in three dimensions under the influence of the potential

$$V = \frac{1}{2}m\omega^2(x^2 + y^2 + z^2).$$

Show that the energy levels have the form $\left(N + \frac{3}{2}\right)\hbar\omega$ where N is a non-negative integer, and find their degeneracies.

3.4 Let D_n^k be the number of ways of writing n as the sum of k non-negative integers n_1, $n_2, \ldots, n_k$. Show that the generating function can be written as

$$\sum_{n=0}^{\infty} D_n^k s^n = \sum_{n_1=0}^{\infty} \sum_{n_2=0}^{\infty} \cdots \sum_{n_k=0}^{\infty} s^{n_1+n_2+\ldots+n_k}.$$

Deduce that the generating function is $(1-s)^{-k}$ and hence or otherwise show that

$$D_n^k = \binom{n+k-1}{n}.$$

Deduce formulae for the degeneracy of isotropic harmonic oscillators in two and three dimensions.

3.5 A particle of mass m moves on the x-axis under the influence of the potential

$$V(x) = \frac{1}{2} m\omega^2 x^2 + \epsilon x.$$

By changing origin, or otherwise, show that the energy levels are

$$\left(N + \frac{1}{2}\right)\hbar\omega - \frac{1}{2}\frac{\epsilon^2}{m\omega^2},$$

for N a non-negative integer.

3.6 Schrödinger's equation for a two-dimensional isotropic oscillator can be written in polar coordinates as

$$-\frac{\hbar^2}{2m}\left[\frac{1}{r}\frac{\partial}{\partial r}\left(r\frac{\partial\psi}{\partial r}\right) + \frac{1}{r^2}\frac{\partial^2\psi}{\partial\theta^2}\right] + \frac{1}{2}m\omega^2 r^2\psi = E\psi.$$

By considering solutions ψ that are separable in polar coordinates verify that the energy levels are of the form $n\hbar\omega$ where n is a positive integer. Find wave functions $\psi(r,\theta)$ for the two lowest energy levels.

3.7° At time $t = 0$ the wave function of a free particle of mass m, moving along the x-axis, is given by

$$\psi(0, x) = a^{-\frac{1}{2}} e^{-|x|/a}.$$

Derive the wave function at a subsequent time t. Calculate the probability that the momentum lies in the range $[-\hbar/a, \hbar/a]$.

4 The hydrogen atom

Herewith it has been demonstrated that the Balmer terms come out correctly from the new quantum mechanics.

WOLFGANG PAULI, *On the hydrogen spectrum*, January 1926

4.1. The structure of atoms

Whilst theoretical physicists struggled to understand the interactions between matter and radiation, new experiments were revealing that atoms had internal structure. In 1897 J.J. Thomson showed that cathode rays consisted of a stream of negatively charged particles much lighter than any known atom. These became known as electrons. It was the 1896 discovery and subsequent investigation of radioactivity which forced the startling realization that, far from being indivisible and indestructible as had been supposed, atoms can disintegrate or be shattered into smaller pieces.

Ernest Rutherford realized that by interposing a thin metal foil in the path of alpha rays emanating from radium, the subatomic particles emitted during radioactive decay could themselves be used to probe the structure of matter on an hitherto inaccessibly small scale. The climax of these experiments with his coworkers, Hans Geiger and Ernest Marsden, came with the discovery that atoms consisted mostly of empty space. This led Rutherford to formulate the popular picture of atoms as miniature solar systems in which negatively charged electrons orbit a positively charged nucleus under the influence of electrostatic rather than gravitational attraction. The nucleus was itself composed of positively charged protons and, as it later transpired, usually some uncharged neutrons, both of these being almost 2000 times more massive as an electron. (Ordinary hydrogen has but a single proton in its nucleus, otherwise there are usually some neutrons as well.) The chemical properties of an atom were determined by its electrons. Since atoms are electrically neutral, unless ionized, the number of protons matched the number of electrons, but the nucleus might contain more or fewer neutrons giving the possibility of chemically identical but physically distinct isotopes.

Although still the popular image of an atom, this picture was soon displaced by the quantum mechanical picture that we shall describe in this chapter. The calculation that more than any other convinced most physicists of the correctness of quantum theory was that giving the spectrum of the hydrogen atom. Niels Bohr had already managed to derive the spectrum using the Einstein–Planck law and Rutherford's model of the atom, but his argument relied on a series of brilliant *ad hoc* assumptions, whose application to other problems often led to false conclusions. In 1925, after some initial scepticism about Heisenberg's theory, Pauli showed how it gave the spectrum for the hydrogen atom without the need for any extra assumptions. It was this more than anything else which convinced most physicists of the correctness of quantum theory. The calculation was a major *tour de force* which occupied Pauli for three weeks, but by the end of the year Schrödinger's new method of wave mechanics had reduced the problem to solving a simple differential equation.

In the simplest case of the hydrogen atom, and also for heavier atoms that have lost all but one electron due to ionization, there is only one electron. Since these two cases are mathematically identical we shall investigate them both together. As in classical mechanics both the electron and the nucleus orbit around their mutual centre of mass. Since the hydrogen nucleus is almost 2000 times heavier than the electron whilst for heavier ions the discrepancy is even more pronounced, the centre of mass is almost at the centre of the nucleus. We shall therefore make the simplifying assumption that the nucleus of the atom is fixed. (We shall return to discuss this in more detail in Section 8.5.)

If the nucleus carries a positive charge Ze and the electron has a charge $-e$ then the electrostatic potential energy is $V = -Ze^2/4\pi\epsilon_0 r$, where r is the distance between the electron and nucleus, and ϵ_0 is the dielectric constant of the vacuum. (The role of $(4\pi\epsilon_0)^{-1}$ in electrostatic theory is much the same as that of the gravitational constant in Newton's theory.) Referred to spherical polar coordinates centred on the nucleus, Schrödinger's equation is thus

$$-\frac{\hbar^2}{2m}\nabla^2\psi - \frac{Ze^2}{4\pi\epsilon_0 r}\psi = E\psi. \tag{4.1}$$

4.2. Central force problems

When dealing with motion under central forces where the potential $V = V(r)$ depends only on the distance r from the centre, it is natural to separate variables in spherical polar coordinates (r, θ, ϕ). When the Laplacian is written in terms of these coordinates the Schrödinger equation becomes

$$-\frac{\hbar^2}{2m}\left[\frac{1}{r}\frac{\partial^2 r\psi}{\partial r^2} + \frac{1}{r^2\sin\theta}\frac{\partial}{\partial\theta}\left(\sin\theta\frac{\partial\psi}{\partial\theta}\right) + \frac{1}{r^2\sin^2\theta}\frac{\partial^2\psi}{\partial\phi^2}\right] + V(r)\psi = E\psi. \tag{4.2}$$

Multiplying by r^2/ψ and substituting $\psi = R(r)\Theta(\theta)\Phi(\phi)$ we obtain

$$-\frac{\hbar^2}{2m}\left[\frac{r}{R}\frac{d^2}{dr^2}(rR) + \frac{1}{\Theta\sin\theta}\frac{d}{d\theta}\left(\sin\theta\frac{d\Theta}{d\theta}\right) + \frac{1}{\Phi\sin^2\theta}\frac{d^2\Phi}{d\phi^2}\right] + r^2V = r^2E. \tag{4.3}$$

The second and third terms in this equation depend only on the angular variables, whilst the others depend only on r, so that each group must be constant. That is, for constant λ we have

$$\frac{1}{\Theta\sin\theta}\frac{d}{d\theta}\left(\sin\theta\frac{d\Theta}{d\theta}\right) + \frac{1}{\Phi\sin^2\theta}\frac{d^2\Phi}{d\phi^2} = -\lambda \tag{4.4}$$

and

$$-\frac{\hbar^2}{2m}\left(\frac{r}{R}\frac{d^2}{dr^2}(rR) - \lambda\right) + r^2V = r^2E. \tag{4.5}$$

On multiplying the angular equation by $\sin^2\theta$, we obtain

$$\frac{\sin\theta}{\Theta}\frac{d}{d\theta}\left(\sin\theta\frac{d\Theta}{d\theta}\right) + \frac{1}{\Phi}\frac{d^2\Phi}{d\phi^2} = -\lambda\sin^2\theta, \tag{4.6}$$

in which the term Φ''/Φ depends only on ϕ and the remaining terms only on θ. We must therefore have $\Phi''/\Phi = -\mu^2$ for some constant μ, and

$$\frac{\sin\theta}{\Theta}\frac{d}{d\theta}\left(\sin\theta\frac{d\Theta}{d\theta}\right) - \mu^2 = -\lambda\sin^2\theta. \tag{4.7}$$

The ϕ equation integrates immediately to give Φ as a linear combination of $\exp(\pm i\mu\phi)$ when $\mu \neq 0$. Since $\Phi(\phi+2\pi) = \Phi(\phi)$ this forces μ to be a real integer. When $\mu = 0$ only the constant solution is periodic. We therefore conclude that we may take $\Phi = \exp(i\mu\phi)$, where μ is any integer.

The θ equation can be rewritten in terms of $c = \cos\theta$ using

$$\sin\theta\frac{d}{d\theta} = \sin\theta\frac{dc}{d\theta}\frac{d}{dc} = -\sin^2\theta\frac{d}{dc} = (c^2-1)\frac{d}{dc}, \tag{4.8}$$

to obtain Legendre's equation

$$(c^2-1)\frac{d}{dc}\left((c^2-1)\frac{d\Theta}{dc}\right) - \mu^2\Theta + \lambda(c^2-1)\Theta = 0. \tag{4.9}$$

We shall not solve this equation in detail here (though it can be solved by standard methods), since it is more readily handled by the algebraic techniques that we shall introduce in Chapter 8. The main point is that the solutions are singular at $c = \pm 1$ unless $\lambda = l(l+1)$ with l an integer greater than or equal to $|\mu|$. When λ takes such a value there is a unique solution P_l^μ that is continuous on the interval $[-1, 1]$. (When $\mu = 0$ this solution is a polynomial, P_l, called a Legendre polynomial.)

Definition 4.2.1. The full angular term $\Theta\Phi = P_l^\mu(\theta)e^{i\mu\phi}$ is called a *spherical harmonic of degree* l and is usually denoted by $Y_l^\mu(\theta, \phi)$.

Theorem 4.2.1. The space of spherical harmonics of degree l has dimension $2l+1$.

Proof. The space is spanned by the $2l+1$ functions Y_l^μ for integral $\mu = -l, -l+1, \ldots, l-1, l$. □

The preceding discussion still leaves the radial equation, which on multiplication by R/r takes the form

$$-\frac{\hbar^2}{2m}\left(\frac{d^2}{dr^2}(rR) - \frac{l(l+1)}{r^2}rR\right) + VrR = ErR. \tag{4.10}$$

This equation can be investigated by the same technique that we used for the harmonic oscillator, starting by examining the behaviour at large distances. For definiteness let us assume that $V(r)$ tends to 0 as $r \to \infty$. (Other potentials, such as the quadratically increasing three-dimensional oscillator potential, can be handled similarly, but have different asymptotic solutions.) In this case the dominant terms in the equation can be written as

$$\frac{d^2(rR)}{dr^2} \sim -\frac{2mE}{\hbar^2}(rR), \tag{4.11}$$

whose solutions are $rR \sim \exp(\pm r\sqrt{-2mE}/\hbar)$.

Now, if E were positive the argument of the exponential would be imaginary and we should have $|rR| = 1$. This would mean that R is not normalizable since

$$\int_{R^3} |R|^2 d^3\mathbf{r} = \int_{R^3} |R|^2 r^2 \sin\theta dr d\theta d\phi = 4\pi \int_0^\infty |rR|^2 dr. \tag{4.12}$$

(Near the origin rR must be less singular than $r^{-\frac{1}{2}}$ or the integral will diverge.) The same argument rules out the case of vanishing E so we conclude that E must be negative. For convenience we write $E = -\hbar^2\kappa^2/2m$, with $\kappa > 0$, so that

$$rR(r) \sim e^{-\kappa r}. \tag{4.13}$$

We shall therefore try to find an exact solution in the form

$$R(r) = \frac{f(r)}{r}e^{-\kappa r}. \tag{4.14}$$

Substitution into Schrödinger's equation yields

$$-\frac{\hbar^2}{2m}\left(f'' - 2\kappa f' + \kappa^2 f - \frac{l(l+1)}{r^2}\right) + Vf = \kappa^2 f, \tag{4.15}$$

or, after simplification,

$$f'' - 2\kappa f' - \frac{l(l+1)}{r^2} - \frac{2mV}{\hbar^2}f = 0. \tag{4.16}$$

For particular potentials, V, this equation can be solved in series.

4.3. The spectrum of the hydrogen atom

We are now ready to solve Schrödinger's equation for a hydrogen-like atom. It is useful to introduce the *Bohr radius*, $a = 4\pi\epsilon_0\hbar^2/me^2$, so that we may write the term $2mV/\hbar^2$ as $-2Z/ar$.

Proposition 4.3.1. The permissible bound state energies for the equation

$$-\frac{\hbar^2}{2m}\nabla^2\psi - \frac{Ze^2}{4\pi\epsilon_0 r}\psi = E\psi$$

are given by

$$E_n = -\frac{1}{2n^2}\cdot\frac{Z^2e^2}{4\pi\epsilon_0 a},$$

for $n = l+1, l+2, \ldots$. The corresponding wave functions take the form

$$\psi_{nlm}(r) = \text{constant}\cdot r^l L_n^l(Zr/a)e^{-Zr/na}Y_l^m(\theta,\phi)$$

where L_n^l is a polynomial of degree $n-l$. In particular, the normalized ground state wave function may be written as

$$\psi_{100}(r) = \left(Z^3/\pi a^3\right)^{\frac{1}{2}} e^{-Zr/a}.$$

Proof. The variables can be separated as in the previous discussion, giving the radial equation

$$\left(f'' - 2\kappa f' - \frac{l(l+1)}{r^2}\right) + \left(\frac{2Z}{ar}\right) f = 0. \tag{4.17}$$

In order to ease the calculation we change the variable to $\rho = Zr/a$. The equation then reduces to

$$\frac{d^2 f}{d\rho^2} - \frac{2a\kappa}{Z}\frac{df}{d\rho} - \frac{l(l+1)}{\rho^2} + \frac{2}{\rho} f = 0. \tag{4.18}$$

We now try the series solution

$$f(\rho) = \sum_{k=0}^{\infty} a_k \rho^{k+c} \tag{4.19}$$

with $a_0 \neq 0$. This gives

$$\sum_{k=0}^{\infty}(k+c)(k+c-1)a_k\rho^{k+c-2} - 2(a\kappa/Z)\sum_{k=0}^{\infty}(k+c)a_k\rho^{k+c-1}$$
$$- l(l+1)\sum_{k=0}^{\infty} a_k\rho^{k+c-2} + 2\sum_{k=0}^{\infty} a_k\rho^{k+c-1} = 0, \tag{4.20}$$

or, collecting terms,

$$\sum_{k=0}^{\infty}(k+c+l)(k+c-l-1)a_k\rho^{k+c-2}$$
$$= 2\sum_{k=0}^{\infty}\left[(k+c)(a\kappa/Z) - 1\right] a_k\rho^{k+c-1}. \tag{4.21}$$

Equating coefficients of ρ^{c-2} we obtain the indicial equation, $c(c-1) = l(l+1)$. This time there is no real choice, for, to ensure that $rR(r)$ gives a normalizable wave function, $|rR|^2$ must have a finite integral near $r = 0$, which is only possible if $2c > -1$, so we must take $c = l + 1$ rather than $c = -l$. (At this point in Schrödinger's original notebook there is the exclamation: 'The devil! It is finite at $r = 0$.') The recurrence relation for the coefficients is therefore

$$(k + 2l + 1)ka_k = 2[(a\kappa/Z)(k + l) - 1]a_{k-1}, \tag{4.22}$$

for $k \geq 1$. Arguing as for the harmonic oscillator we discover that, unless the series terminates, it behaves like $\exp(2\kappa r)$, wiping out the normalizability that we had tried to achieve with the factor of $\exp(-\kappa r)$. We must therefore force the series to terminate. Looking at the recurrrence relation for a_k we see that that will happen provided that $a\kappa = Z/n$ for some $n = (k + l) > l$.

Definition 4.3.1. The positive integer n is called the *principal quantum number.*

By substituting $\kappa = Z/na$, we deduce that the energy is given by

$$E_n = -\frac{\hbar^2\kappa^2}{2m} = -\frac{\hbar^2 Z^2}{2mn^2a^2}, \tag{4.23}$$

which, on using the definition of a, reduces to

$$E_n = -\frac{1}{2n^2} \cdot \frac{Z^2e^2}{4\pi\epsilon_0 a}. \tag{4.24}$$

More succinctly, we may write $E_1 = -Z^2e^2/8\pi\epsilon_0 a$ and $E_n = E_1/n^2$.

In the ground state, where $n = 1$, both l and μ must vanish, and the series for f terminates after the first term to give

$$\psi(r) = a_0 e^{-\kappa r} = a_0 e^{-Zr/a}. \tag{4.25}$$

In order to normalize the ground state wave function we require that

$$\begin{aligned}
1 &= \int_{\mathbf{R}^3} |\psi(r)|^2 r^2 \sin\theta dr d\theta d\phi \\
&= 4\pi \int_0^\infty |r\psi(r)|^2 \, dr \\
&= 4\pi |a_0|^2 \int_0^\infty r^2 e^{-2\kappa r} \, dr \\
&= \pi |a_0|^2 \frac{d^2}{d\kappa^2} \int_0^\infty e^{-2\kappa r} \, dr \\
&= \pi |a_0|^2 \frac{d^2}{d\kappa^2} \left[\frac{1}{2\kappa}\right] \\
&= \frac{\pi |a_0|^2}{\kappa^3} \\
&= \pi a^3 |a_0|^2 / Z^3.
\end{aligned}$$

The normalized ground state wave function may therefore be written as

$$\psi_{100}(r) = \left(Z^3/\pi a^3\right)^{\frac{1}{2}} e^{-Zr/a}. \tag{4.26}$$

In general the wave function ψ_{lmn} with principal quantum number n is the product of $\exp(-Zr/a)Y_l^m(\theta,\phi)$ with a polynomial $L_n^l(Zr/a)$ known as an associated Laguerre polynomial. Figure 4.1 shows the graphs of the first three wave functions. □

Bearing in mind the exponential terms in the wave functions, the Bohr radius can be taken as an estimate of the size of the atom. When the experimental values of ϵ_0 and of e are substituted one obtains the value $a \sim 5 \times 10^{-11}$ metres.

For large values of n the energy E_n converges upwards to 0, the minimum energy needed to escape from the nucleus altogether. (This is the energy corresponding to a parabolic orbit in the classical theory.) Strictly speaking, in the terminology of Definition 2.5.2 we have found the bound state energy levels; once the electron escapes from the nucleus one has to work with scattering states.

The energy difference between the levels with principal numbers j and k is

$$\left(\frac{1}{j^2} - \frac{1}{k^2}\right) E_1, \tag{4.27}$$

and this is the energy available to be radiated away as a photon if the wave function of the electron changes from ψ_j to ψ_k. By Planck's law the photon has frequency

$$\omega_{jk} = \left(\frac{1}{j^2} - \frac{1}{k^2}\right) \frac{E_1}{\hbar}. \tag{4.28}$$

Conversely a photon of this frequency can change the wave function from ψ_k to ψ_j. In other words the atom transmits and absorbs light only at certain well-defined frequencies. The series of frequencies corresponding to $j = 2$ was well known to spectroscopists as the Balmer series. The other series can also be measured to extremely high accuracy and provide a very sensitive verification of the predictions of quantum theory.

The energy $-E_j$ itself also has a physical interpretation as the minimum energy that must be supplied to an electron to enable it to escape from the atom starting from the state labelled by j. This is the ionization energy.

Finally we consider the degeneracy of the energy levels.

Theorem 4.3.2. The energy level E_n for the hydrogen atom has degeneracy n^2.

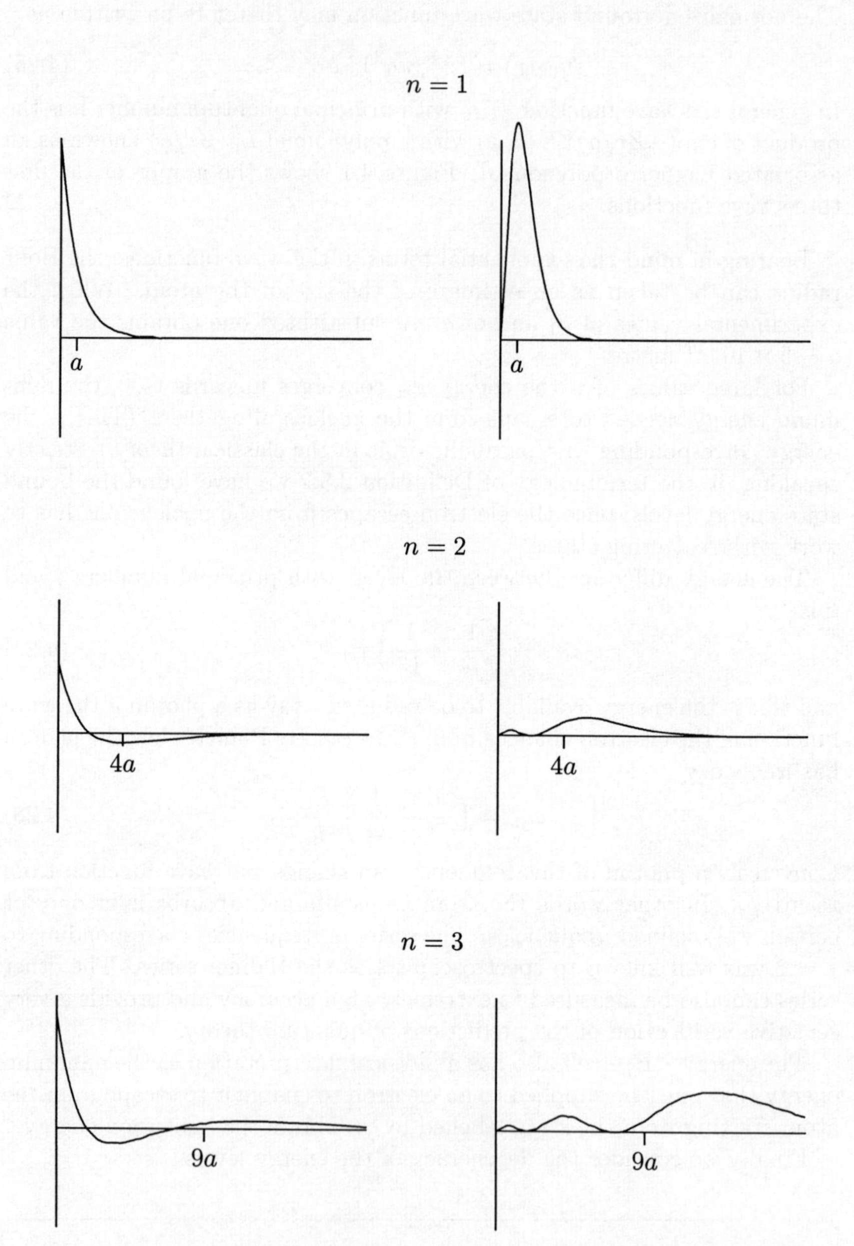

FIGURE 4.1. The wave functions and radial probability density as functions of r for the ground state and first two excited states of a hydrogen-like atom. The scale on the horizontal axis is the radius of the classical orbit that has the same energy.

Proof. We know that the space of spherical harmonics of degree l has dimension $2l+1$. Any degree l that is strictly less than n can give the same energy E_n so that we have a total degeneracy

$$\sum_{l=0}^{n-1}(2l+1) = \sum_{l=0}^{n-1}[(l+1)^2 - l^2] = \sum_{l=1}^{n} l^2 - \sum_{l=0}^{n-1} l^2 = n^2 - 0 = n^2, \quad (4.29)$$

as claimed. □

Exercises

4.1 Find those normalized wave functions for the first and second excited states of the hydrogen-like atom, for which l vanishes.

4.2° In a two-dimensional model of the hydrogen atom Schrödinger's time-independent equation becomes

$$-\frac{\hbar^2}{2m}\left[\frac{1}{r}\frac{\partial}{\partial r}\left(r\frac{\partial\psi}{\partial r}\right) + \frac{1}{r^2}\frac{\partial^2\psi}{\partial\theta^2}\right] - \frac{e^2}{4\pi\epsilon_0 r}\psi = E\psi.$$

By separating the equation in polar coordinates, show that the energy levels are of the form $-\kappa/(2N+1)^2$, for κ a positive constant, and $N = 0, 1, \ldots$. Find the degeneracy of each level.

4.3 Schrödinger's equation for a two-dimensional isotropic oscillator can be written in polar coordinates as

$$-\frac{\hbar^2}{2m}\left[\frac{1}{r}\frac{\partial}{\partial r}\left(r\frac{\partial\psi}{\partial r}\right) + \frac{1}{r^2}\frac{\partial^2\psi}{\partial\theta^2}\right] + \frac{1}{2}m\omega^2 r^2\psi = E\psi.$$

By considering solutions ψ that are separable in polar coordinates verify that the energy levels are of the form $n\hbar\omega$ where n is a positive integer. Find wave functions $\psi(r,\theta)$ for the two lowest energy levels.

4.4 Let $Y(\mathbf{r})$ be a harmonic function on $\mathbf{R}^3$ that is homogeneous of integral degree $l \geq 0$. Show that if $\psi = \phi(r)Y(\mathbf{r})$ is a solution of Schrödinger's equation for the hydrogen atom, then ϕ satisfies the equation

$$-\frac{\hbar^2}{2m}\left[\frac{1}{r}\frac{d^2}{dr^2}(r\phi) + \frac{2l}{r}\frac{d\phi}{dr}\right] - \frac{e^2}{4\pi\epsilon_0 r}\phi = E\phi.$$

By considering the asymptotic forms of solutions and making use of series show that the energy levels for bound states are

$$E_{k,l} = -\left(\frac{e^2}{4\pi\epsilon_0 a}\right)\frac{1}{2(l+k)^2},$$

for k a positive integer. Find a solution of the form $\psi(\mathbf{r}) = z\phi(r)$. [You may use the identities: $\nabla^2(Y\phi) = (\nabla^2 Y)\phi + 2\nabla Y.\nabla\phi + Y(\nabla^2\phi)$ and $\nabla\phi = \phi'(r)\mathbf{r}/r$.]

4.5° A particle of mass m moves in the spherically symmetric potential

$$\frac{\hbar^2\kappa^2}{2m}\left(\frac{a^2}{r} - r\right)^2,$$

where $\kappa > 0$. Write down the time-independent Schrödinger equation and show that if the wave function is written

$$r^{-1}e^{-\kappa r^2/2}f(r)Y_l^m(\theta,\phi)$$

then

$$f'' - 2\kappa r f' + \left(\frac{2mE}{\hbar^2} + 2\kappa^2 a^2 - \kappa - \frac{l(l+1)+\kappa^2 a^4}{r^2}\right)f = 0,$$

where E is the energy eigenvalue. Prove that the energy eigenvalues are given by

$$\frac{\hbar^2}{2m}(4\kappa n + A_l),$$

where $n = 0, 1, 2, \ldots$, and find A_l.

4.6° A particle of mass m moves in three dimensions under the influence of the spherically symmetric potential $\frac{1}{2}m\omega^2 r^2$. Using the results of the previous question, show that the energy levels are given by $E_N = (N + \frac{3}{2})\hbar\omega$, $N = 0, 1, 2, \ldots$, where N is even or odd according as l is even or odd.

4.7° In terms of the parabolic coordinates

$$u = r(1-\cos\theta),$$
$$v = r(1+\cos\theta),$$
$$w = \phi,$$

Schrödinger's equation for the hydrogen atom can be written as

$$-\frac{\hbar^2}{2m}\left\{\frac{4}{u+v}\left[\frac{\partial}{\partial u}\left(u\frac{\partial\psi}{\partial u}\right)+\frac{\partial}{\partial v}\left(v\frac{\partial\psi}{\partial v}\right)\right]+\frac{1}{uv}\frac{\partial^2\psi}{\partial w^2}\right\}$$

$$-\frac{Ze^2}{2\pi\epsilon_0(u+v)}\psi = E\psi.$$

By considering separable solutions $\psi = U(u)V(v)W(w)$ show that the bound states of the hydrogen-like atom have energies

$$E_n = -\frac{1}{2n^2}\cdot\frac{Z^2e^2}{4\pi\epsilon_0 a}$$

for positive integers n. What is the degeneracy of the energy level E_n?

5 Scattering and tunnelling

I am so happy to have escaped the terrible mechanics ... which I never really understood. Now everything is linear, everything can be superposed.

ERWIN SCHRÖDINGER, letter to Willy Wien, 22 February 1926

5.1. Particle beams

Ever since Rutherford's pioneering work, passing alpha rays through a metal foil target, beams of particles have provided a means of probing the fine details of subatomic structure. Modern accelerators hurl particles through a target at velocities only slightly less than that of light. The wavelength of each particle shrinks as its momentum increases and it becomes possible to distinguish features far beyond the resolving power of any microscope. In the simplest cases (such as Rutherford's sedate Edwardian experiments) the particles emerge from their encounter unscathed, and it is by analysing the changes to their momentum that one must build up a picture of the target. To develop some feeling for what happens we shall look at the simplest case in which the particle suffers no deflection. We therefore consider particles moving along a line on which there is some kind of potential barrier.

At large distances from the target the potential is approximately constant: V_L for large negative values of x, say, and V_R for large positive values of x. For simplicity we shall assume that for large enough x the potential is actually constant. Then Schrödinger's equation in these two regions becomes

$$-\frac{\hbar^2}{2m}\frac{d^2\psi_j}{dx^2} + V_j\psi_j = E\psi_j, \tag{5.1}$$

for $j = L$ or R. That is,

$$\frac{d^2\psi_j}{dx^2} = -\frac{2m}{\hbar^2}(E - V_j)\psi_j. \tag{5.2}$$

This has solutions of the form

$$\psi_j = A_j e^{ik_j x} + B_j e^{-ik_j x} \tag{5.3}$$

where $k_j = \sqrt{2m(E - V_j)/\hbar^2}$ and A_j and B_j are constants of integration. (Clearly, k_j is real or imaginary according to whether $E \geq V_j$ or $E < V_j$.

When $E \geq V_j$ we take k_j to be the positive square root of $2m(E - V_j)/\hbar^2$; otherwise we set $k_j = i\kappa_j$ with κ_j positive.)

According to de Broglie's law the term $\exp(ik_L x)$ has momentum $\hbar k_L > 0$, and so represents a wave moving to the right, towards the target. Similarly, the term $\exp(-ik_L x)$ represents a wave moving left, away from the target. To the right of the target these roles are reversed and it is the right moving wave, $\exp(ik_R x)$, which is leaving the target, whilst the left moving wave, $\exp(-ik_R x)$, approaches it. Our main aim in this chapter will be to find the coefficients A_R and B_L of the outgoing waves in terms of the coefficients A_L and B_R of the incident waves. In practice we are very often interested in the situation in which the incident beam is fired at the target from the left. Since no beam is incident from the right (and there are no further targets to cause reflections) the coefficient B_R must vanish, and we have to find A_R and B_L in terms of A_L. (Strictly speaking this interpretation works only when $E \geq V_R$ and k_R is real. However, when $E < V_R$ and $k_R = i\kappa_R$ the wave function takes the form $A_R \exp(-\kappa_R x) + B_R \exp(\kappa_R x)$. This time we have to take $B_R = 0$ to avoid the embarrassment of an exponentially increasing wave function and probability density to the right of the barrier.)

Within the target the potential changes very rapidly over distances so short that it is often convenient to approximate the change by a jump, or discontinuity, in the potential. This raises the question of what happens to the wave function at the jump. We shall impose the following matching condition:

Assumption 5.1.1. The wave function, ψ, and its derivative, ψ', are continuous at a potential jump.

Remark 5.1.1. This is a reasonable requirement; elsewhere this continuity is an automatic consequence of the fact that the wave function must be twice differentiable in order for the Schrödinger equation to make sense.

5.2. Reflection and transmission coefficients

A useful tool in the discussion of beams is the probability current, which in one dimension (Exercise 2.8) is given by

$$j(x,t) = \hbar(\overline{\psi}\psi' - \psi\overline{\psi'})/2mi. \tag{5.4}$$

Proposition 5.2.1. For time-independent problems in one dimension the probability current is a constant.

Proof. The continuity of ψ and ψ' shows that the current is well defined and continuous across a potential jump. Away from discontinuities in the potential we may use Schrödinger's equation, and its consequence, the continuity equation, for the current. For time-independent problems, such as a steady beam, the continuity equation reduces to $j' = 0$, so that the current is a constant. For continuity across potential jumps that constant must be the same in all regions. □

When the wave function takes the special form $A\exp(ikx)+B\exp(-ikx)$ with k real, we may readily compute the probability current to be

$$\begin{aligned} j &= \frac{\hbar}{m}\mathrm{Re}\left[\left(\overline{A}e^{-ikx} + \overline{B}e^{ikx}\right) k \left(Ae^{ikx} - Be^{-ikx}\right)\right] \\ &= \frac{\hbar k}{m}\mathrm{Re}\left[(|A|^2 - |B|^2) + (\overline{B}Ae^{2ikx} - \overline{A}Be^{-2ikx})\right] \\ &= \hbar k \left(|A|^2 - |B|^2\right)/m. \end{aligned} \tag{5.5}$$

(This is the one-dimensional version of Exercise 2.9.)

In practice the potential encountered by a particle in an accelerator as it passes through the target is likely to be very complicated and may have several discontinuities. The constancy of the current nonetheless enables us to keep track of what happens.

Corollary 5.2.2. Suppose that for large negative x the wave function takes the form $A_L\exp(ik_Lx) + B_L\exp(-ik_Lx)$ and for large positive x the form $A_R\exp(ik_Rx) + B_R\exp(-ik_Rx)$, with k_L and k_R positive real numbers. Then

$$|A_L|^2 + \frac{k_R}{k_L}|B_R|^2 = |B_L|^2 + \frac{k_R}{k_L}|A_R|^2.$$

If the beam is incident from the left, so that $B_R = 0$, then

$$|A_L|^2 = |B_L|^2 + \frac{k_R}{k_L}|A_R|^2.$$

Proof. Since the current is constant its values on the extreme left and extreme right must agree. For large negative x it takes the value $\hbar k_L(|A_L|^2 - |B_L|^2)/m$, and for large positive x it takes the value $\hbar k_R(|A_R|^2 - |B_R|^2)/m$. This gives

$$k_L(|A_L|^2 - |B_L|^2) = k_R(|A_R|^2 - |B_R|^2), \tag{5.6}$$

from which the result follows on rearrangement of the terms. □

This result prompts the following definition:

Definition 5.2.1. For a beam incident from the left, the *reflection coefficient* is defined to be $|B_L/A_L|^2$ and the *transmission coefficient* is $(k_R/k_L)|A_R/A_L|^2$.

For beams incident from the left the previous corollary can then be restated as follows:

Corollary 5.2.3. The sum of the reflection and transmission coefficients is 1.

Remark 5.2.1. This result tells us that all the particles in the beam are either reflected or transmitted; none of them gets stuck in the target. (Given the enormous cost of producing the beams in modern accelerators, it would be annoying to lose anything.)

In the commonest examples the potential tends to the same value for large positive and negative values of x, so that $k_L = k_R$ and the transmission coefficient reduces to $|A_R/A_L|^2$.

5.3. Potential jumps

As the simplest possible problem involving a discontinuity, suppose that the potential is given by

$$V(x) = \begin{cases} V_0 & \text{if } x < b \\ V_1 & \text{if } x \geq b, \end{cases} \tag{5.7}$$

where V_0 and V_1 are constant.

As before, we have for $j = 0, 1$

$$\psi_j = A_j e^{ik_j x} + B_j e^{-ik_j x}, \tag{5.8}$$

where $k_j = \sqrt{2m(E - V_j)/\hbar^2}$ and A_j and B_j are constants of integration. There is a single jump so we just have the two equations

$$\begin{aligned} \psi_0(b) &= \psi_1(b) \\ \psi_0'(b) &= \psi_1'(b). \end{aligned} \tag{5.9}$$

On substituting the expressions for ψ_0 and ψ_1 these become

$$\begin{aligned} A_0 e^{ik_0 b} + B_0 e^{-ik_0 b} &= A_1 e^{ik_1 b} + B_1 e^{-ik_1 b} \\ ik_0 A_0 e^{ik_0 b} - ik_0 B_0 e^{-ik_0 b} &= ik_1 A_1 e^{ik_1 b} - ik_1 B_1 e^{-ik_1 b}. \end{aligned} \tag{5.10}$$

The latter equation may be rewritten as

$$A_0 e^{ik_0 b} - B_0 e^{-ik_0 b} = \frac{k_1}{k_0} \left(A_1 e^{ik_1 b} - B_1 e^{-ik_1 b} \right). \tag{5.11}$$

Adding this to the first equation we obtain

$$A_0 = \frac{1}{2k_0} (k_0 + k_1) A_1 e^{i(k_1 - k_0)b} + \frac{1}{2k_0} (k_0 - k_1) B_1 e^{-i(k_0 + k_1)b}. \tag{5.12}$$

Similarly by subtraction we arrive at

$$B_0 = \frac{1}{2k_0} (k_0 - k_1) A_1 e^{i(k_0 + k_1)b} + \frac{1}{2k_0} (k_0 + k_1) B_1 e^{i(k_0 - k_1)b}. \tag{5.13}$$

Proposition 5.3.1. Let $s_{01} = k_0 + k_1$, $d_{01} = k_0 - k_1$,

$$M_{01}(b) = \frac{1}{2k_0} \begin{pmatrix} s_{01} e^{-id_{01}b} & d_{01} e^{-is_{01}b} \\ d_{01} e^{is_{01}b} & s_{01} e^{id_{01}b} \end{pmatrix} \quad \text{and} \quad C_j = \begin{pmatrix} A_j \\ B_j \end{pmatrix},$$

for $j = 0, 1$. Then the waves on either side of the potential jump are related by

$$C_0 = M_{01}(b) C_1.$$

Proof. The previous formulae for A_0 and B_0 may be combined as

$$\begin{pmatrix} A_0 \\ B_0 \end{pmatrix} = \frac{1}{2k_0} \begin{pmatrix} s_{01} e^{-id_{01}b} A_1 + d_{01} e^{-is_{01}b} B_1 \\ d_{01} e^{is_{01}b} A_1 + s_{01} e^{id_{01}b} B_1 \end{pmatrix}, \tag{5.14}$$

from which the result immediately follows. □

5.4. Multiple potential barriers

For a single potential jump the matrix notation that we introduced in the previous section is unnecessarily sophisticated since the problem is easily solved by direct methods. However, as we have already remarked, the particles in a scattering experiment are usually subjected to more than one jump. A slightly more realistic model of the potential barrier encountered in a particle accelerator would have a double potential jump of the form

$$V(x) = \begin{cases} V_0 & \text{if } x < 0, \\ V_1 & \text{if } 0 \leq x < a, \\ V_2 & \text{if } x \geq a, \end{cases} \tag{5.15}$$

where V_0, V_1 and V_2 are constants (Figure 5.1). (The region where the potential is V_1 corresponds to the target.)

We shall suppose that a beam of particles $A_0 \exp(ik_0x)$ is incident on this potential barrier from the left. Some of the beam will be reflected from the barrier, so that for negative x the wave function will have the form

$$\psi_0 = A_0 e^{ik_0x} + B_0 e^{-ik_0x}. \tag{5.16}$$

In the notation of the last section this is related to the wave function in the region $[0, a]$ by

$$C_0 = M_{01}(0)C_1. \tag{5.17}$$

Similarly, extending the notation in the obvious way, we have at $x = a$ the second identity

$$C_1 = M_{12}(a)C_2, \tag{5.18}$$

so that overall

$$C_0 = M_{01}(0)M_{12}(a)C_2. \tag{5.19}$$

The form of the wave function to the right of $x = a$ is $A_2 \exp(ik_2x) + B_2 \exp(-ik_2x)$, but since no beam is incident from the right we must have $B_2 = 0$. That means that

$$C_2 = \begin{pmatrix} A_2 \\ 0 \end{pmatrix} = A_2 \begin{pmatrix} 1 \\ 0 \end{pmatrix}. \tag{5.20}$$

Consequently we have

$$C_0 = A_2 M_{01}(0)M_{12}(a) \begin{pmatrix} 1 \\ 0 \end{pmatrix}. \tag{5.21}$$

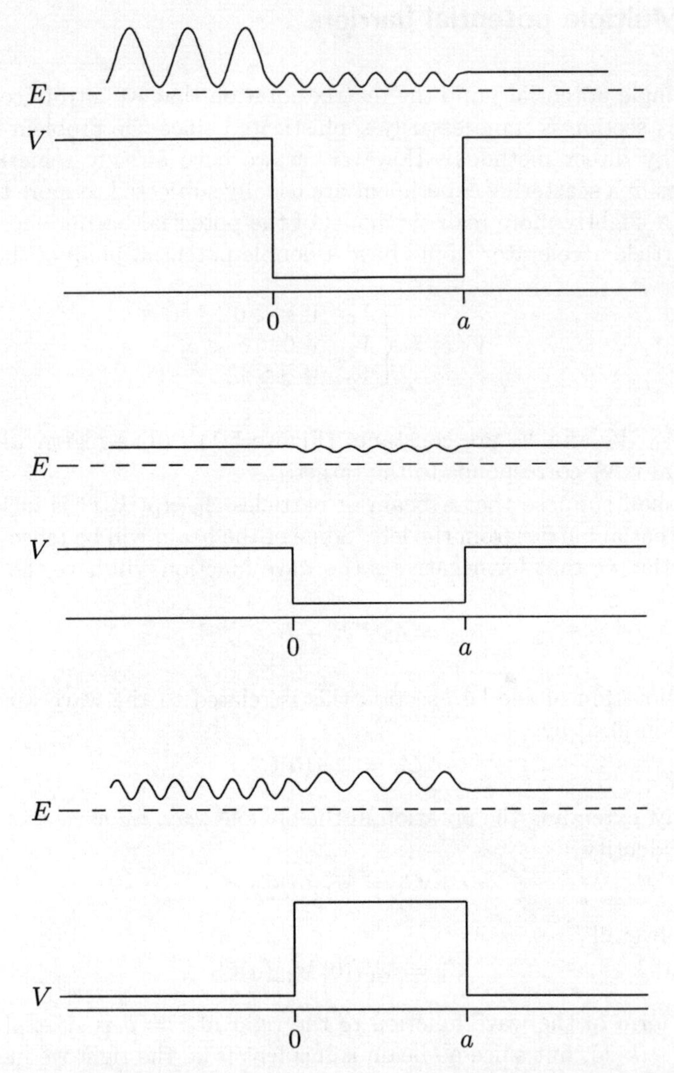

FIGURE 5.1. The probability density of a beam incident from the left on a potential well (top two graphs) or barrier (bottom picture). (The dashed line giving the energy level also serves as the baseline for the graph of probablility density.) To the left of the jumps there is usually interference between the incident and reflected beams, though, as the middle picture shows, for certain sizes of well there may be resonance leading to perfect transmission. To the right there is only a transmitted beam, whose amplitude is constant. In a well the potential energy is lower, so classically the particle would speed up, and so it would be less likely to be found there. For a barrier the reverse would be true.

By multiplying out the various matrices on the right-hand side we can now find expressions for A_2 and B_0 in terms of A_0. In fact we have

$$\begin{aligned}
\begin{pmatrix} A_0 \\ B_0 \end{pmatrix} &= A_2 \frac{1}{4k_0k_1} \begin{pmatrix} s_{01} & d_{01} \\ d_{01} & s_{01} \end{pmatrix} \begin{pmatrix} s_{12}e^{-id_{12}a} & d_{12}e^{-is_{12}a} \\ d_{12}e^{is_{12}a} & s_{12}e^{id_{12}a} \end{pmatrix} \begin{pmatrix} 1 \\ 0 \end{pmatrix} \\
&= \frac{A_2}{4k_0k_1} \begin{pmatrix} s_{01} & d_{01} \\ d_{01} & s_{01} \end{pmatrix} \begin{pmatrix} s_{12}e^{-id_{12}a} \\ d_{12}e^{is_{12}a} \end{pmatrix} \\
&= \frac{A_2}{4k_0k_1} \begin{pmatrix} s_{01}s_{12}e^{-id_{12}a} + d_{01}d_{12}e^{is_{12}a} \\ d_{01}s_{12}e^{-id_{12}a} + s_{01}d_{12}e^{is_{12}a} \end{pmatrix}. \qquad (5.22)
\end{aligned}$$

Proposition 5.4.1. If $E \geq V_1$ so that k_1 is real, then

$$\begin{pmatrix} A_0 \\ B_0 \end{pmatrix} = \frac{A_2}{2k_0k_1} e^{ik_2a} \begin{pmatrix} (k_0 + k_2)k_1 \cos k_1a - i(k_2k_0 + k_1^2) \sin k_1a \\ (k_0 - k_2)k_1 \cos k_1a - i(k_2k_0 - k_1^2) \sin k_1a \end{pmatrix}.$$

If the energy $E \geq \max\{V_0, V_1, V_2\}$ so that all three k are real, the reflection coefficient is given by

$$\left|\frac{B_0}{A_0}\right|^2 = \frac{(k_0 - k_2)^2k_1^2 \cos^2 k_1a + (k_2k_0 - k_1^2)^2 \sin^2 k_1a}{(k_0 + k_2)^2k_1^2 \cos^2 k_1a + (k_2k_0 + k_1^2)^2 \sin^2 k_1a},$$

and the transmission coefficient by

$$\frac{k_2|A_2|^2}{k_0|A_0|^2} = \frac{4k_0k_2k_1^2}{(k_0 + k_2)^2k_1^2 \cos^2 k_1a + (k_2k_0 + k_1^2)^2 \sin^2 k_1a}.$$

Proof. Taking the formula for C_0 and substituting for each s and d in terms of the k_j we obtain

$$C_0 = \frac{A_2}{4k_0k_1} e^{ik_2a} \begin{pmatrix} (k_0 + k_1)(k_1 + k_2)e^{-ik_1a} + (k_0 - k_1)(k_1 - k_2)e^{ik_1a} \\ (k_0 - k_1)(k_1 + k_2)e^{-ik_1a} + (k_0 + k_1)(k_1 - k_2)e^{ik_1a} \end{pmatrix}. \qquad (5.23)$$

When $E \geq V_1$ this can be rewritten in the stated form

$$C_0 = \frac{A_2}{2k_0k_1} e^{ik_2a} \begin{pmatrix} (k_0 + k_2)k_1 \cos k_1a - i(k_2k_0 + k_1^2) \sin k_1a \\ (k_0 - k_2)k_1 \cos k_1a - i(k_2k_0 - k_1^2) \sin k_1a \end{pmatrix}. \qquad (5.24)$$

Taking the ratio of components on the right-hand side we obtain the reflection coefficient:

$$\left|\frac{B_0}{A_0}\right|^2 = \left|\frac{(k_0 - k_2)k_1 \cos k_1 a - i(k_2 k_0 - k_1^2) \sin k_1 a}{(k_0 + k_2)k_1 \cos k_1 a - i(k_2 k_0 + k_1^2) \sin k_1 a}\right|^2 . \tag{5.25}$$

If $E \geq \max\{V_0, V_1, V_2\}$ this reduces further to the form given. Similarly the transmission coefficient may be found by comparison of the first components in the formula for C_0. □

Remark 5.4.1. One may now check directly that their sum is 1, in accordance with Corollary 5.2.3.

5.5. Tunnelling

In many situations of practical significance the middle potential V_1 is greater than the energy of the incident beam, so that $k_1 = i\kappa_1$. The algebraic formula connecting the wave functions on the left- and right-hand sides of the target is still valid for imaginary k_1 and it now gives the following expression for C_0:

$$\frac{A_2}{4k_0\kappa_1} e^{ik_2 a} \begin{pmatrix} (k_0 + i\kappa_1)(\kappa_1 - ik_2)e^{\kappa_1 a} + (k_0 - i\kappa_1)(\kappa_1 + ik_2)e^{-\kappa_1 a} \\ (k_0 - i\kappa_1)(\kappa_1 - ik_2)e^{\kappa_1 a} + (k_0 + i\kappa_1)(\kappa_1 + ik_2)e^{-\kappa_1 a} \end{pmatrix}. \tag{5.26}$$

This simplifies to give

$$\frac{A_2}{2k_0\kappa_1} e^{ik_2 a} \begin{pmatrix} (k_0 + k_2)\kappa_1 \cosh \kappa_1 a + i(\kappa_1^2 - k_0 k_2) \sinh \kappa_1 a \\ (k_0 - k_2)\kappa_1 \cosh \kappa_1 a - i(\kappa_1^2 + k_0 k_2) \sinh \kappa_1 a \end{pmatrix}. \tag{5.27}$$

The reflection and transmission coefficients may be calculated exactly as before. (For simplicity we consider only the case when k_0 and k_2 are real.) The most interesting point is the fact that the transmission coefficient

$$\frac{k_2|A_2|^2}{k_0|A_0|^2} = \frac{4k_0 k_2 \kappa_1^2}{(k_0 + k_2)^2 \kappa_1^2 \cosh^2 \kappa_1 a + (\kappa_1^2 - k_0 k_2)^2 \sinh^2 \kappa_1 a} \tag{5.28}$$

does not vanish.

This example is of far more profound significance than the simplicity of the above calculation might suggest. It lies behind many of the most important quantum devices. The crucial point is that a beam of classical particles of energy E would have insufficient energy to penetrate a barrier of potential $V_1 > E$ at all; they would all be reflected back like a ball

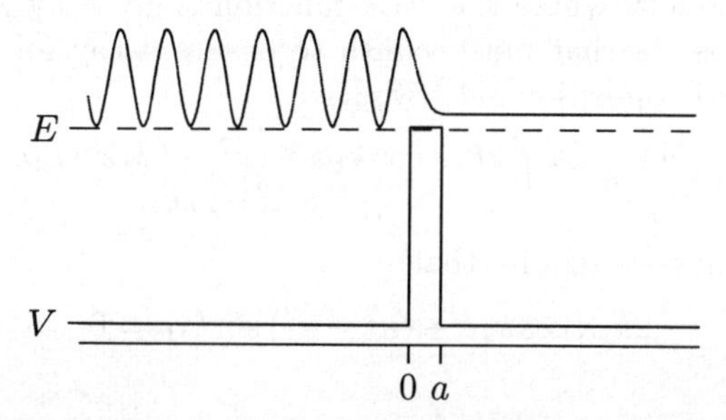

FIGURE 5.2. A beam that is incident on a barrier from the left can tunnel through, even though classically it would lack the energy to traverse the obstacle.

hitting a high wall. Thanks to their wave properties quantum particles can 'tunnel' through a barrier too large for them to surmount (see Figure 5.2).

Devices that exploit this phenomenon are numerous and include the nuclear reactor, the transistor and the superconducting Josephson junction. A recent application occurs in the scanning tunnelling electron microscope (STEM), for which its inventors, Gerd Binnig and Heinrich Rohrer, received the 1986 Nobel Prize in Physics. In this device a fine tungsten needle, its tip sharpened to a point only one atom across, is brought within nanometres of the sample to be investigated (1 nanometre $= 10^{-9}$ m). Electrons can tunnel across the short gap from the sample to the needle producing a detectable current which depends sensitively on the size of the gap. By measuring the current and varying the height of the needle as it is drawn across the sample in a carefully controlled way a computer relief map of the sample can be built up. The device can easily resolve individual atoms and promises to be of enormous use in surface chemistry. More recently it has been used to manipulate individual atoms too.

5.6. Potential wells

This method is not only applicable to problems involving beams. Consider, for example, a particle in a potential well

$$V = \begin{cases} V_0 & \text{if } x \notin [0, a], \\ 0 & \text{if } x \in [0, a], \end{cases} \tag{5.29}$$

where $V_0 > E > 0$. This time k_1 is real and we may write $k_0 = k_2 = i\kappa$, with κ positive. The wave function to the right of the well is $A_2 \exp(-\kappa x) + B_2 \exp(\kappa x)$. Physically one would expect the wave function and probability to decay exponentially in the region of high potential, so we take $B_2 = 0$.

Similarly on the left, where the wave function is given by $A_0 \exp(-\kappa x) + B_0 \exp(\kappa x)$, it is A_0 that must vanish to ensure decay for large negative values of x. By Proposition 5.4.1 we have

$$\begin{pmatrix} 0 \\ B_0 \end{pmatrix} = \frac{A_2}{2k_1\kappa} e^{-\kappa a} \begin{pmatrix} 2k_1\kappa \cos k_1 a + (\kappa^2 - k_1^2) \sin k_1 a \\ (\kappa^2 + k_1^2) \sin k_1 a \end{pmatrix}. \tag{5.30}$$

For consistency we require that

$$2k_1\kappa \cos k_1 a + (\kappa^2 - k_1^2) \sin k_1 a = 0, \tag{5.31}$$

that is

$$(k_1^2 - \kappa^2) \tan k_1 a = 2k_1\kappa. \tag{5.32}$$

Now, we also know that

$$-\frac{\hbar^2\kappa^2}{2m} + V_0 = E = \frac{\hbar^2 k_1^2}{2m}, \tag{5.33}$$

so that κ and k_1 are related by

$$\kappa^2 + k_1^2 = \frac{2m}{\hbar^2} V_0. \tag{5.34}$$

One can now eliminate κ between the equations to obtain a condition on k_1 for solutions to exist. For very large $V_0 \gg E$ we have $\kappa \gg k_1$ so that

$$\tan k_1 a = \frac{2k_1\kappa}{k_1^2 - \kappa^2} \tag{5.35}$$

is very small. In the limit of infinite κ this forces $k_1 a = n\pi$ for some integer n, and energies

$$E_n = \frac{\hbar^2 n^2 \pi^2}{2ma^2}. \tag{5.36}$$

These are precisely the energies of the particle in a square well potential that we derived in Section 2.2. This is reasonable, since the enormous potential jumps at 0 and a should effectively confine the particle within the interval $[0, a]$.

5.7. The scattering matrix

For a complicated succession of potential steps it is more efficient to adopt a purely matrix approach rather than following the progress successively across each barrier. We already know that the coefficients change from left to right across a series of barriers according to the formula

$$C_L = MC_R, \tag{5.37}$$

where M denotes a suitable product of matrices of the form $M_{jk}(b)$. In other words we may capture the overall effect of a succession of potential jumps, no matter how complicated, in a single matrix equation.

Remark 5.7.1. By direct calculation,

$$\det\left(M_{01}(b)\right) = \left(s_{01}^2 - d_{01}^2\right)/4k_0^2 = k_1/k_0. \tag{5.38}$$

Since the outgoing k from one step becomes the input to the next there is a pairwise cancellation of terms to give $\det(M) = k_R/k_L$, where k_R and k_L are the values of k appropriate to the extreme right and left of the barriers. If, as is often the case, the potential is the same at the two extremes, then $\det(M) = 1$.

Although M is the most useful matrix for one-dimensional problems, there is another matrix that generalizes more readily to higher dimensions. The four coefficients A_L, B_L, A_R, B_R appearing in C_L and C_R fall into two pairs: A_L and B_R are associated with waves incident on the barrier, whilst A_R and B_L describe waves leaving the barrier.

Definition 5.7.1. The *scattering matrix* S connects the incoming and outgoing coefficients by the formula

$$\begin{pmatrix} B_L \\ A_R \end{pmatrix} = S \begin{pmatrix} A_L \\ B_R \end{pmatrix}.$$

By comparison with the earlier equation one can find the entries of S in terms of those of M. Indeed if

$$\begin{pmatrix} A_R \\ B_R \end{pmatrix} = \begin{pmatrix} M_{aa} & M_{ab} \\ M_{ba} & M_{bb} \end{pmatrix} \begin{pmatrix} A_L \\ B_L \end{pmatrix}, \tag{5.39}$$

then

$$\begin{pmatrix} B_L \\ A_R \end{pmatrix} = M_{bb}^{-1} \begin{pmatrix} -M_{ba} & 1 \\ \det(M) & M_{ab} \end{pmatrix} \begin{pmatrix} A_L \\ B_R \end{pmatrix}. \tag{5.40}$$

Theorem 5.7.1. If k_L and k_R are real and equal, then the scattering matrix S is unitary, that is $S^*S = 1$.

Proof. If k_L and k_R are real and equal then Corollary 4.2.2 simplifies to give

$$|A_L|^2 + |B_R|^2 = |B_L|^2 + |A_R|^2. \tag{5.41}$$

This can be rewritten as

$$(\overline{A_L} \ \ \overline{B_R})\begin{pmatrix} A_L \\ B_R \end{pmatrix} = (\overline{B_L} \ \ \overline{A_R})\begin{pmatrix} B_L \\ A_R \end{pmatrix} = (\overline{A_L} \ \ \overline{B_R})\, S^* S \begin{pmatrix} A_L \\ B_R \end{pmatrix}. \tag{5.42}$$

Since this is true for all choices of A_L and B_R the matrix S must satisfy $S^*S = 1$, and is therefore unitary. □

Remark 5.7.2. When k_L and k_R are not real the matrix S can be far from unitary. Indeed, if one applies the same mathematical ideas to the case of the finite potential well described in the previous section then we see from the fact that both A_L and B_R vanish that the scattering matrix S must be singular. In fact, considered as a function of the incident momentum k_1, the scattering matrix has poles at the bound state values of k_1. This provides a useful link between the scattering matrix and the bound state energies.

Exercises

5.1 Rederive the results of Proposition 5.4.1 by direct solution of the four matching conditions at the potential jumps.

5.2 Calculate the transmission and reflection coefficients for a beam of particles incident on the following potential barriers from $x = -\infty$:
(a)

$$V(x) = \begin{cases} V_0 & \text{if } x \in [0, a]; \\ 0 & \text{if } x \notin [0, a]. \end{cases}$$

(b)

$$V(x) = \begin{cases} 0 & \text{if } x \in [0, a]; \\ V_0 & \text{if } x \notin [0, a]. \end{cases}$$

If $k^2 = 2mE/\hbar^2$ and $k_0^2 = 2m(E - V_0)/\hbar^2$ are fixed calculate the values of a for which the transmission coefficient has its maximum and minimum values in each case.

5.3 Show that the matrix $M_{01}(b)$ of Proposition 5.3.1 can be written in the form

$$M_{01}(b) = U_0(b)\left(uu^* + \frac{k_1}{k_0} vv^*\right) U_1(b)^*$$

where

$$u = \frac{1}{\sqrt{2}}\begin{pmatrix} 1 \\ 1 \end{pmatrix}, \qquad v = \frac{1}{\sqrt{2}}\begin{pmatrix} 1 \\ -1 \end{pmatrix}$$

are an orthonormal basis of $\mathbf{C}^2$, and

$$U_j(b) = \begin{pmatrix} e^{-ik_jb} & 0 \\ 0 & e^{ik_jb} \end{pmatrix}.$$

5.4° A beam with energy $\hbar^2k^2/2m$ and density $|A|^2$ is incident from large positive values of x, parallel to the x-axis, on a potential barrier of the form

$$V(x) = \begin{cases} 0 & \text{if } x > a, \\ -V_0 & \text{if } 0 < x < a, \\ \infty & \text{if } x < 0, \end{cases}$$

where V_0 is a positive constant. Show that the wave function for $x > a$ can be written as

$$\psi(x) = A\left(e^{-ikx} + e^{i(kx+\phi)}\right)$$

and find $\exp(i\phi)$.

5.5 A particle of mass m moves in a potential

$$V(x) = \begin{cases} V_0 & \text{if } x \notin [0,a], \\ 0 & \text{if } x \in [0,a]. \end{cases}$$

Writing the bound state energies in the form $E = \hbar^2k^2/2m$, derive a formula for k directly from Schrödinger's equation and the continuity conditions. Check that your formula is consistent with equation (5.35).

5.6 A particle of mass m moves in the three-dimensional energy well

$$V(r) = \begin{cases} 0 & \text{if } 0 \le r \le a \\ V_0 & \text{if } r > a. \end{cases}$$

Show that Schrödinger's equation has a continuous solution, $\psi(r)$, with energy $E = \hbar^2k^2/2m < V_0$ provided that

$$k^2 = \frac{2mV_0}{\hbar^2}\sin^2(ka).$$

Show that for very large V_0 this gives energy levels close to those in Exercise 2.4.

5.7 Find the probability current for the wave function $\psi = A\exp(-\kappa x) + B\exp(\kappa x)$ when κ is real.

5.8° Let $V(x) = 0$, $|x| > R$, and the wave function $\psi(x)$ corresponding to particles of momentum $\hbar k$ incident on the potential from $-\infty$ be given by

$$\psi(x) = \begin{cases} e^{ikx} + ae^{-ikx}, & x < -R, \\ be^{ikx}, & x > R. \end{cases}$$

Show that there is another solution

$$\chi(x) = \begin{cases} be^{-ikx}, & x < -R, \\ e^{-ikx} - (b\overline{a}/\overline{b})e^{ikx}, & x > R. \end{cases}$$

Deduce that the reflection and transmission coefficients are the same for particles incident from $-\infty$ as for particles incident from $+\infty$.

5.9 Use the scattering matrix with $B_R = 0$ to show that

$$\begin{pmatrix} A_R \\ B_L \end{pmatrix} = \frac{A_L}{M_{bb}} \begin{pmatrix} M_{ba} \\ \det M \end{pmatrix}.$$

Find the scattering matrix for the square well of Section 5.6. Show that as a function of κ it has poles where equation (5.31) is satisfied.

6 The mathematical structure of quantum theory

> You are only going to spoil Heisenberg's physical ideas with your futile mathematics.
>
> WOLFGANG PAULI to Max Born, 19 July 1925

6.1. States and observables

We have already mentioned that Schrödinger's formulation of quantum mechanics was slightly preceded by a totally different algebraic formulation proposed by Heisenberg. Although Schrödinger's methods were so much simpler that they were generally preferred, both approaches seemed to give the same answers, so it was natural to ask how they were related. This was particularly important because some features of quantum theory were easier to understand in Heisenberg's formalism. In fact, as we have already noted, Schrödinger was able to show that the two methods were essentially equivalent. Soon mathematicians like Hermann Weyl and John von Neumann were able to find a general mathematical scheme that encompassed both approaches, and it is this that we shall now describe.

There are two particularly important concepts in a physical theory: the *states* of the system, which we have so far represented by wave functions, and the *observables*, quantities like position, momentum and energy, that we might want to measure. The theory links these in the form of rules that tell us how to calculate probabilities of events or to find the expectation value of a certain observable when the system is in a given state.

Definition 6.1.1. The states are described by non-zero vectors in a complex inner product space $\mathcal{H}$, and two vectors describe the same state if and only if one is a multiple of the other.

The space $\mathcal{H}$ is chosen to suit the particular problem under consideration, and it is usually subject to some additional technical restrictions that we shall discuss in the next section. Although $\mathcal{H}$ can be finite dimensional as in the matrix treatment of scattering theory, it is more often infinite dimensional. For example, $\mathcal{H}$ is often the vector space of complex-valued wave functions ψ for which

$$\int_{\mathbf{R}^3} |\psi(\mathbf{x})|^2 \, d^3\mathbf{x} < \infty \tag{6.1}$$

(to ensure normalizability), and with the inner product defined by

$$\langle \phi | \psi \rangle = \int_{\mathbf{R}^3} \overline{\phi(\mathbf{x})} \psi(\mathbf{x}) \, d^3\mathbf{x}. \tag{6.2}$$

The fact that we have separated the two vectors by a vertical line rather than a comma accords with the notation used in physics and will serve as a reminder that we are following a physicists' convention that makes inner products linear in the *second* variable and conjugate linear in the first. (Although different from that current in most algebra texts, this convention is gradually gaining ground amongst mathematicians too, and it is almost universal in textbooks on quantum theory. Physicists, following Dirac, often emphasize the inner product structure by writing the vectors inside pointed brackets, for example $|\psi\rangle$, or $\langle\phi|$ for an element of the dual space, but we shall not usually bother with that.)

Definition 6.1.2. A vector ψ is said to be *normalized* if

$$\|\psi\|^2 = \langle \psi | \psi \rangle = 1.$$

In the case of wave functions this accords with the previous definition, equation (2.17). The fact that vectors that differ only by multiples describe the same physical state means that the physics is unaffected when we normalize a vector by multiplying by a suitable constant. Apart from rescaling states and multiplying them by phase factors such as $\exp(-iEt/\hbar)$, the vector space structure permits addition of vectors, which makes it possible to account for superposition and interference effects.

Observables are described by certain kinds of linear transformation. We should first recall that, in algebra, the adjoint of a linear transformation A is defined as the unique linear transformation A^* that satisfies

$$\langle A^* \phi | \psi \rangle = \langle \phi | A \psi \rangle, \tag{6.3}$$

for all ϕ and ψ in $\mathcal{H}$, and that A is called self-adjoint if $A = A^*$, that is if

$$\langle \phi | A \psi \rangle = \langle A^* \phi | \psi \rangle \tag{6.4}$$

for all ϕ and ψ in $\mathcal{H}$. In infinite dimensions this definition is really rather restrictive, so we shall refine it in the next section, once we have looked at its application.

Definition 6.1.3. The *observables* in quantum mechanics are described by self-adjoint linear transformations on $\mathcal{H}$.

Usually one refers to linear transformations in quantum mechanics as operators and to these as self-adjoint operators. Typical examples in Schrödinger's theory are the position and momentum operators:

Definition 6.1.4. The *position operator* $\mathbf{X}$ in three dimensions has components X_j for $j = 1, 2, 3$ that are defined on vectors ψ in $\mathcal{H}$ by

$$(X_j\psi)(\mathbf{x}) = x_j\psi(\mathbf{x});$$

the *momentum operator* $\mathbf{P}$ has components P_j for $j = 1, 2, 3$ that are defined on differentiable ψ by

$$(P_j\psi)(\mathbf{x}) = \frac{\hbar}{i}\frac{\partial\psi}{\partial x_j}(\mathbf{x});$$

each being defined for $j = 1, 2, 3$. In one dimension the position and momentum operators take the simpler form

$$(X\psi)(x) = x\psi(x) \qquad \text{and} \qquad (P\psi)(x) = \frac{\hbar}{i}\frac{d\psi}{dx}.$$

It should be noted that the momentum operator is defined only on functions which are differentiable, and there are also subtler restrictions. For example, P and X can only be applied to those wave functions ψ for which the integrals in

$$\|P\psi\|^2 = \int_{\mathbf{R}} \left|\frac{\hbar}{i}\frac{d\psi}{dx}\right|^2 dx \tag{6.5}$$

and

$$\|X\psi\|^2 = \int_{\mathbf{R}} x^2|\psi(x)|^2\, dx \tag{6.6}$$

are well defined, otherwise their images are not in $\mathcal{H}$. We shall therefore have to allow for operators, A, which are defined only on a subspace.

To see why these operators are self-adjoint we note that

$$\langle X\phi|\psi\rangle = \int_{\mathbf{R}} \overline{x\phi(x)}\psi(x)\, dx = \int_{\mathbf{R}} \overline{\phi(x)}x\psi(x)\, dx = \langle\phi|X\psi\rangle. \tag{6.7}$$

One can show formally that P is self-adjoint, using integration by parts:

$$\begin{aligned}\langle P\phi|\psi\rangle &= \int_{\mathbf{R}} \overline{\frac{\hbar}{i}\frac{d\phi}{dx}}\psi\, dx \\ &= -\frac{\hbar}{i}\int_{\mathbf{R}} \overline{\frac{d\phi}{dx}}\psi\, dx \\ &= -\frac{\hbar}{i}\left\{ \left[\overline{\phi}\psi\right]_{-\infty}^{\infty} - \int_{\mathbf{R}} \overline{\phi}\frac{d\psi}{dx}\, dx \right\} \\ &= -\frac{\hbar}{i}\left[\overline{\phi}\psi\right]_{-\infty}^{\infty} + \int_{\mathbf{R}} \overline{\phi}\frac{\hbar}{i}\frac{d\psi}{dx}\, dx. \end{aligned} \tag{6.8}$$

We already require ϕ and ψ to be both differentiable (so that P is defined) and normalizable, so we expect that $(\overline{\phi}\psi)(x)$ should tend to 0 as $|x| \to \infty$. Then

$$\langle P\phi|\psi\rangle = \int_{\mathbf{R}} \overline{\phi}\frac{\hbar}{i}\frac{d\psi}{dx}\, dx = \langle\phi|P\psi\rangle, \tag{6.9}$$

and P is self-adjoint. (It should be noted that the self-adjointness condition depends on boundary conditions satisfied by wave functions, and this is typical.) The simpler alternative way to deal with momentum is to work with the Fourier transform, and use momentum space wave functions. The Fourier transform is linear and Plancherel's theorem 3.4.1 tells us that it does not matter whether we evaluate inner products using the wave functions ϕ and ψ or their Fourier transforms, because

$$\langle \mathcal{F}\phi|\mathcal{F}\psi\rangle = \langle\phi|\psi\rangle. \tag{6.10}$$

Recalling that a linear transformation that preserves inner products is said to be unitary, we see that the Fourier transform is unitary and respects all the important quantum mechanical structure of $\mathcal{H}$. Proposition 3.4.2 can be reinterpreted as saying that $(\mathcal{F}P\psi)(p) = p\mathcal{F}\psi(p)$, from which the self-adjointness of P follows by the same argument as for X.

Definition 6.1.5. The *Hamiltonian operator*, H, is defined on twice differentiable functions ψ by

$$(H\psi)(\mathbf{x}) = -\frac{\hbar^2}{2m}\nabla^2\psi(\mathbf{x}) + V(\mathbf{x})\psi(\mathbf{x}),$$

or symbolically, by $H = |\mathbf{P}|^2/2m + V(\mathbf{X})$.

In one dimension one takes the Hamiltonian operator $H = P^2/2m + V(X)$. With this definition Schrödinger's time-independent equation can be written as

$$H\psi = E\psi, \tag{6.11}$$

and interpreted as an eigenvalue problem, with the energies as eigenvalues.

Formally H is self-adjoint for

$$H^* = \frac{1}{2m}(P^*)^2 + V(X^*) = \frac{1}{2m}P^2 + V(X) = H. \tag{6.12}$$

However, more careful consideration of its domain of definition shows that H is really only self-adjoint for certain potentials V, because H and H^* may not be defined on the same set of vectors. (Fortunately, for all the potentials that are of interest to us, H really is self-adjoint.)

There is, of course, an important logical distinction between the states and observables themselves and the corresponding mathematical descriptions that we use, but we shall not distinguish them by using different notation and terminology, as it should be clear from the context to which one is referring.

6.2. Some mathematical refinements

In order to be able to handle some of the technical difficulties which have already appeared and to work more confidently in infinite dimensions, it is customary to embellish $\mathcal{H}$ with some extra structure. However, although this is important in clarifying some ideas and making possible a mathematically rigorous treatment of the subject, it does not substantially alter the main physical concepts, and we shall not make much use of it later, so that this section may be omitted on a first reading by those who are prepared to take such matters on trust.

The state space, $\mathcal{H}$, is usually subject to two further constraints, which exploit the fact that the norm enables us to define a distance, $d(\phi, \psi) = \|\phi - \psi\|$, between vectors. The first of these is that it should be *complete*, in the following sense: whenever $\{\psi_n \in \mathcal{H}\}$ is a sequence of vectors, such that, for any positive ϵ, there exists an integer N_ϵ with

$$\|\psi_m - \psi_n\| < \epsilon \tag{6.13}$$

for all $m, n > N_\epsilon$, then there exists a limit vector $\psi \in \mathcal{H}$ such that

$$\|\psi_n - \psi\| \to 0. \tag{6.14}$$

This is a physically sensible condition, since it is often useful to take limits, as, for example, when one wants to sum an infinite Fourier series of the kind we encountered in Section 2.3. (For more detail see Appendix A1.1.) Finite-dimensional spaces are automatically complete, and a general theorem of topology tells us that any metric space can be completed by a procedure analogous to that for constructing the real numbers from the rationals. There is therefore no serious loss of generality in assuming $\mathcal{H}$ to be complete.

The second condition is not strictly necessary, but holds in almost all the examples which are of interest. We recall that a subset S of $\mathcal{H}$ is *dense* if for any vector $\psi \in \mathcal{H}$ and any $\epsilon > 0$ there is a vector $\phi \in S$ such that

$$\|\phi - \psi\| < \epsilon. \tag{6.15}$$

The second requirement is that $\mathcal{H}$ should contain a countable dense subset. This is again automatic in finite dimensions, because we may choose a basis and then take for S the vectors which have rational coordinates.

Definition 6.2.1. A *Hilbert space* is a complex inner product space which is complete and contains a countable dense subset. The state space is assumed to be a Hilbert space.

The discussion of the observables is rather more delicate than that for the states. We have already noted that a linear operator, A, may be defined only on a subspace, $\mathcal{D}(A)$ of $\mathcal{H}$, called the domain of A. We shall assume that this domain is dense in the sense of the above definition. We must also now modify the definition of the adjoint. We take for its domain, $\mathcal{D}(A^*)$, the set of vectors $\psi \in \mathcal{H}$ such that for all $\phi \in \mathcal{D}(A)$ there exists a vector $\Psi \in \mathcal{H}$ with

$$\langle A\phi|\psi\rangle = \langle\phi|\Psi\rangle. \tag{6.16}$$

In fact, Ψ is uniquely determined, since the difference, ξ, of two such vectors would satisfy $\langle\phi|\xi\rangle = 0$, for all ϕ, giving

$$\|\xi\|^2 = \langle\xi - \phi|\xi\rangle + \langle\phi|\xi\rangle = \langle\xi - \phi|\xi\rangle. \tag{6.17}$$

The Cauchy–Schwartz–Bunyakowski inequality would then give

$$\|\xi\|^2 \leq \|\phi - \xi\|\|\xi\|. \tag{6.18}$$

By the density of $\mathcal{D}(A)$, we know that for any $\epsilon > 0$ we can find ϕ such that $\|\phi - \xi\| < \epsilon$. So for all positive ϵ, we have $\|\xi\|^2 \leq \epsilon\|\xi\|$, which forces ξ to vanish. Moreover, since ψ and Ψ both occupy the linear slot in the inner product Ψ depends linearly on ψ, and we may write $\Psi = A^*\psi$, with A^* a uniquely defined linear transformation, which we naturally call the adjoint.

Definition 6.2.2. If $\mathcal{D}(A) = \mathcal{D}(A^*)$ and on this common domain $A = A^*$ then A is said to be *self-adjoint.* This is the sense in which observables in quantum mechanics are required to be self-adjoint.

Often the trickiest part of a problem is to show that a Hamiltonian really is self-adjoint.

This approach to quantum mechanics using Hilbert spaces and self-adjoint operators is not the only possibility. One can also sidestep some of the problems by noting that most of the operators that one wants to use in practice are defined on a common dense domain, $\mathcal{S}$, consisting of the space of all infinitely differentiable wave functions, ψ, for which $X^j P^k \psi$ is normalizable for all positive integers j and k. This is actually too small to be of much use, but it has a very large dual space, $\mathcal{S}'$, called the *tempered distributions*, to which most of the interesting operators can be transposed. However, we shall not avail ourselves of this approach.

The interrelationship between the mathematics and the physics is described in more detail in von Neumann's book *The mathematical foundations of quantum mechanics.* (Indeed, it was in this book that the term 'Hilbert space' was first introduced and some of the important theorems in the area first proved.)

6.3. Statistical aspects of quantum theory

There are mathematical advantages in working directly with expectation values of observables rather than with the probability densities.

Definition 6.3.1. The *expectation value* of an observable A in a state described by the vector ψ in $\mathcal{D}(A)$ is

$$\mathsf{E}_\psi(A) = \frac{\langle\psi|A\psi\rangle}{\|\psi\|^2}.$$

Remark 6.3.1. When ψ is normalized this takes the simpler form $\mathsf{E}_\psi(A) = \langle\psi|A\psi\rangle$. When the state is clear from the context one often abbreviates the notation to $\mathsf{E}_\psi(A) = \langle A\rangle$.

Remark 6.3.2. Although it is only for self-adjoint operators A that this formula has any physical significance it will sometimes be advantageous to use it for general linear transformations.

To see why we adopt this definition consider the case when A is the potential energy operator $V(X)$, and ψ has been normalized. Then

$$\mathsf{E}_\psi(X) = \langle\psi|V(X)\psi\rangle = \int_{\mathbf{R}} \overline{\psi(x)}(V(x)\psi(x))\,dx = \int_{\mathbf{R}} V(x)|\psi(x)|^2\,dx, \tag{6.19}$$

which is the classical formula for the expectation of $V(x)$ when x is randomly distributed with probability density $|\psi(x)|^2$. Thus our assumption is consistent with the previous Definition 2.4.1. Indeed knowing this formula for all potentials V is sufficient to determine that the probability density must be $|\psi(x)|^2$, so that, despite the simplicity of the formula, we have not sacrificed any information by working with expectation values rather than probability densities.

Statistical calculations involving momentum are often greatly simplified by Fourier transformation. For example, Plancherel's theorem gives

$$\begin{aligned}\mathsf{E}_\psi(P) &= \langle\psi|P\psi\rangle \\ &= \int_{\mathbf{R}} \overline{(\mathcal{F}\psi)(p)}(\mathcal{F}P\psi)(p)\,dp \\ &= \int_{\mathbf{R}} p|(\mathcal{F}\psi)(p)|^2\,dp, \end{aligned} \tag{6.20}$$

so that $|(\mathcal{F}\psi)(p)|^2$ plays the role of the probability density for momentum.

Before going on to show that the formula for expectations also determines the probability distribution for the energy we shall consider some of the mathematical properties that make the definition plausible.

Proposition 6.3.1. For every ψ in $\mathcal{H}$ the expectation value E_ψ has the following properties:
(i) $\mathsf{E}_\psi(1) = 1$, where 1 denotes the identity operator on $\mathcal{H}$.
(ii) $\mathsf{E}_\psi(A)$ is real for all self-adjoint operators A.
(iii) $\mathsf{E}_\psi(A) \geq 0$ for all positive operators A.
(iv) $\mathsf{E}_\psi(A)$ depends linearly on A, that is $\mathsf{E}_\psi(\alpha A + \beta B) = \alpha\mathsf{E}_\psi(A) + \beta\mathsf{E}_\psi(B)$ for all complex numbers α and β and all linear operators A and B.

Proof. (i) When A is the identity operator

$$\mathsf{E}_\psi(1) = \frac{\langle\psi|\psi\rangle}{\|\psi\|^2} = 1. \tag{6.21}$$

(ii) Since A is self-adjoint we have

$$\mathsf{E}_\psi(A) = \frac{\langle\psi|A\psi\rangle}{\|\psi\|^2} = \frac{\langle A\psi|\psi\rangle}{\|\psi\|^2} = \frac{\overline{\langle\psi|A\psi\rangle}}{\|\psi\|^2} = \overline{\mathsf{E}_\psi(A)}, \tag{6.22}$$

so that $\mathsf{E}_\psi(A)$ is real.

(iii) We recall that a linear transformation A is said to be positive if

$$\langle\psi|A\psi\rangle \geq 0 \tag{6.23}$$

for all vectors ψ. From this definition it immediately follows that $\mathsf{E}_\psi(A)$ is non-negative. In the most important applications A has the form B^*B, and then one can obtain a more precise result:

$$\|\psi\|^2\mathsf{E}_\psi(B^*B) = \langle\psi|B^*B\psi\rangle = \langle B\psi|B\psi\rangle = \|B\psi\|^2. \tag{6.24}$$

(iv) Finally we note that

$$\begin{aligned}\mathsf{E}_\psi(\alpha A + \beta B) &= \frac{\langle\psi|(\alpha A + \beta B)\psi\rangle}{\|\psi\|^2}\\ &= \alpha\frac{\langle\psi|A\psi\rangle}{\|\psi\|^2} + \beta\frac{\langle\psi|B\psi\rangle}{\|\psi\|^2}\\ &= \alpha\mathsf{E}_\psi(A) + \beta\mathsf{E}_\psi(B),\end{aligned} \tag{6.25}$$

completing the proof. □

We shall often write operators, $c1$, that are constant multiples of the identity just as c. From the expectation values the variance of an observable A can also be calculated as

$$\mathsf{E}_\psi\big((A - \mathsf{E}_\psi(A))^2\big) = \mathsf{E}_\psi(A^2) - \mathsf{E}_\psi(A)^2. \tag{6.26}$$

The left-hand side is clearly positive by (ii) above, which justifies the following definition:

Definition 6.3.2. The *dispersion* of the observable A in the state ψ is given by

$$\Delta_\psi(A) = \left[\mathsf{E}_\psi(A^2) - \mathsf{E}_\psi(A)^2\right]^{\frac{1}{2}}.$$

This is the quantum theoretical analogue of the standard deviation of a classical random variable.

It is natural to ask whether a quantum mechanical observable can ever have a precise value, that is whether the dispersion can vanish.

Proposition 6.3.2. The dispersion of A in the state ψ vanishes if and only if ψ is an eigenvector of A. Moreover, in this case the associated eigenvalue is $\mathsf{E}_\psi(A)$.

Proof. If ψ is an eigenvector of A, with $A\psi = \alpha\psi$, say, then

$$\mathsf{E}_\psi(A) = \frac{\langle\psi|A\psi\rangle}{\|\psi\|^2} = \frac{\langle\psi|\alpha\psi\rangle}{\|\psi\|^2} = \alpha, \tag{6.27}$$

so that

$$A\psi = \mathsf{E}_\psi(A)\psi. \tag{6.28}$$

We also have the identity

$$\frac{\|(A - \mathsf{E}_\psi(A)1)\psi\|^2}{\|\psi\|^2} = \mathsf{E}_\psi((A - \mathsf{E}_\psi(A)1)^2) = \Delta_\psi(A)^2, \tag{6.29}$$

from which it is clear that $\Delta_\psi(A)$ vanishes if and only if $A\psi = \mathsf{E}_\psi(A)\psi$, that is if and only if ψ is an eigenvector of A with eigenvalue $\mathsf{E}_\psi(A)$. The result now follows immediately. □

Remark 6.3.3. As the eigenvalues are given by expectations, Proposition 6.3.1(ii) tells us that they must be real. We also note, for future use, that if ϕ and ψ are both eigenvectors of A with distinct eigenvalues μ and λ, respectively, then

$$\langle A\phi|\psi\rangle - \langle\phi|A\psi\rangle = (\overline{\mu} - \lambda)\langle\phi|\psi\rangle.$$

The left-hand side vanishes for self-adjoint A, and since $\overline{\mu} = \mu \neq \lambda$, we must have $\langle\phi|\psi\rangle = 0$, that is ϕ is orthogonal to ψ.

The eigenvectors of an observable A (also called *eigenfunctions* or *eigenstates* in the Schrödinger formulation) thus play a distinguished role as the states in which A takes a precise value, namely the eigenvalue. Even when the wave function ψ is not itself an eigenvector of A, it may be possible to expand it as a linear combination of orthonormal eigenvectors, ψ_λ,

$$\psi = \sum_\lambda c_\lambda \psi_\lambda. \tag{6.30}$$

(One can even allow infinite sums, provided that the partial sums satisfy the convergence condition of (6.13), since then (6.14) gives a limit.) The coefficients c_λ are as usual given by the formula

$$c_\lambda = \langle \psi_\lambda | \psi \rangle, \tag{6.31}$$

which is obtained by taking the inner product of the expansion formula with the vector ψ_λ and using the orthonormality.

Using the expansion, and supposing that $A\psi_\lambda = \alpha_\lambda \psi_\lambda$, we then have

$$\begin{aligned}
\langle \psi | A\psi \rangle &= \sum_\lambda c_\lambda \langle \psi | A\psi_\lambda \rangle \\
&= \sum_\lambda c_\lambda \alpha_\lambda \langle \psi | \psi_\lambda \rangle \\
&= \sum_\lambda c_\lambda \alpha_\lambda \overline{c_\lambda} \\
&= \sum_\lambda \alpha_\lambda |c_\lambda|^2.
\end{aligned} \tag{6.32}$$

So, if ψ is normalized, we have

$$\mathsf{E}_\psi(A) = \sum_\lambda \alpha_\lambda |c_\lambda|^2. \tag{6.33}$$

Similarly for any polynomial f we obtain

$$\mathsf{E}_\psi(f(A)) = \sum_\lambda f(\alpha_\lambda) |c_\lambda|^2. \tag{6.34}$$

If ψ is not normalized we can replace it by $\psi/\|\psi\|$ to obtain a similar formula with $|c_\lambda|^2$ replaced by $|c_\lambda|^2/\|\psi\|^2$. We have now proved the following result:

Proposition 6.3.3. If it is possible to find an orthonormal basis of $\mathcal{H}$ consisting of eigenvectors of A then the probability that A takes the value α_λ in the state ψ is $|\langle \psi_\lambda | \psi \rangle|^2/\|\psi\|^2$.

We have already observed that Schrödinger's time-independent equation can be interpreted as an eigenvalue equation, and the energies as eigenvalues. The probability of obtaining a given energy E postulated in Assumption 2.6.1 is thus recovered as a special case of Proposition 6.3.3 when $A = H$.

In finite dimensions we know that there is always an orthonormal basis of eigenvectors for any self-adjoint operator A, so that such an expansion is always possible. Unfortunately, in an infinite-dimensional space not even self-adjointness is enough to guarantee that an operator has any eigenvectors, let alone enough to form a basis. For example, if ψ were an eigenvector for the momentum operator P in one dimension with eigenvalue p then we should have

$$P\psi = p\psi. \tag{6.35}$$

Written out explicitly this becomes

$$\frac{d\psi}{dx} = \frac{ip}{\hbar}\psi, \tag{6.36}$$

so that

$$\psi(x) = ce^{ipx/\hbar}, \tag{6.37}$$

for some constant c. We have already seen that this plane wave has infinite norm, so that $\psi \notin \mathcal{H}$. Worse still, if ψ were an eigenvector of X with eigenvalue α we should have

$$(X\psi)(x) = \alpha\psi(x), \tag{6.38}$$

or

$$(x - \alpha)\psi(x) = 0. \tag{6.39}$$

This forces ψ to vanish except on the single point $\{\alpha\}$ and so

$$\|\psi\|^2 = \int_{\mathbf{R}} |\psi(x)|^2 \, dx = 0. \tag{6.40}$$

(Intuitively the area under the graph of a function that vanishes at all but one point is 0; those acquainted with the Lebesgue integral can make this more precise by observing that $\{\alpha\}$ is a null set.) Since eigenvectors are, by definition, non-zero vectors, this shows that X like P has no eigenvectors in $\mathcal{H}$. The fact that neither P nor X can be measured with arbitrary precision will be strengthened in the next chapter into the famous 'Heisenberg uncertainty principle'. (There is, in fact, a generalization of the finite-dimensional spectral theorem valid for self-adjoint operators on

Hilbert spaces, but it is an altogether deeper and more subtle result, and we shall not discuss it further.)

Given a self-adjoint operator, A, in an infinite-dimensional space, it is useful to define its *spectrum* to be the set of complex numbers α for which $A-\alpha$ is not invertible. (The notion of invertible is not quite straightforward either, but our discussion will not be sensitive to the precise details.) Every eigenvalue is in the spectrum, because if $(A-\alpha)\psi = 0$, then $A-\alpha$ is not one–one and so not invertible. In finite dimensions the rank-nullity theorem enables one to show that the converse is also true but in infinite dimensions this breaks down. For example, $(X-\alpha)\psi(x) = (x-\alpha)\psi(x)$ and so any function in the range of $X-\alpha$ must vanish at $x = \alpha$. This means that $X-\alpha$ is not onto and therefore not invertible for any α, so the spectrum of X is the whole of $\mathbf{R}$, whereas we have seen that it has no eigenvalues. The spectrum contains information about the scattering as well as the bound states. It can be shown that the spectrum of A is always a closed subset of $\mathbf{R}$. This can lead to some counterintuitive properties. Consider the two Hamiltonians $H_j = P^2/2m + \frac{1}{2}m\omega_j^2 x^2$ for $j = 1, 2$. Their difference $H_1 - H_2$ has eigenvalues given by the energy differences, $(n_1+\frac{1}{2})\hbar\omega_1 - (n_2+\frac{1}{2})\hbar\omega_2$. A theorem of Kronecker says that if ω_1/ω_2 is irrational this set is dense in $\mathbf{R}$. The spectrum, which must contain these values and is also closed, must be the whole of $\mathbf{R}$.

Example 6.3.1. We conclude this section by calculating the mean and dispersion of the momentum in the n-th eigenstate of the one-dimensional square well. We recall that

$$\psi_n(x) = \sqrt{2/a}\sin(n\pi x/a) \tag{6.41}$$

for $x \in [0, a]$ and vanishes elsewhere. Now for any real normalized wave function ψ we have

$$\begin{aligned}\mathsf{E}_\psi(P) &= \langle\psi|P\psi_n\rangle \\ &= \frac{\hbar}{i}\int_0^a \psi_n(x)\psi'(x)\,dx \\ &= \frac{\hbar}{2i}[\psi(x)^2]_0^a \\ &= 0,\end{aligned} \tag{6.42}$$

which we should expect anyway by symmetry.

This tells us that the dispersion is given by

$$\Delta_\psi(P)^2 = \mathsf{E}_\psi(P^2) - \mathsf{E}_\psi(P)^2 = \mathsf{E}_{\psi_n}(P^2). \tag{6.43}$$

Since the Hamiltonian within the well is given by $H = P^2/2m$ this can also be expressed as

$$\Delta_\psi(P)^2 = 2m\mathsf{E}_\psi(H), \tag{6.44}$$

and, taking $\psi = \psi_n$ which is an eigenfunction of H with value E_n, we obtain

$$\Delta_{\psi_n}(P)^2 = 2m\mathsf{E}_{\psi_n}(H) = 2mE_n = \frac{n^2\pi^2\hbar^2}{a^2}. \tag{6.45}$$

6.4. Time evolution in quantum theory

So far our discussion of the abstract mathematical structure of quantum mechanics has not concerned itself with the dynamical questions of how the systems change in time. In order to address this deficiency we must allow the vector describing the state to depend on time. Let us write ψ_t for the state at time t.

Assumption 6.4.1. The vector ψ_t satisfies the abstract Schrödinger equation

$$i\hbar\frac{d}{dt}\psi_t = H\psi_t,$$

where H is the relevant Hamiltonian operator.

Remark 6.4.1. When ψ is a wave function in one or three dimensions then this clearly reduces to the differential equation introduced in Section 2.1.

Theorem 6.4.1. The abstract Schrödinger equation has the formal solution

$$\psi_t = \exp\left(-\frac{i}{\hbar}\int_0^t H\,dt\right)\psi_0.$$

Proof. Assuming (as is true in this case) that the exponential of an operator can be defined and has the usual properties, then

$$\frac{d}{dt}\left(\exp\left(\frac{i}{\hbar}\int_0^t H\,dt\right)\psi_t\right) = \frac{d}{dt}\left(\exp\left(\frac{i}{\hbar}\int_0^t H\,dt\right)\right)\psi_t$$

$$+\exp\left(\frac{i}{\hbar}\int_0^t H\,dt\right)\frac{d\psi_t}{dt}$$
$$=\exp\left(\frac{i}{\hbar}\int_0^t H\,dt\right)\left[\frac{i}{\hbar}H\psi_t+\frac{d\psi_t}{dt}\right] \quad (6.46)$$

By Schrödinger's equation the expression in square brackets vanishes and so $\exp(i\int_0^t H\,dt/\hbar)\psi_t$ is a constant. Since when $t=0$ it is ψ_0, we obtain the formal solution given above. □

In terms of $U_t=\exp(-i\int_0^t H\,dt/\hbar)$ we may write $\psi_t=U_t\psi_0$, and the abstract form of Schrödinger's equation can be recovered by differentiation. When H is constant this simplifies further to give $U_t=\exp(-iHt/\hbar)$. In this case, since H is self-adjoint we have

$$U_t^*=e^{itH^*/\hbar}=e^{itH/\hbar}=U_{-t}. \quad (6.47)$$

As with normal exponentials U_{-t} is the inverse of U_t, so that

$$U_t^*U_t=U_{-t}U_t=U_0=1=U_tU_t^*, \quad (6.48)$$

which shows that U_t is unitary. This means that

$$\|\psi_t\|^2=\|U_t\psi_0\|^2=\|\psi_0\|^2, \quad (6.49)$$

and confirms in the abstract setting the finding of Proposition 2.5.2, that the time evolution does not change the normalization of ψ. In fact, with the appropriate technical assumptions, the above arguments can all be made rigorous.

6.5. Measurements in quantum theory

Measurement occupies a much more important place in quantum mechanics than it did in the older theories. In classical physics it was assumed that, by taking sufficient care, disturbances to the system during a measurement could be kept below any given level of tolerance. In quantum theory that is certainly not the case.

Suppose, for example, that one measures an observable A and finds a precise value α for it. It could have started in any state ψ for which the probability $|\langle\psi|\psi_\alpha\rangle|^2$ for obtaining this value is not zero. However, after the measurement one knows the exact value so there is zero dispersion, and, by Proposition 6.3.2, the state must be described by an eigenvector ψ_α corresponding to α. The effect of the measurement has been to change the wave function from ψ to ψ_α, or some multiple of it.

Of course, the measurement may only determine a range of possibilities rather than a precise value. In that case there are a number of possible eigenvectors, and the most that can be asserted after the measurement is that the vector representing the state should lie in the subspace that they span. One simple way of ensuring this is to make the following projection postulate, due to von Neumann and refined by Gerhart Lüders.

Assumption 6.5.1. Let Q be the orthogonal projection onto the subspace spanned by the possible eigenvectors of A consistent with the outcome of the measurement. Then after the measurement the state vector has changed from ψ to $Q\psi$.

Unlike the continuous time evolution described by Schrödinger's equation, a measurement usually diminishes the norm of the wave function, since $\|Q\psi\|^2$ defines the probability of the outcome.

Example 6.5.1. Suppose that α is a non-degenerate eigenvalue of A and that ψ_α is a normalized eigenvector associated with it. The projection, P_α^A, onto the one-dimensional space that it spans, is then given by

$$P_\alpha^A \psi = \langle \psi_\alpha | \psi \rangle \psi_\alpha, \tag{6.50}$$

and this will represent the effect on the state ψ of measuring the value α. Thus the projection postulate not only asserts that ψ has changed to a multiple of ψ_α, but also tells us what multiple. In fact, that multiple is determined by

$$\|P_\alpha^A \psi\|^2 = \|\langle \psi_\alpha | \psi \rangle \psi_\alpha\|^2 = |\langle \psi_\alpha | \psi \rangle|^2, \tag{6.51}$$

so that, if ψ is normalized, then $\|P_\alpha^A \psi\|^2$ is just the probability of measuring the value α. This argument can be extended to show that for more general projections one gets just the probability of obtaining a value in the given range.

Definition 6.5.1. The *transition probability* between two normalized vectors ϕ and ψ in $\mathcal{H}$ is

$$\tau(\phi, \psi) = |\langle \phi | \psi \rangle|^2.$$

We can then write the probability of measuring the non-degenerate eigenvalue α as $\tau(\psi, \psi_\alpha)$. From the earlier discussion this is also the probability that the state will change from ψ to ψ_α during the measurement, which is the reason for calling it a transition probability. It can also be used during ordinary time evolution: for example, the probability that after time t the system is still in its original state ψ_0 is $\tau(\psi_t, \psi_0) = |\langle\psi_t|\psi_0\rangle|^2$. It is worth noting that τ is symmetric so that the probability of a change from ψ to ϕ during a measurement is the same as the probability of a transition from ϕ to ψ.

Exercises

6.1 Show that if A is a linear operator on $\mathcal{H}$ such that $\langle\psi|A\psi\rangle/\|\psi\|^2$ is real for all states ψ then A must be self-adjoint.
[*Hint*: Consider vectors $\psi + c\phi$ for $c \in \mathbf{C}$.]

6.2 Let (r, θ, ϕ) be spherical polar coordinates for a particle in $\mathbf{R}^3$. Show that $-i\hbar\partial/\partial r$ is not a self-adjoint operator on the wave functions, but $-i\hbar(\partial/\partial r + 1/r)$ is self-adjoint.

6.3 Let $\mathcal{P}$ be the *parity operator* defined on wave functions on $\mathbf{R}$ by

$$(\mathcal{P}\psi)(x) = \psi(-x).$$

Show that $\mathcal{P}^2 = 1$ and deduce that the eigenvalues of $\mathcal{P}$ are ± 1. Characterize the eigenvectors in terms of even and odd wave functions.

6.4 Show that the operator $\mathcal{P}$ of the previous question satisfies $\mathcal{P}P = -P\mathcal{P}$, and $\mathcal{P}V(X) = V(-X)\mathcal{P}$, for any function V. Deduce that, when the potential, V, is an even function $\mathcal{P}$ commutes with the Hamiltonian, H. Hence or otherwise show that, in this case, it is sufficient to consider eigenvectors of H that are odd or even functions. If the system is in a non-degenerate energy state show that the expectation value of the position operator is 0.

6.5° A quantum mechanical system with only three independent states is described by $\mathcal{H} = \mathbf{C}^3$. The Hamiltonian operator is

$$H = \hbar\omega \begin{pmatrix} 1 & 2 & 0 \\ 2 & 0 & 2 \\ 0 & 2 & -1 \end{pmatrix}.$$

Show that the eigenvalues of H are $3\hbar\omega$, 0 and $-3\hbar\omega$, and find the corresponding eigenvectors.

At time $t = 0$ the system is in the state

$$\psi_0 = \begin{pmatrix} 1 \\ 0 \\ 0 \end{pmatrix}.$$

Find the Schrödinger state vector ψ_t at a subsequent time t. Let p_1, p_2 and p_3 denote the probabilities of observing the system in the states $\begin{pmatrix} 1 \\ 0 \\ 0 \end{pmatrix}$, $\begin{pmatrix} 0 \\ 1 \\ 0 \end{pmatrix}$, $\begin{pmatrix} 0 \\ 0 \\ 1 \end{pmatrix}$, respectively. Show that $0 \leq p_2 \leq \frac{1}{2}$.

6.6° A two-state quantum system in a magnetic field **B** has the Hamiltonian operator

$$H = -\frac{e\hbar}{2\mu}\begin{pmatrix} B_3 & B_1 - iB_2 \\ B_1 + iB_2 & -B_3 \end{pmatrix},$$

where μ is a constant. At time $t = 0$ the system is in the state described by the vector $\begin{pmatrix} 1 \\ 0 \end{pmatrix}$. Show that the probability of its being in a state $\begin{pmatrix} 0 \\ 1 \end{pmatrix}$ at a time t is

$$\frac{|B|^2 - B_3^2}{|B|^2}\sin^2\left(\frac{e|B|t}{2\mu}\right),$$

where $|B|^2 = B_1^2 + B_2^2 + B_3^2$.

6.7 Show that if ψ_t satisfies Schrödinger's equation and the observable A does not depend on t then

$$i\hbar\frac{d}{dt}\langle\psi_t|A\psi_t\rangle = \langle\psi_t|(AH - HA)\psi_t\rangle.$$

6.8° A hydrogen-like atom with nuclear charge Ze is initially in its ground state when radioactive decay suddenly reduces the nuclear charge to $Z'e$. Show that the probability that a subsequent measurement of energy will find it in the new ground state is $[2\sqrt{ZZ'}/(Z + Z')]^6$. [The ground state wave function is given in Proposition 4.3.1. The formula $\int_0^\infty r^n \exp(-\lambda r)\,dr = n!\lambda^{-(n+1)}$ may be used without proof.]

7 The commutation relations

In my paper the fact that *XY* was not equal to *YX* was very disagreeable to me. I felt that this was the only point of difficulty with the whole scheme.

WERNER HEISENBERG, recalling his paper of July 1925

7.1. The commutation relations

Our results so far, in particular Section 3.4, strongly suggest that although it is possible to find formalisms in which either momentum or position are easy to handle, one cannot simultaneously manage both. The underlying reason for this is that the operators P and X do not commute. In fact they are related by

$$\begin{aligned}(PX\psi)(x) &= \frac{\hbar}{i}\frac{d}{dx}(x\psi(x)) \\ &= \frac{\hbar}{i}\psi(x) + \frac{\hbar}{i}x\frac{d\psi}{dx}(x) \\ &= \frac{\hbar}{i}\psi(x) + (XP\psi)(x).\end{aligned} \tag{7.1}$$

In other words,

$$PX = \frac{\hbar}{i}1 + XP. \tag{7.2}$$

Definition 7.1.1. The *commutator*, $[A, B]$, of two operators A and B is defined by

$$[A, B] = AB - BA.$$

In this notation, we have shown that $[P, X] = -i\hbar 1$. In three dimensions a similar calculation shows that $[P_j, X_k] = -i\hbar\delta_{jk}1$. On the other hand multiplications by coordinates commute with each other, and so do partial differentiations (at least for the sort of well-behaved wave functions that we are using). These results can be summarized as follows:

Theorem 7.1.1. (The canonical commutation relations) In one dimension the position and momentum operators are related by

$$[P, X] = \frac{\hbar}{i} 1.$$

In higher dimensions the relations are

$$[P_j, P_k] = 0, \qquad [P_j, X_k] = \frac{\hbar}{i} \delta_{jk} 1, \qquad [X_j, X_k] = 0.$$

These commutation relations strongly resemble the classical Poisson bracket relations for generalized coordinates and momenta. Recalling that the Poisson bracket is defined by

$$\{f, g\} = \frac{\partial f}{\partial p_j} \frac{\partial g}{\partial x_j} - \frac{\partial f}{\partial x_j} \frac{\partial g}{\partial p_j}, \tag{7.3}$$

we have

$$\{p_j, x_k\} = \delta_{jk}, \tag{7.4}$$

$$\{p_j, p_k\} = 0 = \{x_j, x_k\}. \tag{7.5}$$

The following properties of the commutator also parallel those of the Poisson bracket.

Proposition 7.1.2. For all operators A, B and C the commutator satisfies the following identities:
(i) $[A, B] = -[B, A]$;
(ii) $[A, B]$ is linear in both A and B;
(iii) $[A, BC] = B[A, C] + [A, B]C$;
(iv) (the Jacobi identity) $[A, [B, C]] + [B, [C, A]] + [C, [A, B]] = 0$.

Proof. (i) $[A, B] = AB - BA = -(BA - AB) = -[B, A]$.
(ii) For any complex numbers β and γ,

$$\begin{aligned} [A, \beta B + \gamma C] &= A(\beta B + \gamma C) - (\beta B + \gamma C)A \\ &= \beta(AB - BA) + \gamma(AC - CA) \\ &= \beta[A, B] + \gamma[A, C]. \end{aligned} \tag{7.6}$$

Linearity in the first variable follows similarly, or by use of (i).

(iii)

$$\begin{aligned}[A, BC] &= ABC - BCA \\ &= (AB - BA)C + B(AC - CA) \\ &= B[A, C] + [A, B]C. \end{aligned} \tag{7.7}$$

(iv) Finally:

$$\begin{aligned}[A, [B, C]] &= [A, BC] - [A, CB] \\ &= B[A, C] + [A, B]C - C[B, A] - [A, C]B \\ &= [B, [A, C]] + [[A, B], C] \\ &= -[B, [C, A]] - [C, [A, B]], \end{aligned} \tag{7.8}$$

from which Jacobi's identity follows on rearrangement of the terms. □

Remark 7.1.1. The similarity with Poisson brackets led Dirac to suggest that each function, f, on the classical phase space should be replaced in quantum theory by an operator, $Q(f)$, in such a way that for any pair of functions f and g one had

$$[Q(f), Q(g)] = -i\hbar Q(\{f, g\}). \tag{7.9}$$

This idea has served as the inspiration for most of the mathematical investigations of quantization. It is, however, now known that this identity cannot be satisfied for all functions simultaneously without violating some of the other conditions of quantum theory (see Exercise 7.19). This can lead to ambiguities when one has to decide whether the quantum analogue of a classical observable such as px^2 should be PX^2, X^2P, XPX or some combination of these. Fortunately one can arrange that

$$[Q(f), Q(g)] + i\hbar(Q\{f, g\}) = O(\hbar^3), \tag{7.10}$$

so that for most practical purposes any differences between the answers obtained is likely to be small. In any case, physicists now have sufficient experience of quantum mechanics that they no longer have to start by considering the corresponding classical system first.

7.2. Heisenberg's uncertainty principle

The non-commutativity of P and X has some profound consequences that we shall now start to investigate.

Lemma 7.2.1. Let A, B and C be self-adjoint operators such that $[A, B] = iC$. Then
(i) For all real t, $(A - itB)^*(A - itB) = A^2 + tC + t^2B^2$.
(ii) For any normalized vector $\psi \in \mathcal{H}$,

$$\|(A - itB)\psi\|^2 = \mathsf{E}_\psi(A^2) + t\mathsf{E}_\psi(C) + t^2\mathsf{E}_\psi(B^2).$$

(iii) For any vector $\psi \in \mathcal{H}$, $\mathsf{E}_\psi(A^2)\mathsf{E}_\psi(B^2) \geq \frac{1}{4}\mathsf{E}_\psi(C)^2$.
(iv) There is equality in (iii) if and only if there exists a real number t such that $(A - itB)\psi = 0$ or, equivalently, $(A^2 + t^2B^2)\psi = -tC\psi$.

Proof. (i) The adjoint of $(A - itB)$ is $(A + itB)$ so that we have

$$\begin{aligned}(A - itB)^*(A - itB) &= (A + itB)(A - itB)\\ &= A^2 - it(AB - BA) + t^2B^2\\ &= A^2 + tC + t^2B^2.\end{aligned} \tag{7.11}$$

(ii) For convenience we shall work with a normalized vector ψ. Then by 6.3.1(iv) we see that

$$\begin{aligned}\mathsf{E}_\psi(A^2) + t\mathsf{E}_\psi(C) + t^2\mathsf{E}_\psi(B^2) &= \mathsf{E}_\psi(A^2 + tC + t^2B^2)\\ &= \mathsf{E}_\psi((A - itB)^*(A - itB))\\ &= \|(A - itB)\psi\|^2.\end{aligned} \tag{7.12}$$

(iii) From (ii) it is clear that $\mathsf{E}_\psi(A^2) + t\mathsf{E}_\psi(C) + t^2\mathsf{E}_\psi(B^2) \geq 0$.

The quadratic expression $\mathsf{E}_\psi(A^2) + t\mathsf{E}_\psi(C) + t^2\mathsf{E}_\psi(B^2)$ is non-negative, so its discriminant,

$$\mathsf{E}_\psi(C)^2 - 4\mathsf{E}_\psi(A^2)\mathsf{E}_\psi(B^2), \tag{7.13}$$

must be non-positive, and this gives the inequality (iii).

(iv) Inequality occurs in (iii) when the discriminant vanishes, which is equivalent to the quadratic having a repeated real root t. On the other hand, since

$$\mathsf{E}_\psi(A^2) + tE_\psi(C) + t^2\mathsf{E}_\psi(B^2) = \|(A - itB)\psi\|^2, \tag{7.14}$$

t is a root of the quadratic if and only if $(A - itB)\psi = 0$. By (i) this same value of t also satisfies

$$(A^2 + t^2B^2)\psi = -tC\psi. \qquad \square$$

Corollary 7.2.2. (Heisenberg's uncertainty principle) The dispersions of the position and momentum are related by

$$\Delta_\psi(P)\Delta_\psi(X) \geq \tfrac{1}{2}\hbar.$$

The lower bound is achieved if and only if

$$\psi(x) = \exp\left(-\frac{t}{2\hbar}(x-\mu)^2 + \alpha\right)$$

for some positive real constant t and complex constants μ and α.

Proof. Set $A = P - \mathsf{E}_\psi(P)$ and $B = X - \mathsf{E}_\psi(X)$. Then, since multiples of the identity commute with all operators, we have

$$[A, B] = [P, X] - \mathsf{E}_\psi(X)[P, 1] - \mathsf{E}_\psi(P)[1, X - \mathsf{E}_\psi(X)1] = [P, X] = \frac{\hbar}{i}1, \tag{7.15}$$

so that $C = -\hbar 1$. Using Definition 6.3.2, the lemma then gives

$$\Delta_\psi(P)^2\Delta_\psi(X)^2 = \mathsf{E}_\psi(A^2)\mathsf{E}_\psi(B^2) \geq \tfrac{1}{4}\hbar^2, \tag{7.16}$$

from which the desired inequality now follows on taking positive square roots.

For equality, we need a real t such that $A\psi = itB\psi$, or

$$P\psi = \mathsf{E}_\psi(P)\psi + it[X - \mathsf{E}_\psi(X)1]\psi. \tag{7.17}$$

For convenience we write $\lambda = \mathsf{E}_\psi(P) - it\mathsf{E}_\psi(X)$, so that

$$\frac{\hbar}{i}\frac{d\psi}{dx} = (\lambda + itx)\psi(x). \tag{7.18}$$

Integrating we obtain

$$\ln(\psi(x)) = \frac{i}{\hbar}\left(\frac{itx^2}{2} + \lambda x\right) + \beta, \tag{7.19}$$

for some constant β, so that

$$\psi(x) = \exp\left(-\frac{tx^2}{2\hbar} + \frac{i\lambda}{\hbar}x + \beta\right). \tag{7.20}$$

Finally we note that ψ will not be normalizable unless t is positive. Setting $\mu = i\lambda/t$ and $\alpha = \beta - \lambda^2/2\hbar t$, we arrive at the stated form of wave function. □

Remark 7.2.1. Heisenberg's uncertainty principle tells us that, in any state ψ, there is a lower bound to the product of the dispersions of P and X, so that greater precision in a measurement of position can only be bought at the expense of less precision in the measurement of momentum and vice versa. Soon after he had discovered this result, Heisenberg suggested the following physical model to enhance its plausibility. If we wish to determine the position of a particle very accurately using, for example, some form of microscope then we must work with light (or other radiation) of a correspondingly short wavelength, since distances cannot be resolved to better than about half a wavelength: $\Delta_\psi(X) \sim \frac{1}{2}(\text{wavelength})$. On the other hand photons carry a momentum $\hbar/(\text{wavelength})$. Since a photon must collide with the particle that we are observing if we are to see it at all, an unknown fraction of this momentum may be transferred to the particle giving $\Delta_\psi(P) \sim \hbar/(\text{wavelength}) \sim \frac{1}{2}\hbar/\Delta_\psi(X)$. Although this example provides a rather striking image, it does rather suggest that the uncertainty in P arises only after our observation of X, whereas the uncertainty principle actually refers to dispersions in the same state at the same time.

Definition 7.2.1. The states ψ for which $\Delta_\psi(P)\Delta_\psi(X) = \frac{1}{2}\hbar$ are called *minimal uncertainty states.*

The ground state of the harmonic oscillator,

$$\psi_0(x) = \left(\frac{m\omega}{\pi\hbar}\right)^{\frac{1}{4}} e^{-m\omega x^2/2\hbar}, \tag{7.21}$$

is a minimal uncertainty state obtained by taking $t = m\omega$ and $\mu = 0$. This is no coincidence, because if $\mu = \mathsf{E}_\psi(X) + i\mathsf{E}_\psi(P)/t$ vanishes the final equivalence in Lemma 7.2.1 tells us that

$$(P^2 + m^2\omega^2 X^2)\psi = m\omega \mathsf{E}_\psi(\hbar 1)\psi = m\omega\hbar\psi, \tag{7.22}$$

or

$$\left(\frac{P^2}{2m} + \frac{1}{2}m\omega^2 X^2\right)\psi = \frac{1}{2}\hbar\omega\psi. \tag{7.23}$$

7.3. The time–energy uncertainty relation

In the theory of relativity, space and time are linked, and so are momentum and energy. One would therefore expect a relativistic theory to have an uncertainty principle for energy and time, as well as for position and momentum. There are indeed such relations, well known in areas, such as radar signal processing, that study time-dependent wave phenomena, but they have a slightly different aspect in quantum mechanics. In part this stems from the fact that we cannot simply define a 'time operator' to be multiplication by t. (At any given time this would just be multiplication by a constant and could be normalized away.) However, the time–energy uncertainty reveals itself in other ways. For example, the time evolution of an energy eigenstate is given by multiplication by $\exp(-iEt/\hbar)$. Since the physical state is unchanged by such multiplications, that state lasts for ever. This is a special case of a more general relationship between the lifetime of a state and the precision with which its energy can be known.

Lemma 7.3.1. Let H be the Hamiltonian of a system and ψ_t denote the state at time t, where initially $\psi_0 = \psi$, and assume that $\langle\psi|\psi_t\rangle$ is continuously twice differentiable in near $t = 0$. Then for small t we have

$$\langle\psi|\psi_t\rangle = \exp\left(-\frac{it}{\hbar}\mathsf{E}_\psi(H) - \frac{t^2}{2\hbar^2}\Delta_\psi(H)^2\right)[1 + o(t^2)],$$

where $o(t^2)$ denotes a term that tends to 0 faster than t^2 as $t \to 0$.

Proof. We may as well assume that ψ is normalized, and we shall drop the suffix ψ from E and Δ. We first note that

$$\begin{aligned} i\hbar\frac{d}{dt}\exp(it\mathsf{E}(H)/\hbar)\,\langle\psi|\psi_t\rangle &= \exp(it\mathsf{E}(H)/\hbar)\,[\langle\psi|H\psi_t\rangle - \mathsf{E}(H)\langle\psi|\psi_t\rangle] \\ &= \exp(it\mathsf{E}(H)/\hbar)\,\langle\psi|\,[H - \mathsf{E}(H)]\,\psi_t\rangle. \end{aligned} \tag{7.24}$$

This vanishes when $t = 0$, since $\mathsf{E}(H) = \langle\psi|H\psi\rangle$. Differentiating again, we obtain

$$-\hbar^2\frac{d^2}{dt^2}\exp(it\mathsf{E}(H)/\hbar)\,\langle\psi|\psi_t\rangle = \exp(it\mathsf{E}(H)/\hbar)\,\langle\psi|\,[H - \mathsf{E}(H)]^2\,\psi_t\rangle, \tag{7.25}$$

which reduces to

$$\langle\psi|\,[H - \mathsf{E}(H)]^2\,\psi\rangle = \Delta(H)^2 \tag{7.26}$$

at $t = 0$. From these two results the product rule can be applied to show that the first and second derivatives of

$$\exp\left(\frac{it}{\hbar}\mathsf{E}(H) + \frac{t^2}{2\hbar^2}\Delta(H)^2\right)\langle\psi|\psi_t\rangle \tag{7.27}$$

both vanish at $t = 0$. Since this function takes the value 1 at $t = 0$, Taylor's theorem gives the identity

$$\exp\left(\frac{it}{\hbar}\mathsf{E}(H) + \frac{t^2}{2\hbar^2}\Delta(H)^2\right)\langle\psi|\psi_t\rangle = 1 + o(t^2), \tag{7.28}$$

from which the result follows. □

Corollary 7.3.2. The probability that the state ψ will not change within a short time t under the evolution given by H is

$$\exp\left(-\left(t\Delta_\psi(H)/\hbar\right)^2\right)\left[1 + o\left(t^2\right)\right].$$

Proof. Using the lemma we see that the probability that there will be no change of state is

$$|\langle\psi|\psi_t\rangle|^2 = \exp\left(-\frac{t^2}{\hbar^2}\Delta_\psi(H)^2\right)[1 + o(t^2)].$$ □

The exponential term remains bigger than $\frac{1}{2}$ unless

$$t\Delta_\psi(H)\hbar > \sqrt{\ln(2)},$$

and this shows that the timescale on which the state changes is inversely proportional to the variance of the energy.

7.4. Simultaneous measurability

It is not only position and momentum or energy and time which are governed by an uncertainty principle; there are many other such pairs of *complementary* observables to which Lemma 7.2.1 can be applied. One can even apply it in the case of commuting observables A and B with $C = 0$. However, in that case it gives rise to the trivial inequality

$$\Delta_{\psi}(A)\Delta_{\psi}(B) \geq 0, \tag{7.29}$$

and it is not usually possible to achieve the lower bound. That happens only if one of the two dispersions vanishes, and by Proposition 6.3.2 that can only occur when ψ is an eigenvector of the relevant operator. Ideally we should like ψ to be an eigenvector of both A and B so that both observables could be measured precisely. However, we have already seen that in infinite-dimensional spaces even self-adjoint operators need have no eigenvectors at all. Nonetheless, the following generalization of a well-known finite-dimensional theorem (proved in Appendix A1) provides a useful condition that is sufficient to ensure a good supply of states that are simultaneously eigenvectors of two commuting observables.

Proposition 7.4.1. Let A and B be self-adjoint operators on the inner product space $\mathcal{H}$, let $\mathcal{H}_A$ and $\mathcal{H}_B$ denote the subspaces spanned by eigenvectors of A and B, respectively, and let $\mathcal{H}_{A,B}$ denote the span of the vectors that are simultaneously eigenvectors for both A and B. If $AB = BA$ then $\mathcal{H}_{A,B} = \mathcal{H}_A \cap \mathcal{H}_B$.

This has the following immediate corollary:

Corollary 7.4.2. If $AB = BA$ and $\mathcal{H}$ admits an orthonormal basis of eigenvectors for A then $\mathcal{H}_{A,B} = \mathcal{H}_B$.

Proof. Since $\mathcal{H}$ admits an orthonormal basis of eigenvectors for A we have $\mathcal{H}_A = \mathcal{H}$, so that

$$\mathcal{H}_{A,B} = \mathcal{H} \cap \mathcal{H}_B = \mathcal{H}_B. \qquad \square$$

This result means that instead of looking for vectors that are eigenvectors of B alone we may as well look for vectors that are simultaneously eigenvectors of A and of B since every B-eigenvector is in the span of these.

This inspires the following definition:

Definition 7.4.1. The observables corresponding to commuting operators are said to be *compatible* or *simultaneously measurable.*

7.5. The harmonic oscillator revisited

The last sections have emphasized the algebraic structure of quantum theory, so it is natural to ask whether one can solve real quantum mechanical problems without resorting to differential equations at all. In fact, for many elementary systems this is the case.

As an illustration we look at the harmonic oscillator. The Hamiltonian operator is

$$H = \frac{1}{2m}P^2 + \frac{1}{2}m\omega^2 X^2, \tag{7.30}$$

so that we should like to find E and ψ satisfying

$$\left(\frac{1}{2m}P^2 + \frac{1}{2}m\omega^2 X^2\right)\psi = E\psi, \tag{7.31}$$

where P and X are self-adjoint operators such that $[P, X] = -i\hbar 1$.

As remarked at the end of Section 7.2, the ground state of the harmonic oscillator is a minimal uncertainty state with $t = m\omega$ and $\mu = 0$. In that case $\mathsf{E}_\psi(P)$ and $\mathsf{E}_\psi(X)$ vanish so that $A = P$, $B = X$, and the operator $A - itB$ simplifies to $P - im\omega X$.

Now, in the classical Hamiltonian treatment of the oscillator $p \pm im\omega x$ has particularly simple equations of motion (see Exercise 7.18), so this suggests that we would do well to study the operator

$$a_- = P - im\omega X \tag{7.32}$$

and its adjoint

$$a_+ = a_-^* = P + im\omega X. \tag{7.33}$$

Lemma 7.5.1. The operators a_- and a_+ satisfy the equations

$$a_\mp^* a_\mp = a_\pm a_\mp = 2m\left(H \mp \tfrac{1}{2}\hbar\omega\right).$$

Moreover, for any normalized vector $\psi \in \mathcal{H}$

$$\|a_\mp \psi\|^2 = 2m\mathsf{E}_\psi\left(H \mp \tfrac{1}{2}\hbar\omega\right).$$

Proof. Lemma 7.2.1(i) gives

$$\begin{aligned}(P \mp im\omega X)^*(P \mp im\omega X) &= P^2 + m^2\omega^2 X^2 \mp m\omega\hbar 1 \\ &= 2m\left(H \mp \tfrac{1}{2}\hbar\omega\right). \end{aligned} \tag{7.34}$$

The formula for $\|a_\mp \psi\|^2$ likewise follows from the second part of the same lemma. □

Corollary 7.5.2. Let ψ be an eigenvector of H such that $H\psi = E\psi$. Then
(i) $\|a_\mp \psi\|^2 = 2m\left(E \mp \frac{1}{2}\hbar\omega\right)\|\psi\|^2$;
(ii) $E \geq \frac{1}{2}\hbar\omega$;
(iii) $E = \frac{1}{2}\hbar\omega$ if and only if $\psi(x) = \psi_0(x) = A\exp\bigl(-m\omega x^2/2\hbar\bigr)$.

Proof. (i) Since for ψ an eigenvector we have $\mathsf{E}_\psi(H) = E$, part (i) follows from the preceding lemma.

(ii) From the first part it is clear that

$$E \mp \tfrac{1}{2}\hbar\omega \geq 0, \tag{7.35}$$

so that

$$E \geq \pm\tfrac{1}{2}\hbar\omega. \tag{7.36}$$

The sharper inequality now gives the result.

(iii) Bearing in mind the identity (i), we see that equality occurs if and only if $\|a_-\psi\|$ vanishes, that is if and only if

$$(P - im\omega X)\,\psi = 0. \tag{7.37}$$

This is precisely the equation that we solved in Corollary 7.2.2 to get the minimal uncertainty states. Substituting the known values $t = m\omega$ and $\mu = 0$ now gives the result. □

We are thus able to obtain the ground state energy and wave function quite painlessly by purely algebraic techniques, and we shall now go on to consider the higher energy levels. The first step is to show that the operators a_+ and a_- respectively raise or lower the eigenvalue of eigenvectors of H by $\hbar\omega$.

Lemma 7.5.3. The operators $a_\pm$ satisfy the equations

$$Ha_\pm = a_\pm \left(H \pm \hbar\omega\right),$$

or equivalently, $[H, a_\pm] = \pm\hbar\omega a_\pm$.

Proof. According to Lemma 7.5.1

$$2m\left(H \mp \tfrac{1}{2}\hbar\omega\right)a_\pm = (a_\pm a_\mp)\,a_\pm = a_\pm\,(a_\mp a_\pm) = a_\pm 2m\left(H \pm \tfrac{1}{2}\hbar\omega\right). \tag{7.38}$$

Simplifying we obtain

$$Ha_\pm = a_\pm \left(H \pm \hbar\omega\right), \tag{7.39}$$

from which the stated commutation relation immediately follows. □

Corollary 7.5.4. If $H\psi = E\psi$, then for $N = 1, 2, \ldots$,

$$Ha_\pm^N\psi = (E \pm N\hbar\omega)a_\pm^N\psi.$$

Moreover, if $\psi \neq 0$ then $a_+^N\psi \neq 0$, and $a_-^N\psi$ vanishes if and only if E takes one of the values $\frac{1}{2}\hbar\omega$, $\frac{3}{2}\hbar\omega$, $\frac{5}{2}\hbar\omega, \ldots, (N - \frac{1}{2})\hbar\omega$.

Proof. Since the lemma already furnishes the case of $N = 1$, let us assume inductively that the result is true for N. Then, by applying the operators of the lemma to $a_\pm^N\psi$, we obtain

$$\begin{aligned} Ha_\pm^{N+1}\psi &= a_\pm(H \pm \hbar\omega)a_\pm^N\psi \\ &= a_\pm\left[(E \pm N\hbar\omega) \pm \hbar\omega\right]a_\pm^N\psi \\ &= \left[E \pm (N+1)\hbar\omega\right]a_\pm^{N+1}\psi, \end{aligned} \tag{7.40}$$

as required for the inductive step. Moreover, by the inductive hypothesis $a_+^N\psi$ does not vanish, and by Corollary 7.5.2(i) and (ii)

$$\|a_+^{N+1}\psi\|^2 = 2m\left(E + \tfrac{1}{2}\hbar\omega\right)\|a_+^N\psi\|^2 \geq 2m\hbar\omega\|a_+^N\psi\|^2 \neq 0. \tag{7.41}$$

If $a_-^{N+1}\psi$ vanishes then $a_-^N\psi$ is in the kernel of a_-, and so, by Corollary 7.5.2(iii), it is either 0 or it is a multiple of ψ_0 and its energy, $E - N\hbar\omega$, must be $\frac{1}{2}\hbar\omega$. In the latter case the energy is therefore $E = (N + \frac{1}{2})\hbar\omega$, whilst in the former case the inductive hypothesis tells us that E takes one of the values $\frac{1}{2}\hbar\omega$, $\frac{3}{2}\hbar\omega, \ldots, (N - \frac{1}{2})\hbar\omega$. □

Theorem 7.5.5. The set of eigenvalues of $H = P^2/2m + \frac{1}{2}m\omega^2X^2$ is

$$\left\{\left(N + \tfrac{1}{2}\right)\hbar\omega : N = 0, 1, 2, \ldots\right\}.$$

The eigenvectors corresponding to the eigenvalue $\left(N + \frac{1}{2}\right)\hbar\omega$ are the multiples of

$$\psi_N = a_+^N\psi_0,$$

where

$$\psi_0(x) = \left(\frac{m\omega}{\pi\hbar}\right)^{\frac{1}{4}} e^{-m\omega x^2/2\hbar}.$$

Proof. We know that ψ_0 is an eigenvector of H with eigenvalue $\frac{1}{2}\hbar\omega$, so Corollary 7.5.4 tells us that $a_+^N\psi_0$ is an eigenvector of H with value $(N + \frac{1}{2})\hbar\omega$. Conversely if ψ is an eigenvector with value E then $a_-^n\psi$ is either 0 or an eigenvector with eigenvalue $E - n\hbar\omega$. For n larger than $E/\hbar\omega$, $E - n\hbar\omega$ is negative and so excluded as an admissible eigenvalue by Corollary 7.5.2(ii). We therefore conclude that $a_-^n\psi$ must vanish and, by Corollary 7.5.4, E takes a value of the required form.

Finally, if ψ is any eigenvector with eigenvalue $(N + \frac{1}{2})\hbar\omega$ then $a_-^N\psi$ is an eigenvector of eigenvalue $\frac{1}{2}\hbar\omega$, and so is a multiple of ψ_0. Since the same applies to ψ_N we conclude that $a_-^N\psi = \lambda a_-^N\psi_N$ for some constant λ. But now, unless it vanishes, $\psi - \lambda\psi_N$ is an eigenvector with eigenvalue $(N + \frac{1}{2})\hbar\omega$ that is outside the range of energies for which $a_-^N(\psi - \lambda\psi_N)$ can vanish according to Corollary 7.5.4. To avoid a contradiction we must have $\psi = \lambda\psi_N$, showing that the eigenvalues are non-degenerate and that every eigenvector is a multiple of the appropriate ψ_N. □

7.6. Uniqueness of the commutation relations

Looking back at the arguments of the last section the only place in which any use was made of the explicit form of the operators P and X was to establish that the $E = \frac{1}{2}\hbar\omega$ energy level was spanned by a single wave function ψ_0, and derive the formula for ψ_0. Elsewhere we used only the commutation relations between P and X. We could therefore have replaced the explicit use of P and X by the algebraic assumption that there exists an eigenvector Ω for the energy level $E = \frac{1}{2}\hbar\omega$, which is unique up to multiples. Equivalently, by virtue of Corollary 7.5.2(iii), we could have assumed that $\ker a_-$ is a one-dimensional space spanned by Ω.

In fact this extra assumption is sufficient to determine the algebraic structure completely, as we can then map the abstract space $\mathcal{H}$ across to the usual space of wave functions by

$$\sum_{n=0}^{\infty} c_n \left(\frac{i}{\hbar} a_+\right)^n \Omega \to \sum_{n=0}^{\infty} c_n \psi_n. \tag{7.42}$$

In other words when $\ker(P - im\omega X)$ is one dimensional it is only possible to satisfy the commutation relations on a space that is isomorphic to Schrödinger's space of wave functions. Under mild technical assumptions the above argument can be made rigorous and leads to a result known as the Stone–von Neumann uniqueness theorem. It brings the assurance that there is nothing in the detailed theory of wave functions that cannot, in principle, also be achieved by the algebraic approach, since they work in isomorphic spaces.

7.7. A generating function for the oscillator wave functions

The algebraic approach to the harmonic oscillator described in the last section greatly simplifies the task of finding the wave functions. It enables one to find the ground state by solving a first-order differential equation rather than the second-order Schrödinger equation, and then provides the explicit formula $\psi_N = (ia_+/\hbar)^N \psi_0$ for the wave functions corresponding to any higher energy level. That formula can be simplified still further by some elementary calculations.

Lemma 7.7.1. The operator a_+ can be written as

$$\frac{i}{\hbar} a_+ = e^{m\omega x^2/2\hbar} \frac{d}{dx} e^{-m\omega x^2/2\hbar}.$$

Proof. We first note that

$$\frac{i}{\hbar}a_+ = \frac{i}{\hbar}(P + im\omega X) = \left(\frac{d}{dx} - \frac{m\omega}{\hbar}x\right). \tag{7.43}$$

Now, for any differentiable function ϕ we have

$$\left(\frac{d}{dx} - \frac{m\omega}{\hbar}x\right)\phi = e^{m\omega x^2/2\hbar}\frac{d}{dx}\left(e^{-m\omega x^2/2\hbar}\phi\right). \tag{7.44}$$

Combining these we deduce the operator identity

$$\frac{i}{\hbar}a_+ = e^{m\omega x^2/2\hbar}\frac{d}{dx}e^{-m\omega x^2/2\hbar}, \tag{7.45}$$

as required. □

Corollary 7.7.2. For $N = 0, 1, 2, \ldots$ we have the identity

$$\psi_N(x) = \left(\frac{m\omega}{\pi\hbar}\right)^{\frac{1}{4}} e^{m\omega x^2/2\hbar}\frac{d^N}{dx^N}e^{-m\omega x^2/2\hbar}\psi_0.$$

Proof. This follows from a simple calculation using the identity in the preceding lemma:

$$\begin{aligned}
\psi_N(x) &= \left(\frac{i}{\hbar}a_+\right)^N \psi_0 \\
&= \left(e^{m\omega x^2/2\hbar}\frac{d}{dx}e^{-m\omega x^2/2\hbar}\right)^N \psi_0 \\
&= e^{m\omega x^2/2\hbar}\frac{d^N}{dx^N}e^{-m\omega x^2/2\hbar}\psi_0 \\
&= \left(\frac{m\omega}{2\hbar}\right)^{\frac{1}{4}} e^{m\omega x^2/2\hbar}\frac{d^N}{dx^N}e^{-m\omega x^2/\hbar}.
\end{aligned}$$

□

Rather than dealing with the individual wave functions it is more efficient to combine them.

Definition 7.7.1. The *generating function* for the harmonic oscillator wave functions is defined by

$$G(s,x) = \sum_{N=0}^{\infty} \frac{s^N}{N!} \psi_N(x).$$

Using our earlier formula for ψ_N we can obtain an explicit formula for G.

Theorem 7.7.3. For all real values of s and x the generating function is given by

$$G(s,x) = \left(\frac{m\omega}{\pi\hbar}\right)^{\frac{1}{4}} \exp\left[-\frac{m\omega}{\hbar}\left(s^2 + 2sx + \frac{1}{2}x^2\right)\right].$$

Proof. By the definition of G and the explicit formula for ψ_N we have

$$\begin{aligned} G(s,x) &= \sum_{n=0}^{\infty} \frac{s^N}{N!} \psi_N(x) \\ &= \left(\frac{m\omega}{\pi\hbar}\right)^{\frac{1}{4}} e^{m\omega x^2/2\hbar} \sum_{n=0}^{\infty} \frac{s^N}{N!} \frac{d^N}{dx^N}\left(e^{-m\omega x^2/\hbar}\right). \end{aligned} \tag{7.46}$$

By applying Taylor's theorem to the analytic function $\exp(-m\omega x^2/\hbar)$ we may sum the series to obtain

$$\begin{aligned} G(s,x) &= \left(\frac{m\omega}{\pi\hbar}\right)^{\frac{1}{4}} e^{m\omega x^2/2\hbar} e^{-m\omega(x+s)^2/\hbar} \\ &= \left(\frac{m\omega}{\pi\hbar}\right)^{\frac{1}{4}} \exp\left[-\frac{m\omega}{\hbar}\left(s^2 + 2sx + \frac{1}{2}x^2\right)\right]. \end{aligned}$$

□

The generating function provides a very economical way of calculating normalizations and expectations, such as

$$\begin{aligned} \sum_{M,N=0}^{\infty} \frac{s^N}{N!} \frac{t^M}{M!} \langle \psi_N | \psi_M \rangle &= \int_{\mathbf{R}} \overline{G(s,x)} G(t,x)\, dx \\ &= \left(\frac{m\omega}{\pi\hbar}\right)^{\frac{1}{2}} \int_{\mathbf{R}} \exp\left(-\frac{m\omega}{\hbar}(s^2 + 2sx + t^2 + 2tx + x^2)\right) dx \\ &= \left(\frac{m\omega}{\pi\hbar}\right)^{\frac{1}{2}} e^{2m\omega st/\hbar} \int_{\mathbf{R}} \exp\left(-\frac{m\omega}{\hbar}(x+s+t)^2\right) dx. \end{aligned} \tag{7.47}$$

On substituting $u = x + s + t$ and doing the integration we find that

$$\sum_{M,N=0}^{\infty} \frac{s^N}{N!}\frac{t^M}{M!}\langle\psi_N|\psi_M\rangle = e^{2m\omega st/\hbar} = \sum_{N=0}^{\infty} \frac{1}{N!}\left(\frac{2m\omega st}{\hbar}\right)^N. \tag{7.48}$$

On comparing coefficients we see that

$$\langle\psi_N|\psi_M\rangle = \delta_{MN} N! \left(\frac{2m\omega}{\hbar}\right)^N, \tag{7.49}$$

confirming the orthogonality of the eigenvectors and providing the appropriate normalization.

We may also use the generating function to prove the following useful result:

Theorem 7.7.4. If the wave function ψ is orthogonal to ψ_N for all $N \geq 0$ then $\psi = 0$.

Proof. Since $G(s, x)$ is defined by a power series it can be extended to complex values of s, and we have

$$\begin{aligned} G(a + i\alpha, x) &= \left(\frac{m\omega}{\pi\hbar}\right)^{\frac{1}{4}} \exp\left[-\frac{m\omega}{\hbar}\left(\frac{1}{2}x^2 + 2(a + i\alpha)x + (a + i\alpha)^2\right)\right] \\ &= \exp\left(-\frac{m\omega}{\hbar}\left[\alpha(x + a) - \alpha^2\right]\right) G(a, x). \end{aligned} \tag{7.50}$$

Equation (3.39) for the inversion of the Fourier transform, (3.38), can be applied to a product of functions to give

$$F(x, x)\psi(x) = \frac{1}{2\pi\hbar}\int_{\mathbf{R}^2} e^{ip(x-y)/\hbar} F(x, y)\psi(y)\, dydp. \tag{7.51}$$

Setting $F(x, y) = G(a, x)G(a, y)$ and substituting $p = 2m\omega\alpha$ then gives

$$\begin{aligned} G(a, x)^2\psi(x) &= \frac{m\omega}{\pi\hbar}\int_{\mathbf{R}^2} e^{2im\omega\alpha(x-y)/\hbar} G(a, x)G(a, y)\psi(y)\, dyd\alpha, \\ &= \frac{m\omega}{\pi\hbar}\int_{\mathbf{R}^2} e^{-2m\omega\alpha^2/\hbar} G(a + i\alpha, x)\overline{G(a + i\alpha, y)}\psi(y)\, dyd\alpha, \\ &= \frac{m\omega}{\pi\hbar}\int_{\mathbf{R}} e^{-2m\omega\alpha^2/\hbar} G(a + i\alpha, x)\langle G(a + i\alpha)|\psi\rangle\, d\alpha. \end{aligned} \tag{7.52}$$

If $\langle\psi_N|\psi\rangle$ vanishes for all N then so does

$$\langle G(a+i\alpha)|\psi\rangle = \sum \frac{(a-i\alpha)^N}{N!}\langle\psi_N|\psi\rangle, \tag{7.53}$$

which means that $G(a,x)^2\psi(x) = 0$, and the result follows immediately. □

This result tells us that there are enough harmonic oscillator wave functions to span the space (in the sense of the definition given in Appendix A1). This can be seen still more explicitly in Exercie 7.20.

7.8*. Coherent states

The algebraic approach to the harmonic oscillator may seem rather abstract, but for some applications it is actually closer to the physics than a differential equation.

Laser light is usually approximately monochromatic; that is, it has a precisely defined frequency, ω. Each component, ϕ, of the electromagnetic field therefore satisfies the equation

$$\frac{\partial^2\phi}{\partial t^2} = -\omega^2\phi, \tag{7.54}$$

so that ϕ satisfies the same equation as a classical harmonic oscillator. Knowing that we are simply dealing with an oscillator it is easy to quantize the light wave, as, according to Planck and Einstein, we must. The energy levels will then be $(N+\frac{1}{2})\hbar\omega$. The ground state energy $\frac{1}{2}\hbar\omega$ is then interpreted as the energy of the vacuum before the beam has been turned on. Each successive level adds an energy $\hbar\omega$, that is precisely the energy of one extra photon. Thus the energy level $(N+\frac{1}{2})\hbar\omega$ is interpreted as having N photons added to the vacuum. Taking a vector ψ and forming $a_+\psi$ raises the energy by $\hbar\omega$, and so can be regarded as the creation of one photon. Similarly a_-, which lowers the energy, can be thought of as annihilating one photon. The characterization of the ground state as the vector satisfying $a_-\psi = 0$ corresponds to the physical idea that the vacuum contains no photons for a_- to annihilate.

This physical picture leads to the following terminology:

Definition 7.8.1. The operator a_+ is called a *creation operator.* Similarly a_- is called an *annihilation operator.*

The detailed theory of lasers has to explain the way in which the beams are generated and that involves a careful consideration of the interaction between matter and radiation that lies beyond the scope of this book. (The principles were first recognized by Einstein in 1916.) However, the beams themselves can often be described by the state $C(s)$ obtained by normalizing the generating function $G(s)$. We have already shown in the preceding section that

$$\langle G(s)|G(t)\rangle = e^{2m\omega st/\hbar}, \tag{7.55}$$

so that

$$\|G(s)\|^2 = e^{2m\omega s^2/\hbar}, \tag{7.56}$$

and we may take

$$C(s) = e^{-m\omega s^2/\hbar}G(s) = e^{-m\omega s^2/\hbar}\sum_{n=0}^{\infty}\frac{s^n}{n!}\psi_n. \tag{7.57}$$

Definition 7.8.2. The state described by $C(s)$ is known as a *coherent state.*

The name derives from the fact that $C(s)$ can be used to describe coherent light, although in that case the parameter m would no longer be interpreted as a mass.

Proposition 7.8.1. Coherent states satisfy

$$C(s,x) = \left(\frac{m\omega}{\pi\hbar}\right)^{\frac{1}{4}}\exp\left(-\frac{m\omega}{2\hbar}(x+2s)^2\right);$$

and

$$\langle C(s)|C(t)\rangle = \exp\left(-\frac{m\omega}{\hbar}|s-t|^2\right).$$

Proof. The first assertion follows from the definition since

$$C(s,x) = e^{-m\omega s^2/\hbar}G(s,x) = \exp\left(-\frac{m\omega}{2\hbar}(x+2s)^2\right). \tag{7.58}$$

The second follows from

$$\begin{aligned}\langle C(s)|C(t)\rangle &= \exp\left(-\frac{m\omega}{\hbar}(s^2+t^2)\right)\langle G(s)|G(t)\rangle \\ &= \exp\left(-\frac{m\omega}{\hbar}(s^2+t^2-2st)\right), \end{aligned} \tag{7.59}$$

which gives the stated result. □

As the exponential of a quadratic function of x, $C(s)$ is a minimum uncertainty state. (If we allow complex values of s then Corollary 7.2.2 tells us that every minimal uncertainty state is a scalar multiple of a coherent state.)

Being a sum over different ψ_n, $C(s)$ describes a state with no specific number of photons. The probability of finding n photons is the same as the probability that the energy is $(n+\frac{1}{2})\hbar\omega$. Bearing in mind that ψ_n is not normalized this is

$$\begin{aligned}\frac{|\langle\psi_n|C(s)\rangle|^2}{\|\psi_n\|^2} &= e^{-2m\omega s^2/\hbar}\frac{|\langle\psi_n|G(s)\rangle|^2}{\|\psi_n\|^2} \\ &= e^{-2m\omega s^2/\hbar}\frac{|s|^{2n}\|\psi_n\|^2}{(n!)^2}. \end{aligned} \tag{7.60}$$

We showed in the preceding section that

$$\|\psi_n\|^2 = n!\left(\frac{2m\omega}{\hbar}\right)^n, \tag{7.61}$$

so the probability reduces to

$$\frac{1}{n!}\left(\frac{2m\omega s^2}{\hbar}\right)^n e^{-2m\omega s^2/\hbar}, \tag{7.62}$$

that is the mean number of photons follows a Poisson distribution with mean $2m\omega s^2/\hbar$.

7.9*. Squeezed states

It is quite easy to show that both position and momentum have zero expectation values in the ground state, ψ_0, of the harmonic oscillator, and the energy is split evenly between potential and kinetic energy, so that

$$\mathsf{E}_{\psi_0}(P^2/2m) = \tfrac{1}{4}\hbar\omega = \mathsf{E}_{\psi_0}\left(\tfrac{1}{2}m\omega^2X^2\right), \tag{7.63}$$

(see Exercises 7.6 and 7.7). Combining these facts we see that

$$\Delta_{\psi_0}(P)^2 = \mathsf{E}_{\psi_0}(P^2) = \tfrac{1}{2}m\hbar\omega = \Delta_{\psi_0}(m\omega X)^2, \tag{7.64}$$

so that the ground state not only achieves the minimum permitted by the uncertainty principle, but it distributes that uncertainty evenly between the position and momentum. When we take the ground state Ψ_ζ of an oscillator with a different frequency, $\zeta\omega$, we similarly obtain

$$\Delta_{\Psi_\zeta}(P)^2 = \tfrac{1}{2}m\hbar\zeta\omega = \Delta_{\Psi_\zeta}\left((m\zeta\omega X)^2\right)^2, \tag{7.65}$$

whence it follows that $\Delta_{\Psi_\zeta}(P)^2 = \frac{1}{2}m\hbar\zeta\omega$ and $\Delta_{\Psi_\zeta}(m\omega X)^2 = \frac{1}{2}m\hbar\zeta^{-1}\omega$. By choosing $\zeta < 1$ we may reduce the dispersion of P at the expense of increasing that of $m\omega X$, whilst retaining the minimal uncertainty overall. It is possible to prepare such states of light in the laboratory, and they are known as *squeezed states*, with squeezing parameter $r = -\frac{1}{2}\ln\zeta$. (It is also possible to work with complex values of ζ, but then one can no longer interpret $\zeta\omega$ as a frequency, the expression for r becomes more complicated, and the uncertainty is not minimal.) The state Ψ_ζ is not an eigenstate of the physical Hamiltonian, $P^2/2m + \frac{1}{2}m\omega^2X^2$, and so it changes with time. Initially the dispersion of P increases whilst that of $m\omega X$ diminishes until the above values are reversed, and then the changes continue in a periodic cycle. Nonetheless this is entirely predictable and it provides a way of sidestepping some of the effects of the uncertainty principle. This is particularly useful when transmitting signals with small numbers of photons, so that the uncertainty limits are important.

It is quite easy to calculate the expected number of photons in the squeezed state. Bearing in mind that the energy is $(n+\frac{1}{2})\hbar\omega$, we make the following definition:

Definition 7.9.1. The *number operator* is defined to be

$$N = (\hbar\omega)^{-1}(P^2/2m + m\omega^2X^2/2) - \tfrac{1}{2}.$$

Using our earlier expressions for the kinetic and potential energy, the expectation value of N is given by

$$\begin{aligned}(\hbar\omega)^{-1}\mathsf{E}(P^2/2m + m\omega^2X^2/2) - \tfrac{1}{2} &= \tfrac{1}{4}\left(\zeta + \zeta^{-1}\right) - \tfrac{1}{2}\\ &= \left(\frac{\zeta^{\frac{1}{2}} - \zeta^{-\frac{1}{2}}}{2}\right)^2.\end{aligned} \tag{7.66}$$

Unless $\zeta = 1$, this is positive showing that the squeezed states do contain photons. By a more careful analysis one can find the complete probability distribution for N.

Exercises

7.1 Let A and B be self-adjoint operators. Show that their commutator satisfies

$$[A, B]^* = -[A, B].$$

Hence or otherwise show that $-i[A, B]$ is self-adjoint.

7.2 Let X_1, X_2, X_3 and P_1, P_2, P_3 be the position and momentum observables for a particle moving in three dimensions. Define the angular momentum observables $L_1 = (X_2P_3 - X_3P_2)$, $L_2 = (X_3P_1 - X_1P_3)$ and $L_3 = (X_1P_2 - X_2P_1)$ and show that

$$[L_1, L_2] = i\hbar L_3.$$

7.3° A particle of mass m moves along the x-axis under the influence of a potential $V(x) = \frac{1}{2}m\omega^2x^2$. Show that, if $T = P^2/2m$ is its kinetic energy, then

$$\mathsf{E}_\psi(T)\mathsf{E}_\psi(V) \geq \left(\frac{\hbar\omega}{4}\right)^2.$$

7.4 Show that for a differentiable function, f,

(i) $[P, f(X)] = -i\hbar f'(X)$;

(ii) $[XP, f(X)] = -i\hbar X f'(X)$.

By considering Fourier transforms deduce that, for any differentiable function, g,

(iii) $[X, g(P)] = i\hbar g'(P)$;

(iv) $[XP, g(P)] = i\hbar P g'(P)$.

7.5 By taking $g(P) = \exp(-iaP/\hbar)$ in part (iii) of the previous exercise, or otherwise, show that

$$e^{iaP/\hbar}Xe^{-iaP/\hbar} = X + a.$$

Deduce that for any function f that can be expanded in a power series,

$$e^{iaP/\hbar}f(X)e^{-iaP/\hbar} = f(X + a).$$

Hence, or otherwise, deduce Weyl's form of the commutation relations:

$$e^{iaP/\hbar}e^{ibX/\hbar} = e^{iba/\hbar}e^{ibX/\hbar}e^{iaP/\hbar}.$$

7.6 Let ψ be an eigenvector with energy E for the harmonic oscillator Hamiltonian $H = P^2/2m + \frac{1}{2}m\omega^2X^2$. By considering $\mathsf{E}(a_\pm)$ and $\mathsf{E}_\psi(a_\pm^2)$ or otherwise, show that $\mathsf{E}_\psi(X)$ and $\mathsf{E}_\psi(P)$ vanish and that

$$\mathsf{E}_\psi(T) = \mathsf{E}_\psi(V),$$

where $T = P^2/2m$. Deduce that $\Delta_\psi(P)^2 = mE$, and find the corresponding expression for $\Delta_\psi(X)$. Verify the formulae for the expectation values of T and V directly for the eigenstates of the one-dimensional harmonic oscillator.
[The eigenstates ψ_n corresponding to the energy levels $(n + \frac{1}{2})\hbar\omega$ of the harmonic oscillator satisfy

$$\left(\frac{m\omega}{\hbar}\right)^{\frac{1}{2}} X\psi_n = \left(\frac{n}{2}\right)^{\frac{1}{2}} \psi_{n-1} + \left(\frac{n+1}{2}\right)^{\frac{1}{2}} \psi_{n+1}.]$$

7.7 Show that if $H\psi = E\psi$ then for any operator A

$$\langle\psi|[H, A]\psi\rangle = 0.$$

Suppose that $H = T + V$ where $T = P^2/2m$ and $V = kX^N$ with k a complex number.

(i) By taking $A = X$ show that $\mathsf{E}_\psi(P) = 0$.
(ii) By taking $A = XP$ derive the *virial theorem*:

$$2\mathsf{E}_\psi(T) = N\mathsf{E}_\psi(V).$$

(iii) Deduce that $\mathsf{E}_\psi(T) = NE/(N+2)$, and find $\Delta_\psi(P)$.
(iv) Show that the hydrogen atom potential is homogeneous of degree -1 and deduce that any bound state energy E is negative.
(v) When $V = \frac{1}{2}m\omega^2X^2$ show that

$$\Delta_\psi(P)\Delta_\psi(X) = \frac{E}{\omega}.$$

7.8 At time $t = 0$ a quantum mechanical system is in a state for which both the expected values of position and momentum are 0. The Hamiltonian operator for the system is $H = P^2/2m$. Prove that at a subsequent time t, $\Delta_\psi(P)$ has the same value as at $t = 0$, but that

$$\frac{d}{dt}\left((\Delta_\psi(X))^2\right) = \frac{1}{m}\mathsf{E}_\psi(XP + PX).$$

Hence, or otherwise, show that $\Delta_\psi(X)^2$ increases quadratically with t.
[*Hint* The results of Exercise 6.7 may be used.]

7.9° An operator a satisfies $[a, a^*] = 1$ where a^* is the adjoint of a. If the operator $N = a^*a$ show that N has for eigenvalues the set of non-negative integers. Show also that if each eigenvalue k is non-degenerate then it is possible to choose the corresponding normalized eigenvectors ψ_k in such a way that

$$a^*\psi_k = \sqrt{k+1}\psi_{k+1}.$$

Hence, or otherwise, derive the energy levels of the Hamiltonian $H = P^2/2m + \frac{1}{2}m\omega^2X^2$. Derive the ground state and first excited state as functions of x.

7.10° The operator a satisfies the relations

$$a^2 = 0, \qquad a^*a + aa^* = 1.$$

The operator N is defined by $N = a^*a$. Find $[N, a]$ and $[N, a^*]$, and show that N is a self-adjoint projection, that is $N^2 = N = N^*$. Obtain a matrix representation for a, a^* and N in which N is diagonal.

7.11 A charged particle moving in the plane perpendicular to a magnetic field B has Hamiltonian

$$H = \frac{1}{2m}\left[\left(P_1 + \frac{1}{2}eX_2B\right)^2 + \left(P_2 - \frac{1}{2}eX_1B\right)^2\right].$$

Show that the set of energy eigenvalues is

$$\left\{\left(n + \frac{1}{2}\right)\frac{|eB|\hbar}{m} : n = 0, 1, 2, \ldots\right\}.$$

7.12 Use the generating function

$$G(s, x) = \left(\frac{m\omega}{\pi\hbar}\right)^{\frac{1}{4}} \exp\left[-\frac{m\omega}{\hbar}\left(s^2 + 2sx + \frac{1}{2}x^2\right)\right]$$

to find the normalized wave functions for the first and second excited states of the harmonic oscillator.

7.13 Show that

$$a_-C(s) = (2im\omega s)C(s).$$

7.14 Show that $\langle\psi_M|X\psi_N\rangle$ vanishes unless $|M - N| = 1$, and for all $N \geq 0$ find $\langle\psi_N|X\psi_N\rangle$ and $\langle\psi_N|X^2\psi_N\rangle$

(i) by using generating functions,

(ii) by writing $X = (a_+ - a_-)/2im\omega$.

7.15 Let $G_t(s,x) = \exp(-\frac{1}{2}i\omega t)G(\exp(-i\omega t)s, x)$. Show that

$$i\hbar\frac{\partial G_t}{\partial t} = \left(-\frac{\hbar^2}{2m}\frac{\partial^2}{\partial x^2} + \frac{1}{2}m\omega^2 x^2\right) G_t.$$

At time $t = 0$ the wave function for the harmonic oscillator is

$$\psi(0,x) = \left(\frac{m\omega}{\pi\hbar}\right)^{\frac{1}{4}} \exp\left(-\frac{m\omega}{2\hbar}(x-a)^2\right).$$

Find the wave function at time t.

7.16 A particle moves on the x-axis in a potential, V, that is periodic of period a. (That is, $V(x+a) = V(x)$ for all $x \in \mathbf{R}$.) Let T_a denote the operator defined by

$$(T_a\psi)(x) = \psi(x+a).$$

Show that T_a commutes with the Hamiltonian operator H.

Show also that $T_a\psi = \lambda\psi$ if and only if $\lambda^{-x/a}\psi(x)$ is periodic with period a. Deduce that the energy eigenstates of the form $\psi(x) = \exp(ikx/a)\phi(x)$, with ϕ periodic of period a, span the space of all energy eigenstates.

7.17 Show using Lemma 7.3.1 that

$$\langle\psi|\psi_t\rangle = [1 + o(t^2)] \int e^{-itE/\hbar} \exp\left(-\frac{[E - \mathsf{E}_\psi(H)]^2}{2\Delta_\psi(H)^2}\right) dE.$$

[This shows that for small times s the behaviour is though the energy were normally distributed with mean $\mathsf{E}_\psi(H)$ and variance $\Delta_\psi(H)^2$.]

7.18 Show that in *classical* Hamiltonian mechanics the equations of motion for a one-dimensional harmonic oscillator with potential $\frac{1}{2}m\omega^2x^2$ can be written in complex form as

$$\frac{d}{dt}(p + im\omega x) = i\omega(p + im\omega x).$$

Deduce that

$$[p(t) + im\omega x(t)] = e^{i\omega t}[p(0) + im\omega x(0)],$$

and find an expression for the position at time t in terms of the initial values, $p(0)$ and $x(0)$.

7.19 Suppose that there is a map, $\mathcal{Q}$, from functions of p and x to operators, which satisfies $[\mathcal{Q}(f), \mathcal{Q}(g)] = -i\hbar\mathcal{Q}(\{f, g\})$, and also that $\mathcal{Q}(p) = P$ and $\mathcal{Q}(x) = X$.

(i) Show that $\mathcal{Q}(x^k) - X^k$ and $\mathcal{Q}(p^k) - P^k$ commute with P and X, for any $k > 1$.

(ii) Show that $12\{p^3, x^3\} = \{\{p^3, x^2\}, \{x^3, p^2\}\}$.

(iii) Show that $-12\hbar^2[P^3, X^3] \neq [[P^3, X^2], [X^3, P^2]]$.

(iv) Deduce the Groenewald–van Hove theorem that $\mathcal{Q}$ cannot exist.

[You may assume that the only operators which commute with P and X are multiples of the identity. The results of Exercise 7.4 may be useful.]

7.20 Show that for any wave function ψ, one has

$$\psi(x) = \frac{m\omega}{\pi\hbar} \int_{\mathbf{R}^2} e^{-2m\omega(a^2+\alpha^2)/\hbar} G(a + i\alpha, x) \langle G(a + i\alpha)|\psi\rangle \, d\alpha da.$$

8 Angular momentum

> If therefore the constant *h* of Planck has ... an atomic significance, it may mean that the angular momentum of an atom can only rise or fall by discrete amounts when an electron leaves or returns.
>
> JOHN NICHOLSON, in *The constitution of the solar corona*, 1912

8.1. Angular momentum in quantum mechanics

In many of the systems studied in classical mechanics angular momentum is far more useful than ordinary linear momentum. For example, in the study of planetary orbits, or indeed any motion under a central force, the conservation of angular momentum plays a crucial role. We shall therefore turn our attention to the description of angular momentum in quantum mechanics.

The classical angular momentum of a particle is described by the vector $\mathbf{x} \times \mathbf{p}$. This suggests that quantum mechanical angular momentum should be described by three operator components:

Definition 8.1.1. The quantum mechanical *orbital angular momentum* vector, $\mathbf{L}$, has for components the operators

$$L_1 = X_2P_3 - X_3P_2, \qquad L_2 = X_3P_1 - X_1P_3, \qquad L_3 = X_1P_2 - X_2P_1.$$

By use of the alternating symbol

$$\epsilon_{jkl} = \begin{cases} 1 & \text{if } (jkl) \text{ is a cyclic permutation of } (123) \\ -1 & \text{if } (jkl) \text{ is a cylic permutation of } (213) \\ 0 & \text{in all other cases,} \end{cases} \tag{8.1}$$

and the summation convention (that we sum over any repeated index over the values 1, 2 and 3), we may also write this as

$$L_j = \epsilon_{jkl}X_kP_l. \tag{8.2}$$

The fundamental algebraic relations for the angular momentum are summarized in the following result.

Proposition 8.1.1. The angular momentum satisfies the following commutation relations:

(i) $[L_j, P_k] = i\hbar\epsilon_{jkl}P_l$;

(ii) $[L_j, X_k] = i\hbar\epsilon_{jkl}X_l$;

(iii) $[L_j, L_k] = i\hbar\epsilon_{jkl}L_l$.

Proof. (i) Using Proposition 7.1.2(iii) together with the commutation relations, 7.1.1, we have

$$[X_sP_t, P_k] = [X_s, P_k]P_t + X_s[P_t, P_k] = i\hbar\delta_{ks}P_t. \tag{8.3}$$

So

$$[L_j, P_k] = \epsilon_{jsl}[X_sP_l, P_k] = i\hbar\epsilon_{jsl}\delta_{ks}P_l = i\hbar\epsilon_{jkl}P_l. \tag{8.4}$$

(ii) The formula for $[L_j, X_k]$ is obtained similarly.
(iii) Finally,

$$\begin{aligned}[L_1, L_2] &= [L_1, X_3P_1 - X_1P_3] \\ &= X_3[L_1, P_1] + [L_1, X_3]P_1 - X_1[L_1, P_3] - [L_1, X_1]P_3 \\ &= 0 - i\hbar X_2P_1 + i\hbar X_1P_2 - 0 \\ &= i\hbar L_3, \end{aligned} \tag{8.5}$$

and the other non-trivial commutation relations between the components of $\mathbf{L}$ follow by permutations of the indices. □

It is useful now to abstract from the above result the following more general idea:

Definition 8.1.2. Any three operators J_1, J_2, and J_3 that satisfy the commutation relations

$$[J_j, J_k] = i\hbar\epsilon_{jkl}J_l$$

are called *angular momentum operators.*

Most of the following analysis works equally for orbital and general angular momentum operators. The commutation relations can be cast into a useful coordinate-free form by introducing the angular momentum operator in the direction of the vector $\mathbf{a}$

$$\mathbf{a}.\mathbf{J} = a_1 J_1 + a_2 J_2 + a_3 J_3. \tag{8.6}$$

Then we have in place of (iii):

Proposition 8.1.2.

$$[\mathbf{a}.\mathbf{J}, \mathbf{b}.\mathbf{J}] = i\hbar(\mathbf{a} \times \mathbf{b}).\mathbf{J}.$$

Proof. The derivation of this is left as a simple exercise. □

8.2. Ladder operators

The fact that the three components of angular momentum do not commute with each other is at first sight somewhat disconcerting, for this means that they cannot simultaneously be measured precisely, and yet we expected them to be constants of the motion in central force problems. To overcome this difficulty we need another related operator.

Definition 8.2.1. The *total angular momentum* is defined by

$$J^2 = J_1^2 + J_2^2 + J_3^2.$$

This is more tractable as the following result shows.

Proposition 8.2.1. Suppose that $[J_j, A_k] = i\hbar\epsilon_{jkl}A_l$, $[J_j, B_k] = i\hbar\epsilon_{jkl}B_l$, and that $\mathbf{A}.\mathbf{B} = A_1B_1 + A_2B_2 + A_3B_3$. Then

$$[J_j, \mathbf{A}.\mathbf{B}] = 0$$

for $j = 1, 2, 3$.

Proof. With the summation convention we may write $\mathbf{A.B}$ as $A_k B_k$ to obtain

$$\begin{aligned}
[J_j, \mathbf{A.B}] &= [J_j, A_k B_k] \\
&= A_k [J_j, B_k] + [J_j, A_k] B_k \\
&= A_k i\hbar \epsilon_{jkl} B_l + i\hbar \epsilon_{jkl} A_l B_k \\
&= i\hbar \epsilon_{jkl} \left(A_k B_l + A_l B_k \right) \\
&= 0,
\end{aligned} \tag{8.7}$$

where the last line follows from the fact that ϵ_{jkl} is antisymmetric in k and l whilst the bracketed term is symmetric. □

Corollary 8.2.2. Each of the operators X^2, P^2, $\mathbf{X.P}$, and L^2 commutes with L_j, and J^2 commutes with J_j, for $j = 1, 2, 3$.

This result, which is an immediate consequence of the last two propositions, means that we can find simultaneous eigenvectors of J^2 and J_3. Once J_3 has been given an eigenvalue the relation $[J_1, J_2] = i\hbar J_3$ will closely resemble the commutation relation between X and P, and this suggests that it might be useful to introduce some more operators:

Definition 8.2.2. The *ladder operators* $J_\pm$ are defined by

$$J_\pm = J_1 \pm iJ_2.$$

Proposition 8.2.3. The ladder operators satisfy the following relations:
(i) $[J^2, J_\pm] = 0$;
(ii) $J_\mp J_\pm = J^2 - \left(J_3^2 \pm \hbar J_3\right)$;
(iii) $[J_+, J_-] = 2\hbar J_3$;
(iv) $J_3 J_\pm = J_\pm (J_3 \pm \hbar)$ or, equivalently, $[J_3, J_\pm] = \pm\hbar J_\pm$.

Proof. (i) The first part is a trivial consequence of Corollary 8.2.2.

(ii) Applying Lemma 7.2.1 with $A = J_1$, $B = J_2$, $C = -\hbar J_3$, and $t = \mp 1$, we have

$$J_\mp J_\pm = J_1^2 + J_2^2 \mp \hbar J_3$$
$$= J^2 - \left(J_3^2 \pm \hbar J_3\right). \tag{8.8}$$

(iii) This part follows immediately from the previous one:

$$[J_+, J_-] = \left[J^2 - \left(J_3^2 - \hbar J_3\right)\right] - \left[J^2 - \left(J_3^2 + \hbar J_3\right)\right]$$
$$= 2\hbar J_3. \tag{8.9}$$

(iv) Finally, we have by direct calculation:

$$[J_3, J_\pm] = [J_3, J_1 \pm iJ_2]$$
$$= i\hbar J_2 \pm \hbar J_1$$
$$= \pm\hbar(J_1 \pm i\hbar J_2)$$
$$= \pm\hbar J_\pm, \tag{8.10}$$

which completes the proof. □

Corollary 8.2.4. Let ψ be a common eigenvector of J^2 and J_3, satisfying $J^2\psi = \lambda\hbar^2\psi$ and $J_3\psi = m\hbar\psi$. Then
(i) $J^2 J_\pm\psi = \lambda\hbar^2 J_\pm\psi$;
(ii) $J_3 J_\pm\psi = (m \pm 1)\hbar J_\pm\psi$;
(iii) $\|J_\pm\psi\|^2 = [\lambda - m(m \pm 1)]\hbar^2\|\psi\|^2$;
(iv) $\lambda \geq m(m \pm 1)$ and $\lambda = m(m \pm 1)$ if and only if $J_\pm\psi = 0$.

Proof. We know from Corollary 8.2.2 that J^2 and J_3 commute and so have common eigenvectors. We know also from part (i) of the preceding proposition that J^2 commutes with $J_\pm$, from which the first part of the corollary follows immediately.

Applying both sides of the operator identity 8.2.3(iv) to ψ we obtain the second identity.

Since J_+ and J_- are adjoints we have

$$\|J_\pm\psi\|^2 = \langle\psi|J_\mp J_\pm\psi\rangle, \tag{8.11}$$

and from Proposition 8.2.3(ii) we deduce that

$$\begin{aligned}\|J_{\pm}\psi\|^2 &= \langle\psi|\left[J^2 - \left(J_3^2 \pm \hbar J_3\right)\right]\psi\rangle \\ &= [\lambda - m(m \pm 1)]\hbar^2\|\psi\|^2. \end{aligned} \tag{8.12}$$

The inequality relating λ and m follows immediately, together with the condition for equality. □

8.3. Representations of the angular momentum operators

We have now assembled the same kind of information about the angular momentum operators that Corollary 7.5.2 supplied for the creation and annihilation operators and we can now follow the strategy of Section 7.5. Just as with the energy of the harmonic oscillator, there are barriers that prevent us from changing the eigenvalues of J_3 too much, but this time the inequality $\lambda \geq m(m \pm 1)$ provides both upper and lower bounds, blocking the action of both J_+ and J_-. As with the oscillator, this is the key to discovering those J_3 eigenvalues.

Theorem 8.3.1. The eigenvalues of J^2 have the form $j(j+1)\hbar^2$ for $j = 0, \frac{1}{2}, 1, \frac{3}{2}, 2, \frac{5}{2}, \ldots$.

For each choice of j the eigenvalues of J_3 are $m\hbar$ for $m = -j$, $-j+1, \ldots, j-1, j$. The degeneracy of each eigenvalue is the same as that of $j\hbar$.

Proof. The inequality provided by Corollary 8.2.4(iv) is particularly transparent when written in the form $\lambda + \frac{1}{4} \geq (m \pm \frac{1}{2})^2$, and clearly shows that for a given λ, the eigenvalue m is bounded above and below. Since, by repeated use of Corollary 8.2.4(ii), $J_{\pm}^N\psi$ either vanishes or is another eigenvector with eigenvalue $(m \pm N)$, it is clear that for some non-negative integers p and q both $J_+^{p+1}\psi$ and $J_-^{q+1}\psi$ must vanish. We may as well assume that these are the least integers with this property, so that $J_+^p\psi \neq 0$ is an eigenvector of J_3 with eigenvalue $(m+p)\hbar$ and the identity $J_+(J_+^p\psi) = 0$ is the condition required in 7.2.4(iv) for the equality $\lambda = (m+p)(m+p+1)$. Similarly, from $J_-(J_-^q\psi) = 0$ we deduce that $\lambda = (m-q)(m-q+1)$.

Combining these we see that $(m+p)(m+p+1) = (m-q)(m-q-1)$, which can be rearranged as

$$2m(p+q+1) = (q-p)(p+q+1). \tag{8.13}$$

Since $p+q+1$ is positive we deduce that $m = (q-p)/2$. We may therefore write $m+p=j$ where $j = (q+p)/2$ and deduce that

$$\lambda = (m+p)(m+p+1) = j(j+1). \tag{8.14}$$

It is immediate from our definition that j is half of a non-negative integer, and that $m = j-p$ must differ from it by an integer. To see that each such value of m does occur in an eigenvalue of J_3, we first note that $\psi_j = J_+^p \psi$ provides an eigenvector with eigenvalue $(m+p)\hbar = j\hbar$, and then that $J_-^{j-m}\psi_j$ is an eigenvector with eigenvalue $m\hbar$. One may check the degeneracies as for the harmonic oscillator. □

Corollary 8.3.2. If there are no degeneracies then the eigenspace on which J^2 takes the eigenvalue $j(j+1)\hbar^2$ is $2j+1$ dimensional with a basis of the form $\{\psi_m \in \mathcal{H} : 0 \le j-m \le 2j\}$ such that

$$\begin{aligned} J_3\psi_m &= m\hbar\psi_m, \\ J_\pm\psi_m &= \sqrt{(j\mp m)(j\pm m+1)}\hbar\psi_{m\pm 1}. \end{aligned}$$

Proof. According to Corollary 8.2.4(iii), when $J^2\psi = j(j+1)\hbar^2\psi$ and $J_3\psi = m\hbar\psi$, we have

$$\|J_-\psi\|^2 = \Big[j(j+1)-m(m-1)\Big]\hbar^2\|\psi\|^2 = (j+m)(j-m+1)\hbar^2\|\psi\|^2. \tag{8.15}$$

Let ψ_j be a normalized common eigenvector for which the J_3 eigenvalue is $j\hbar$. We may iteratively define

$$\psi_{j-q} = [q(2j-q+1)]^{-\frac{1}{2}}\hbar^{-1}J_-\psi_{j-q+1}, \tag{8.16}$$

for $q = 1, 2, \dots, 2j$, to obtain a sequence of normalized eigenvectors having each of the permissible eigenvalues for J_3. Applying J_+ to this defining relation and using Proposition 8.2.3(ii) we obtain

$$\begin{aligned} J_+\psi_{j-q} &= [q(2j-q+1)]^{-\frac{1}{2}}\hbar^{-1}(J^2 - J_3^2 + \hbar J_3)\psi_{j-q+1} \\ &= [q(2j-q+1)]^{\frac{1}{2}}\hbar\psi_{j-q+1}, \end{aligned} \tag{8.17}$$

from which the stated expression for the action of J_+ follows. The action of J_- is, by definition, as asserted. □

Spectacular demonstrations of this quantization are now available. At temperatures below 2.19 K liquid helium becomes a 'superfluid' and starts to exhibit quantum phenomena on a macroscopic scale. If a small bucket of the fluid is set spinning, vortices can be formed, each of which carries one quantum $\hbar$ of angular momentum.

We shall now show how Corollary 8.3.2 can be used to get explicit operators satisfying the commutation relations for angular momentum.

Example 8.3.1. When $j = 0$ then $2j + 1 = 1$ and the space is spanned by a single vector ψ_0 such that

$$J_3\psi_0 = 0, \qquad J_\pm\psi_0 = 0, \qquad J^2\psi_0 = 0. \tag{8.18}$$

It is easy to verify that these operators do satisfy the angular momentum commutation relations, but for obvious reasons this is sometimes known as the *trivial representation* of those relations.

Example 8.3.2. The next permissible value is $j = \frac{1}{2}$. For $j = \frac{1}{2}$ the space is spanned by two vectors $\psi_\pm = \psi_{\pm\frac{1}{2}}$, such that

$$J_3\psi_\pm = \pm\tfrac{1}{2}\hbar\psi_\pm \tag{8.19}$$

$$J_+\psi_+ = 0 = J_-\psi_- \tag{8.20}$$

$$J_+\psi_- = \hbar\psi_+ \qquad J_-\psi_+ = \hbar\psi_-. \tag{8.21}$$

With respect to the basis ψ_+, ψ_- we have the matrices

$$J_3 = \frac{1}{2}\hbar\begin{pmatrix} 1 & 0 \\ 0 & -1 \end{pmatrix}, \qquad J_+ = \hbar\begin{pmatrix} 0 & 1 \\ 0 & 0 \end{pmatrix}, \qquad J_- = \hbar\begin{pmatrix} 0 & 0 \\ 1 & 0 \end{pmatrix}. \tag{8.22}$$

Using the identities $J_1 = \frac{1}{2}(J_+ + J_-)$, and $J_2 = -\frac{1}{2}i(J_+ - J_-)$ we derive

$$J_1 = \frac{1}{2}\hbar\begin{pmatrix} 0 & 1 \\ 1 & 0 \end{pmatrix}, \qquad J_2 = \frac{1}{2}\hbar\begin{pmatrix} 0 & -i \\ i & 0 \end{pmatrix}. \tag{8.23}$$

Definition 8.3.1. The matrix representation of J_1, J_2 and J_3 given above is called the *spin representation* of angular momentum. The three matrices

$$\sigma_1 = \begin{pmatrix} 0 & 1 \\ 1 & 0 \end{pmatrix}, \qquad \sigma_2 = \begin{pmatrix} 0 & -i \\ i & 0 \end{pmatrix}, \qquad \sigma_3 = \begin{pmatrix} 1 & 0 \\ 0 & -1 \end{pmatrix},$$

are called the *Pauli spin matrices.*

Using the Pauli spin matrices we may write the spin representation in the form $J_k = \frac{1}{2}\hbar\sigma_k$ for $k = 1, 2, 3$, and

$$\mathbf{a.J} = \frac{1}{2}\hbar\mathbf{a}.\boldsymbol{\sigma} = \frac{1}{2}\hbar\begin{pmatrix} a_3 & a_1 - ia_2 \\ a_1 + ia_2 & -a_3 \end{pmatrix}. \tag{8.24}$$

Example 8.3.3. When $j = 1$ we have a space spanned by $2j + 1 = 3$ vectors, $\psi_+ = \psi_{+1}$, ψ_0, and $\psi_- = \psi_{-1}$. Each of these is an eigenvector of J^2 with eigenvalue $2\hbar^2$. According to our formulae

$$\begin{array}{ccc} J_3\psi_+ = \hbar\psi_+ & J_3\psi_0 = 0 & J_3\psi_- = -\hbar\psi_- \\ J_-\psi_+ = \sqrt{2}\hbar\psi_0 & J_-\psi_0 = \sqrt{2}\hbar\psi_- & J_-\psi_- = 0 \\ J_+\psi_+ = 0 & J_+\psi_0 = \sqrt{2}\hbar\psi_+ & J_+\psi_- = \sqrt{2}\hbar\psi_0. \end{array} \tag{8.25}$$

With respect to the basis ψ_+,ψ_0, ψ_-, the angular momentum operators are represented by the following matrices:

$$J_+ = \hbar\begin{pmatrix} 0 & \sqrt{2} & 0 \\ 0 & 0 & \sqrt{2} \\ 0 & 0 & 0 \end{pmatrix}, \qquad J_- = \hbar\begin{pmatrix} 0 & 0 & 0 \\ \sqrt{2} & 0 & 0 \\ 0 & \sqrt{2} & 0 \end{pmatrix},$$

$$J_3 = \hbar\begin{pmatrix} 1 & 0 & 0 \\ 0 & 0 & 0 \\ 0 & 0 & -1 \end{pmatrix}.$$

In this form one can readily check that the commutation relations for the J are satisfied. In fact, we shall show in the next section that this is just a heavily disguised version of ordinary three-dimensional space.

8.4. Orbital angular momentum and spherical harmonics

Theorem 8.3.1 not only tells us which eigenvalues of J^2 and J_3 are possible, but, as the examples show, its corollary also tells us how all the operators act on a particular basis of vectors. However, although this works well in an abstract inner product space, it is not always possible to realize the same operators on a space of wave functions.

Proposition 8.4.1. For orbital angular momentum the parameters j and m in Theorem 8.3.1 must both be integers.

Proof. To understand the distinction between orbital angular momentum and the abstract algebraic situation that we studied in the previous sections we need to study the action of the angular momentum operators on wave functions in greater detail. In terms of spherical polar coordinates (r, θ, ϕ) the position vector is $\mathbf{x} = (r \sin\theta \cos\phi, r \sin\theta \sin\phi, r \cos\theta)$. According to the chain rule

$$\frac{\partial}{\partial \phi} = \frac{\partial x_1}{\partial \phi}\frac{\partial}{\partial x_1} + \frac{\partial x_2}{\partial \phi}\frac{\partial}{\partial x_2} + \frac{\partial x_3}{\partial \phi}\frac{\partial}{\partial x_3} = -x_2\frac{\partial}{\partial x_1} + x_1\frac{\partial}{\partial x_2}, \tag{8.26}$$

so that

$$\frac{\hbar}{i}\frac{\partial}{\partial \phi} = X_1 P_2 - X_2 P_1 = L_3. \tag{8.27}$$

Now $L_3\psi_m = m\hbar\psi_m$ can be rewritten as

$$\frac{\partial \psi_m}{\partial \phi} = im\psi_m, \tag{8.28}$$

which integrates to

$$\psi_m(r, \theta, \phi) = e^{im\phi}\psi_m(r, \theta, 0). \tag{8.29}$$

On the other hand we want

$$\psi_m(r, \theta, \phi) = \psi_m(r, \theta, \phi + 2\pi) = e^{2\pi i m}\psi_m(r, \theta, \phi). \tag{8.30}$$

Unless m is integral this forces ψ_m to vanish identically, so that the integrality of m follows immediately. Since j is just the maximum value of m, it must also be integral. □

We must now try to reconcile this result with the existence of the spin representation for $j = \frac{1}{2}$ which we explicitly constructed in Example 8.3.2. We have just seen that in general a rotation through 2π multiplies ψ_m by a factor of $\exp(2\pi i m)$, which is -1 when j is an odd multiple of $\frac{1}{2}$. In the abstract theory the vectors ψ_m and $-\psi_m$ are regarded as defining the same state, and so there is nothing wrong with introducing a minus sign when we rotate through 2π. Wave functions are more restrictive, however, since we need to assign to them a value at each point. Experiments have made it clear that the electron and many other subatomic particles found in nature have, in addition to any orbital angular momentum, an intrinsic angular momentum or *spin* given by $j = \frac{1}{2}$. It is the sum of the orbital and spin angular momentum which is conserved, and not the orbital angular momentum on its own. In situations where the spin matters (for example, when an electron enters a magnetic field) we cannot simply use ordinary wave functions. What one in fact does is to combine the spin and orbital angular momenta by taking two-component wave functions, or *spinors*, whose values lie in the two-dimensional spin representation space.

It is also true that for any integer l there is a $(2l+1)$-dimensional subspace of wave functions on which the angular momentum operators act in the way described in Corollary 8.3.2. This can be shown by explicit solution of the equations

$$\begin{aligned} L^2\psi_m &= l(l+1)\hbar^2\psi_m \\ L_3\psi_m &= m\hbar\psi_m, \end{aligned} \tag{8.31}$$

in polar coordinates, which amounts to the analysis of Section 4.2. However, there is an alternative approach that identifies the angular momentum wave functions as particular solutions of Laplace's equation. This exploits the operator analogue of the vector identity

$$|\mathbf{x}|^2|\mathbf{p}|^2 - (\mathbf{x}.\mathbf{p})^2 = |\mathbf{x}\times\mathbf{p}|^2. \tag{8.32}$$

Proposition 8.4.2. Let $D = X_1P_1 + X_2P_2 + X_3P_3$. Then

$$L^2 = X^2P^2 - D\,(D - i\hbar)\,.$$

Proof. We first note that, using Proposition 8.1.1, we have

$$\begin{aligned} L_1^2 &= L_1\,(X_2P_3 - X_3P_2) \\ &= (X_2L_1 + i\hbar X_3)\,P_3 - (X_3L_1 - i\hbar X_2)\,P_2 \\ &= X_2L_1P_3 - X_3L_1P_2 + i\hbar\,(X_2P_2 + X_3P_3)\,. \end{aligned} \tag{8.33}$$

Adding this to the analogous equations for L_2^2 and L_3^2 we get

$$L^2 = \epsilon_{jkl}X_jL_lP_k + 2i\hbar D. \tag{8.34}$$

The canonical commutation relations (Theorem 7.1.1) also give

$$P_jX_k - X_jP_k = (X_kP_j - X_jP_k) - i\hbar\delta_{jk} = -\epsilon_{jkl}L_l - i\hbar\delta_{jk}. \tag{8.35}$$

Multiplying on the left by X_j, on the right by P_k and summing over j and k, this gives

$$D^2 - X^2P^2 = -\epsilon_{jkl}X_jL_lP_k - i\hbar D. \tag{8.36}$$

Adding this to the last expression for L^2 yields

$$L^2 + D^2 - X^2P^2 = i\hbar D, \tag{8.37}$$

which reduces to the stated relationship. □

Corollary 8.4.3. The Laplace operator may be written as

$$\Delta = \frac{1}{r}\frac{\partial^2}{\partial r^2}r - \frac{1}{\hbar^2 r^2}L^2.$$

Proof. We have $P^2 = -\hbar^2\Delta$, $X^2 = r^2$, and

$$D = \frac{\hbar}{i}\mathbf{x}.\nabla = \frac{\hbar}{i}r\frac{\partial}{\partial r}, \tag{8.38}$$

so that

$$L^2 = -\hbar^2\left[r^2\Delta - r\frac{\partial}{\partial r}\left(r\frac{\partial}{\partial r} + 1\right)\right]. \tag{8.39}$$

Rearranging, this gives

$$\Delta = \frac{1}{r}\frac{\partial}{\partial r}\left(r\frac{\partial}{\partial r} + 1\right) - \frac{1}{\hbar^2 r^2}L^2, \tag{8.40}$$

which then simplifies to give the stated result. □

Theorem 8.4.4. Let F be a fixed function of $r \in [0, \infty)$ that satisfies

$$\int_0^\infty r^{2(l+1)}|F(r)|^2\,dr < \infty,$$

and let V_l be the space of functions of the form $F(r)Y(\mathbf{x})$ with Y a polynomial that is both harmonic and also homogeneous of degree l. Then V_l is a $(2l+1)$-dimensional space on which L_1, L_2, and L_3 act with total angular momentum $l(l+1)\hbar^2$.

Proof. The condition on F is included just to ensure normalizability. Euler's theorem on homogeneous functions tells us that

$$DY = \frac{\hbar}{i}\left(x_1\frac{\partial Y}{\partial x_1} + x_2\frac{\partial Y}{\partial x_2} + x_3\frac{\partial Y}{\partial x_3}\right) = -il\hbar Y. \tag{8.41}$$

Together with $P^2Y = -\hbar^2\Delta Y = 0$, this gives

$$\begin{aligned} L^2Y &= X^2P^2Y - D(D - i\hbar)Y \\ &= -il\hbar(-il\hbar - i\hbar)Y = l(l+1)\hbar^2 Y. \end{aligned} \tag{8.42}$$

Thus L^2 has the correct value.

We have already seen that $L_3 = -i\hbar\partial/\partial\phi$, which commutes with Δ and consequently preserves the space of harmonic functions. It similarly preserves the degree of homogeneity and so maps V_l to itself. By symmetry the same applies to L_1 and L_2.

Now a homogeneous polynomial of degree l in three variables has $(l+1)(l+2)/2$ coefficients and so that is the dimension of the space of such polynomials. The two differentiations of the Laplace operator lower the degree to $l-2$, so that Laplace's equation imposes $l(l-1)/2$ constraints on the coefficients of the resulting polynomial ΔY. Thus the space V_l has dimension

$$\frac{(l+1)(l+2)}{2} - \frac{l(l-1)}{2} = \frac{4l+2}{2} = 2l+1. \tag{8.43}$$

(Alternatively one can show that the eigenvalue $l\hbar$ for L_3 is non-degenerate. This is best done by using Corollary 8.2.4 to identify these eigenvectors with vectors killed by L_+.) □

Example 8.4.1. Polynomials of degree 0 are just constant and so automatically harmonic. They give the one-dimensional space and trivial representation discussed in Example 8.3.1.

Example 8.4.2. Polynomials of degree 1 can be written in the form

$$Y_{\mathbf{a}}(\mathbf{x}) = \mathbf{a}.\mathbf{x} = a_1x_1 + a_2x_2 + a_3x_3, \tag{8.44}$$

where the components a_1, a_2, and a_3 may be complex. These are also automatically harmonic and provide the three-dimensional space described in Example 8.3.3. We can readily find the basis used there by noting that

$$\begin{aligned} L_3Y_{\mathbf{a}} &= \frac{\hbar}{i}\left(x_1\frac{\partial}{\partial x_2} - x_2\frac{\partial}{\partial x_1}\right)(a_1x_1 + a_2x_2 + a_3x_3) \\ &= \frac{\hbar}{i}(x_1a_2 - x_2a_1). \end{aligned} \tag{8.45}$$

Thus $L_3(\mathbf{a}.\mathbf{x}) = 0$ if and only if $a_1 = a_2 = 0$. The polynomial $\psi_0(\mathbf{x}) = x_3$ is therefore an eigenvector of L_3 with value 0. Similarly

$$\psi_\pm(\mathbf{x}) = \frac{1}{\sqrt{2}}(x_1 \pm ix_2) \tag{8.46}$$

satisfies

$$L_3\psi_\pm = \pm\hbar\psi_\pm. \tag{8.47}$$

Example 8.4.3. There are six independent polynomials of degree 2, the quadratics x_1^2, x_2^2, x_3^2, x_2x_3, x_3x_1, and x_1x_2, and this time Laplace's equation imposes a genuine constraint, since, for example, x_1^2 is certainly not harmonic. Each of the quadratics $x_j^2 - x_k^2$ and x_jx_k with $j < k$ satisfies Laplace's equation, but they are related by the identity

$$x_1^2 - x_3^2 = (x_1^2 - x_2^2) + (x_2^2 - x_3^2). \tag{8.48}$$

There are therefore just five independent quadratics on which the orbital angular momentum operators act, and forming a basis for the five-dimensional space when $l = 2$.

8.5. Centre of mass coordinates

We can now return to central force problems and furnish a more complete analysis than was possible in Chapter 4. We start by considering the general problem of two particles with masses m_1 and m_2 interacting through a force that depends only on the distance between them. More precisely, writing $\mathbf{r}_1$ and $\mathbf{r}_2$ for the positions of the two particles, we assume that the interaction is described by a potential $V(|\mathbf{r}_1 - \mathbf{r}_2|)$. The wave function Ψ for this system is a function of both $\mathbf{r}_1$ and $\mathbf{r}_2$, and it satisfies the Schrödinger equation

$$-\frac{\hbar^2}{2m_1}\Delta_1\Psi - \frac{\hbar^2}{2m_2}\Delta_2\Psi + V(|\mathbf{r}_1 - \mathbf{r}_2|)\Psi = E_t\Psi, \tag{8.49}$$

where Δ_j is the Laplacian for the j-th particle and E_t is the total energy.

To handle this system we adopt a trick from classical mechanics and change to centre of mass coordinates: that is, we introduce the total mass $M = m_1 + m_2$, the centre of mass position vector $\mathbf{R} = (m_1\mathbf{r}_1 + m_2\mathbf{r}_2)/M$, the difference vector $\mathbf{r} = \mathbf{r}_1 - \mathbf{r}_2$, and the reduced mass

$$m = \left(\frac{1}{m_1} + \frac{1}{m_2}\right)^{-1} = \frac{m_1m_2}{M}. \tag{8.50}$$

It is then a routine exercise in partial differentiation (see Exercise 8.1) to check that

$$\frac{1}{m_1}\Delta_1 + \frac{1}{m_2}\Delta_2 = \frac{1}{M}\Delta_R + \frac{1}{m}\Delta, \tag{8.51}$$

where Δ_R and Δ denote the Laplacian operators for the coordinates of $\mathbf{R}$ and of $\mathbf{r}$, respectively, so that Schrödinger's equation can be written as

$$-\frac{\hbar^2}{2M}\Delta_R\Psi - \frac{\hbar^2}{2m}\Delta\Psi + V(r)\Psi = E_t\Psi. \tag{8.52}$$

Proposition 8.5.1. The equation

$$-\frac{\hbar^2}{2M}\Delta_R\Psi - \frac{\hbar^2}{2m}\Delta\Psi + V(r)\Psi = E_t\Psi$$

has separable solutions of the form $\Psi = \phi(\mathbf{R})\psi(\mathbf{r})$, where ϕ and ψ satisfy the equations

$$-\frac{\hbar^2}{2M}\Delta_R\phi = E_R\phi$$

and

$$-\frac{\hbar^2}{2m}\Delta\psi + V(r)\psi = E\psi,$$

with $E_R + E = E_t$.

Proof. This straightforward substitution is left for the reader. □

The first equation simply represents the force-free motion of the centre of mass, whilst the second gives the motion of a particle of mass m in a potential V with fixed centre. This provides retrospective justification for the fixed nucleus assumption imposed in Section 4.1. The only change is that m should be regarded as the reduced mass of the system rather than the electron mass. However, the mass, m_2, of a hydrogen nucleus is around 1836 times the electron mass, m_1, so that

$$\begin{aligned} m &\sim \left(\frac{1}{m_1} + \frac{1}{1836m_1}\right)^{-1} \\ &= \left(\frac{1837}{1836m_1}\right)^{-1} \\ &= \frac{1836}{1837}m_1, \end{aligned} \tag{8.53}$$

which differs from m_1 by only 0.05%.

Remark 8.5.1. There are other situations in which the masses of the two particles are comparable and the centre of mass system makes a significant difference. For example, it is possible to have an electron orbit not a proton as in the hydrogen atom but a positron, a particle of the same mass as an electron but oppositely charged, forming what is called positronium. There are also bound atom-like states of oppositely charged quarks, the most basic known constituents of nuclear matter.

8.6. Angular momentum states

Once we have reduced to centre of mass coordinates to obtain the equation

$$-\frac{\hbar^2}{2m}\nabla^2\psi + V(r)\psi \tag{8.54}$$

we can exploit the connection between the Laplacian and the angular momentum. Corollary 8.4.3 tells us that Schrödinger's equation may be rewritten as

$$-\frac{\hbar^2}{2m}\frac{1}{r}\frac{\partial^2}{\partial r^2}(r\psi) + \frac{1}{2mr^2}L^2\psi + V(r)\psi = E\psi. \tag{8.55}$$

By Corollary 8.2.2 L^2 and L_3 commute with P^2 and with r^2 and so commute with the Hamiltonian operator,

$$H = \frac{P^2}{2m} + V(r). \tag{8.56}$$

We may therefore choose a wave function ψ that is simultaneously a wave function for H, L^2, and L_3. If we take ψ satisfying $L^2\psi = l(l+1)\hbar^2\psi$ then the Schrödinger equation reduces to

$$\frac{\hbar^2}{2m}\left(-\frac{1}{r}\frac{\partial^2}{\partial r^2}(r\psi) + \frac{l(l+1)}{r^2}\psi\right) + V(r)\psi = E\psi, \tag{8.57}$$

which is precisely the radial equation (4.9) for separable solutions, Providing retrospective justification for this. The analysis of the hydrogen atom then proceeds exactly as in Chapter 4.

8.7. An algebraic solution of the hydrogen atom

Pauli's solution of the hydrogen atom problem in quantum mechanics exploited a classical technique due to Hamilton, and, in component form,

Lagrange, who had shown that the vector, $\mathbf{A} = (mK)^{-1}(\mathbf{r} \times \mathbf{p}) \times \mathbf{p} + \mathbf{r}/r$, is constant during classical motion under the potential $-K/r$. This vector, which points towards the centre of attraction from the point of closest approach, and whose magnitude is just the eccentricity of the orbit, determines the orbit. (It is sometimes called the Runge–Lenz vector, after two rediscoverers.) In quantum mechanics one can similarly define

$$\mathbf{A} = \frac{1}{2mK}(\mathbf{L} \times \mathbf{P} - \mathbf{P} \times \mathbf{L}) + \frac{1}{X}\mathbf{X}, \tag{8.58}$$

which commutes with the Hamiltonian, satisfies

$$\mathbf{A}.\mathbf{L} = 0, \tag{8.59}$$

and, by use of Proposition 8.1.1 and Corollary 8.2.2, is seen to satisfy

$$[L_j, A_k] = i\hbar\epsilon_{jkl}A_l. \tag{8.60}$$

After a good deal of strenuous algebra, working from the commutation relations, it is also possible to show that

$$[A_j.A_k] = i\hbar(-2H/mK^2)\epsilon_{jkl}L_l \tag{8.61}$$

and

$$A^2 = (2H/mK^2)\left(L^2 + \hbar^2\right) + 1. \tag{8.62}$$

For solutions of Schrödinger's equation $H\psi = E\psi$, one introduces the new sets of operators

$$\mathbf{J}^{\pm} = \frac{1}{2}\left[\mathbf{L} \pm (-2E/mK^2)^{\frac{1}{2}}\mathbf{A}\right], \tag{8.63}$$

which, using equations (8.60, (8.61), are easily seen to commute with each other, and to satisfy

$$[J_j^{\pm}, J_k^{\pm}] = i\hbar\epsilon_{jkl}J_l^{\pm}. \tag{8.64}$$

They therefore provide two commuting sets of angular momentum operators, and we shall be able to choose wave functions satisfying

$$J^{\pm 2}\psi = j_{\pm}\left(j_{\pm} + 1\right)\hbar^2\psi. \tag{8.65}$$

Taking into account equations (8.59) and (8.62), we also have $J^{+2} = J^{-2}$, which means that $j_+ = j_-$, and

$$2\left(J^{+2} + J^{-2}\right) + \hbar^2 = -mK^2/2E, \tag{8.66}$$

which gives

$$\left[4j_+(j_+ + 1) + 1\right]\hbar^2 = -mK^2/2E. \tag{8.67}$$

From this we immediately deduce that

$$E = -mK^2\hbar^2/2(2j_+ + 1)^2, \tag{8.68}$$

in agreement with Proposition 4.3.1.

8.8. The spin representation

Although orbital angular momentum has to be integral we have already seen that there are operators that obey the angular momentum commutation relations with half integral j, for example the Pauli spin matrices with $j = \frac{1}{2}$. The spin matrices have many fascinating properties, some of which are summarized in the following result.

Theorem 8.8.1. Let $\boldsymbol{\sigma} = (\sigma_1, \sigma_2, \sigma_3)$ denote the Pauli spin matrices and let

$$\boldsymbol{\sigma}.\mathbf{a} = \mathbf{a}.\boldsymbol{\sigma} = a_1\sigma_1 + a_2\sigma_2 + a_3\sigma_3.$$

Then
(i) for any $\mathbf{a} \in \mathbf{R}^3$ the matrix $\mathbf{a}.\boldsymbol{\sigma}$ is a self-adjoint matrix with trace 0 and every tracefree self-adjoint matrix is of this form;
(ii) for all $\mathbf{a}, \mathbf{b} \in \mathbf{R}^3$ we have

$$(\mathbf{a}.\boldsymbol{\sigma})(\mathbf{b}.\boldsymbol{\sigma}) = (\mathbf{a}.\mathbf{b})1 + i(\mathbf{a} \times \mathbf{b}).\boldsymbol{\sigma}.$$

Proof. (i) Equation (8.24) gives the expression

$$\mathbf{a}.\boldsymbol{\sigma} = \begin{pmatrix} a_3 & a_1 - ia_2 \\ a_1 + ia_2 & -a_3 \end{pmatrix}, \tag{8.69}$$

which is obviously self-adjoint (hermitian) and has trace 0. Conversely, if X is any 2×2 self-adjoint matrix then it is easy to check that

$$X = \frac{1}{2}\left[\mathrm{tr}(X)1 + \mathrm{tr}(X\sigma_1)\sigma_1 + \mathrm{tr}(X\sigma_2)\sigma_2 + \mathrm{tr}(X\sigma_3)\sigma_3\right]. \tag{8.70}$$

So if $\mathrm{tr}(X) = 0$ then we have the required form.

For the second part, it is easy to calculate from Definition 8.3.1 that the Pauli spin matrices satisfy

$$\sigma_j^2 = 1, \tag{8.71}$$

$$\sigma_1\sigma_2 = i\sigma_3, \qquad \sigma_2\sigma_3 = i\sigma_1, \qquad \sigma_3\sigma_1 = i\sigma_2, \tag{8.72}$$

and similarly

$$\sigma_2\sigma_1 = -i\sigma_3, \qquad \sigma_3\sigma_2 = -i\sigma_1, \qquad \sigma_1\sigma_3 = -i\sigma_2. \tag{8.73}$$

From these relations part (ii) is easily checked. (It may also be checked by a direct calculation of the left-hand side, see Exercise 8.11.) □

Equations (8.72), (8.73) immediately give what are sometimes called the anticommutation relations for spin:

$$\sigma_j\sigma_k + \sigma_k\sigma_j = 2\delta_{jk}. \tag{8.74}$$

These can be combined with the commutation relations, in the following corollary:

Corollary 8.8.2.

$$[\mathbf{a}.\boldsymbol{\sigma}, \mathbf{b}.\boldsymbol{\sigma}] = 2i(\mathbf{a} \times \mathbf{b}).\boldsymbol{\sigma}$$

$$(\mathbf{a}.\boldsymbol{\sigma})(\mathbf{b}.\boldsymbol{\sigma}) + (\mathbf{b}.\boldsymbol{\sigma})(\mathbf{a}.\boldsymbol{\sigma}) = 2(\mathbf{a}.\mathbf{b}).$$

Proof. These can be derived by adding and subtracting the formulae for $(\mathbf{a}.\boldsymbol{\sigma})(\mathbf{b}.\boldsymbol{\sigma})$ and $(\mathbf{b}.\boldsymbol{\sigma})(\mathbf{a}.\boldsymbol{\sigma})$ given in the preceding theorem. □

8.9. Historical notes

The quantization of angular momentum was first suggested by J.W. Nicholson in 1912 in an early attempt to apply quantum theory to the atom. Unfortunately he worked with J.J. Thomson's 'plum pudding' model of the atom as a ball of positive charge in which the electrons were embedded, which Rutherford's experimental results were just showing to be untenable. However, a year later Niels Bohr combined the idea of quantizing angular momentum with Rutherford's picture of the atom as a miniature solar system to provide a highly successful theory of atomic spectra.

The Pauli spin matrices had in fact appeared in mathematics long before Pauli's rediscovery of them in the context of quantum theory. In view of their abstract derivation it may seem somewhat surprising that they have turned out to be useful in nature. However, it then transpired that they could be used to describe the intrinsic spin of the electron. This is the context in which they were investigated by Pauli, who was trying to understand certain puzzling features of the spectra of atoms. There is a certain irony in the fact that, although Pauli introduced the idea of a 'non-classical two-valuedness', he initially dismissed the suggestions of Kronig, Goudsmit and Uhlenbeck that this had anything to do with electron spin. Later Stern and Gerlach found direct experimental confirmation of the idea when they showed that after passing through a highly inhomogeneous magnetic field a beam of electrons splits into two beams corresponding to the two possible spin states of the electrons.

Exercises

8.1 Prove equation (8.51) and Proposition 8.5.1.

8.2 By comparing the formula of Corollary 8.4.3 with the expression for Δ in spherical polar coordinates (r, θ, ϕ), or otherwise, show that L^2 can be written as

$$L^2\psi = -\frac{\hbar^2}{\sin\theta}\frac{\partial}{\partial\theta}\left(\sin\theta\frac{\partial\psi}{\partial\theta}\right) + \frac{1}{\sin^2\theta}\frac{\partial^2\psi}{\partial\phi^2}.$$

Hence, or otherwise, show that there exist separable solutions of the equations

$$L^2\psi = l(l+1)\hbar^2\psi$$
$$L_3\psi = m\hbar\psi$$

of the form

$$\psi = e^{im\phi}P_l^m(\cos\theta).$$

Find expressions for P_1^0 and P_2^0 in the form of power series.

8.3° The operators L_1, L_2, and L_3 satisfy the angular momentum commutation relations and $L_\pm = L_1 \pm iL_2$, and the vectors ψ_m satisfy

$$L^2\psi_m = l(l+1)\hbar^2\psi_m \qquad \text{and} \qquad L_3\psi_m = m\hbar\psi_m.$$

By considering $\langle\psi_m|L_+^2\psi_m\rangle$, or otherwise, show that

$$\langle\psi_m|L_1^2\psi_m\rangle = \langle\psi_m|L_2^2\psi_m\rangle$$

and find the value in terms of l and m when ψ_m is normalized. Show also that

$$\langle\psi_m|L_1L_2\psi_m\rangle = \tfrac{1}{2}im\hbar^2.$$

8.4° The Hamiltonian for a particle moving in a magnetic field is given by

$$H = \frac{\mathbf{P}^2}{2M} + \frac{\mu}{r}L_3,$$

where μ is a positive constant, and L_3 is the third component of the orbital angular momentum operator. By considering a simultaneous eigenvector $\phi_{E,l,m}$ of H, $\mathbf{L}^2$, and L_3 with corresponding eigenvalues E, $l(l+1)\hbar^2$, and $m\hbar$, respectively, prove that there exists a non-negative integer N such that

$$E = -\frac{1}{2}M\mu^2\left(\frac{m}{l+N+1}\right)^2.$$

What are the possible values of m and l?

8.5° The components of the orbital angular momentum operator are given in terms of the spherical polar angles θ and ϕ by

$$L_1 = i\hbar\left(\sin\phi\frac{\partial}{\partial\theta} + \cos\phi\cot\theta\frac{\partial}{\partial\phi}\right),$$

$$L_2 = i\hbar\left(-\cos\phi\frac{\partial}{\partial\theta} + \sin\phi\cot\theta\frac{\partial}{\partial\phi}\right),$$

$$L_3 = -i\hbar\frac{\partial}{\partial\phi}.$$

Show that

$$L_+ = \hbar e^{i\phi}\left(\frac{\partial}{\partial\theta} + i\cot\theta\frac{\partial}{\partial\phi}\right)$$

and determine the simultaneous eigenfunctions of L^2 and L_3 in terms of θ and ϕ when the eigenvalue of L^2 is $2\hbar^2$.

8.6° A particle of mass m moves in the spherically symmetric potential

$$V(r) = -\frac{\hbar^2}{2m}\left(\frac{k^2}{r} - \frac{\alpha^2}{r^2}\right).$$

Show that the energy levels are given by

$$E_{nl} = -\frac{\hbar^2k^4}{8m}\left\{n + \frac{1}{2} + \left[\left(l+\frac{1}{2}\right)^2 + \alpha^2\right]^{\frac{1}{2}}\right\}^{-2}$$

where l is the orbital angular momentum quantum number and $n = 0, 1, 2, 3, \ldots$.

8.7° Show that the components of the orbital angular momentum operator $\mathbf{L} = \mathbf{X} \times \mathbf{P}$ and the components of $\mathbf{X}$ satisfy

$$\mathbf{L}.\mathbf{X} = 0 = \mathbf{X}.\mathbf{L}$$

and derive the commutation relation

$$[L^2, [L^2, \mathbf{X}]] = 2\hbar^2(L^2\mathbf{X} + \mathbf{X}L^2).$$

Deduce that for all ϕ and ψ

$$\langle L^4\phi|\psi\rangle - \langle L^2\phi|L^2\psi\rangle + \langle\phi|L^4\psi\rangle = 2\hbar^2(\langle L^2\phi|X\psi\rangle + \langle\phi|XL^2\psi\rangle).$$

A system has eigenstates ψ_{lm} such that

$$L^2\psi_{lm} = l(l+1)\hbar^2\psi_{lm},$$
$$L_3\psi_{lm} = m\hbar\psi_{lm}.$$

Show that the matrix element $\langle\psi_{l',m'}|\psi_{lm}\rangle$ vanishes unless

$$(l - l' + 1)(l - l' - 1)(l + l')(l + l' + 2) = 0.$$

8.8° The Hamiltonian for a rigid body rotating freely about the origin is given by

$$H = \frac{1}{2}\left(\frac{L_1^2}{A_1} + \frac{L_2^2}{A_2} + \frac{L_3^2}{A_3}\right),$$

where A_1, A_2, and A_3 are the principal moments of inertia. If the orthonormal vectors ψ_{lm} satisfy

$$L^2\psi_{lm} = l(l+1)\hbar^2\psi_{lm}, \qquad L_3\psi_{lm} = m\hbar\psi_{lm},$$

show that $\langle\psi_{ln}|H\psi_{lm}\rangle$ is given by

$$\frac{\hbar^2}{4}\left(\frac{1}{A_1} + \frac{1}{A_2}\right)[l(l+1) - m^2] + \frac{\hbar^2 m^2}{2A_3}$$

if $n = m$, by

$$\frac{\hbar^2}{8}\left(\frac{1}{A_1} - \frac{1}{A_2}\right)\{[l(l+1) - m(m \pm 1)][l(l+1) - (m \pm 1)(m \pm 2)]\}^{\frac{1}{2}}$$

if $n = m \pm 2$, and that all the other matrix elements of H vanish. Deduce that the energy levels for the rigid rotator for which $l = 2$ are

$$\frac{\hbar^2}{2}\left(\frac{1}{A_2} + \frac{1}{A_3}\right), \qquad \frac{\hbar^2}{2}\left(\frac{1}{A_3} + \frac{1}{A_1}\right), \qquad \frac{\hbar^2}{2}\left(\frac{1}{A_1} + \frac{1}{A_2}\right).$$

Find a general formula for the energy levels when $A_1 = A_2$.

8.9° The Hamiltonian for a kinematically symmetric rigid rotator is given by

$$H = \tfrac{1}{2}A(J_1^2 + J_2^2) + \tfrac{1}{2}CJ_3^2,$$

where the self-adjoint operators J_1, J_2, and J_3 satisfy the angular momentum commutation relations. Prove that the eigenvalues of the Hamiltonian are

$$\tfrac{1}{2}\left[Aj(j+1) + (C-A)m^2\right]\hbar^2$$

with $m = -j, -j+1, \ldots, j$ and $j = 0, \frac{1}{2}, 1, \ldots$.

8.10° A charged particle moving in a uniform magnetic field is confined to the plane $x_3 = 0$ at right angles to the field. The Hamiltonian is given by

$$H = \frac{1}{2\mu}\left[(P_1^2 + P_2^2) + \beta^2(X_1^2 + X_2^2) + 2\beta(X_1P_2 - X_2P_1)\right],$$

where P_1 and P_2 are the momenta conjugate to the coordinates X_1 and X_2 (respectively), μ is the mass, and β a positive constant. It can be shown, and you may assume, that the component $L_3 = X_1P_2 - X_2P_1$ of the angular momentum commutes with H. By considering solutions of the time-independent Schrödinger equation of the form $\exp(im\theta)\exp(-\beta r^2/2\hbar)f(r)$ where r and θ are the usual plane polar coordinates in the x_1x_2-plane, or otherwise, show that the energy levels are $\hbar\beta(2N+1)/\mu$, where $N = 0, 1, 2, \ldots$. In establishing that the energy levels are of the above form, where, if anywhere, did you use the fact that L_3 and H commute?
[You may assume that

$$\frac{\partial^2}{\partial x_1^2} + \frac{\partial^2}{\partial x_2^2} = \frac{1}{r}\frac{\partial}{\partial r}r\frac{\partial}{\partial r} + \frac{1}{r^2}\frac{\partial^2}{\partial \theta^2} \quad \text{and} \quad x_1\frac{\partial}{\partial x_2} - x_2\frac{\partial}{\partial x_1} = \frac{\partial}{\partial \theta}.]$$

8.11 Prove the result of 8.8.1(ii), which asserts that

$$(\mathbf{a}.\boldsymbol{\sigma})(\mathbf{b}.\boldsymbol{\sigma}) = (\mathbf{a}.\mathbf{b})1 + i(\mathbf{a} \times \mathbf{b}.\boldsymbol{\sigma}).$$

Let $\mathbf{a}.\boldsymbol{\sigma} = a_1\sigma_1 + a_2\sigma_2 + a_3\sigma_3$, where σ_1, σ_2, and σ_3 are the Pauli spin matrices. Show that

$$e^{i\mathbf{a}.\boldsymbol{\sigma}} = \cos|a| + i\mathbf{a}.\boldsymbol{\sigma}\frac{\sin|a|}{|a|}.$$

8.12° The spin operator $\mathbf{S}$ is self-adjoint and satisfies the commutation relations

$$[S_1, S_2] = i\hbar S_3, \quad [S_2, S_3] = i\hbar S_1, \quad [S_3, S_1] = i\hbar S_2,$$

and $\mathbf{S}^2 = \frac{3}{4}\hbar^2$. Show that the eigenvalues of S_3 are $\pm\hbar$.

Assuming that the eigenvalues of S_3 are non-degenerate show that by a suitable choice of phase

$$A\phi_+ = \hbar\phi_- \quad \text{and} \quad A^*\phi_- = \hbar\phi_+,$$

where $A = S_1 - iS_2$ and $\phi_\pm$ are the eigenvectors of S_3 corresponding to the eigenvalues $\pm\frac{1}{2}\hbar$.

Two spin $\frac{1}{2}$ particles with spin operators $\mathbf{S}(1)$ and $\mathbf{S}(2)$ are coupled so that the Hamiltonian is

$$k\mathbf{S}(1).\mathbf{S}(2).$$

Show that

$$\mathbf{S}(1).\mathbf{S}(2) = \frac{A^*(1)A(2) + A(1)A^*(2)}{2} + S_3(1)S_3(2).$$

Hence, or otherwise, obtain the energy eigenvalues.

8.13° Suppose that K_1, K_2, and K_3 satisfy the commutation relations

$$[K_1, K_2] = i\hbar(\alpha^2 K_3 - \beta), \qquad [K_2, K_3] = i\hbar K_1, \qquad [K_3, K_1] = i\hbar K_2.$$

Show that $J_1 = \alpha^{-1}K_1$, $J_2 = \alpha^{-1}K_2$, $J_3 = K_3 - \beta/\alpha^2$ satisfy the relations for angular momentum. Hence, or otherwise, show that there are solutions in which $K_1^2 + K_2^2 - 2\beta K_3 + \alpha^2 K_3^2 + (\beta/\alpha)^2$ takes the value $j(j+1)\hbar^2\alpha^2$ and K_3 has the eigenvalues $m\hbar + \beta/\alpha^2$ with m in the same range as before. Show that when $\beta = 1$, $\alpha = [(J + \frac{1}{2})\hbar]^{-\frac{1}{2}}$ and $j = J$, the eigenvalues of K_3 are of the form $(n + \frac{1}{2})\hbar$ with $0 \leq n \leq 2J$, whilst the value of $K_1^2 + K_2^2 - 2K_3 + [(J + \frac{1}{2})\hbar]^{-1}K_3^2$ is $-\hbar/(4J + 2)$. What happens to the commutation relations and to these values in the limit as $J \to \infty$?

8.14 Let M_+, M_-, and H be operators satisfying the commutation relations

$$[H, M_\pm] = \pm M_\pm, \qquad [M_+, M_-] = f(H),$$

where f is a function that can be expanded in a convergent power series. Show that $M_\pm$ maps an eigenvector of $\ker(H - \mu)$ to $\ker(H -$

$\mu \mp 1)$.

Show (without worrying about convergence questions) that for any function g that can be expanded in a power series $M_{\pm}g(H) = g(H \pm 1)M_{\pm}$. Hence, or otherwise, show that $M_+M_- + M_-M_+ + g(H)$ commutes with M_+, M_-, and H provided that

$$[f(H \pm 1) + f(H)] \pm [g(H \pm 1) - g(H)] = 0.$$

Show that the following choices are possible:

(i) $f(H) = \alpha H + \beta 1$,
$g(H) = \alpha H^2 + \beta H$;

(ii) $f(H) = \sinh(2\beta H)/\sinh(2\beta)$,
$g(H) = [\cosh(2\beta H) - 1]/[\cosh(2\beta) - 1]$.

By mimicking the argument for angular momentum show that in each of these cases there are operators on a $(2l + 1)$-dimensional space satisfying the given relations.

To what physical systems do the cases $\alpha = 0$ and $\beta = 0$ in (i) correspond?

9* Symmetry in quantum theory

It has been rumoured that the *group pest* is gradually being cut out of quantum mechanics. This is certainly not true as far as the rotation and Lorentz groups are concerned.

H. Weyl, *Group theory and quantum mechanics*, 1930

9.1. Group representations

Physical systems often have some kind of obvious symmetry, such as the rotational symmetry of the central force field about a nucleus. Mathematically this means that the system is invariant under the action of a symmetry group G. (The main ideas of group theory are reviewed in Appendix A1.) Such a physical symmetry must also be present in quantum theory in a way that is compatible with the other important structures, namely the vector spaces and inner products. The simplest way in which this can happen is for G to act on the inner product space $\mathcal{H}$ by unitary transformations, since these respect both the linear structure and the inner product. This suggests the following definition:

Definition 9.1.1. A *unitary representation* of a group G on an inner product space $\mathcal{H}$ is a homomorphism from G to the group of unitary operators on $\mathcal{H}$.

In quantum mechanics we assume that $\mathcal{H} \neq \{0\}$. Since there should be no risk of confusion we shall usually abbreviate the terminology and refer simply to a *representation* of G. If we write U for the homomorphism then for each x in G we have a unitary operator $U(x)$ such that

$$U(xy) = U(x)U(y) \tag{9.1}$$

for all x and y in G. It is worth noting that by combining the unitarity and the group homomorphism property we have

$$U(x)^* = U(x)^{-1} = U(x^{-1}). \tag{9.2}$$

The definition is really justified by the wealth of examples of representations appearing in quantum theory. We shall give a few of them here and some more in the exercises at the end of the chapter.

Example 9.1.1. First a general mathematical example: every group G has a *trivial representation* on the one dimensional space $\mathbf{C}$, which maps each group element to the identity operator, that is

$$U(x) = 1 \tag{9.3}$$

for all x in G.

Example 9.1.2. If A is a rotation of $\mathbf{R}^3$ we can define an operator $U(A)$ on ordinary wave functions by

$$(U(A)\psi)(\mathbf{r}) = \psi(A^{-1}\mathbf{r}). \tag{9.4}$$

Since rotations preserve volumes the integral defining the inner product will be unchanged, so that $U(A)$ is unitary. For any rotations A and B and any wave function ψ we also have

$$\begin{aligned}(U(A)U(B)\psi)(\mathbf{r}) &= (U(B)\psi)(A^{-1}\mathbf{r}) \\ &= \psi(B^{-1}A^{-1}\mathbf{r}) \\ &= \psi((AB)^{-1}\mathbf{r}) = (U(AB)\psi)(\mathbf{r}),\end{aligned} \tag{9.5}$$

which shows that U is a representation of the rotation group.

Example 9.1.3. This argument can be extended to give a representation for all orthogonal transformations. One particularly interesting case occurs when one takes the two-element group consisting of ± 1. This is generated by the *parity operator*, $\mathcal{P}$, corresponding to the orthogonal transformation -1 (cf. Exercise 6.3). Since $\mathcal{P}^2 = 1$, the parity operator has eigenvalues ± 1. The eigenvectors satisfy

$$\psi(-\mathbf{r}) = (\mathcal{P}\psi)(\mathbf{r}) = \pm\psi(\mathbf{r}) \tag{9.6}$$

so that they are odd or even functions, depending on the sign of the eigenvalue. The kinetic energy of a particle, since it depends on second derivatives, is insensitive to sign changes. For a particle moving in an even potential $V(\mathbf{r}) = V(-\mathbf{r})$ the Hamiltonian therefore commutes with $\mathcal{P}$. The eigenfunctions of H may therefore be chosen to be eigenfunctions of $\mathcal{P}$ as well, that is odd or even functions.

Example 9.1.4. If $\mathbf{a}$ is an element of $\mathbf{R}^3$ we can similarly define

$$(U(\mathbf{a})\psi)(\mathbf{r}) = \psi(\mathbf{r} - \mathbf{a}) \tag{9.7}$$

to obtain a unitary representation of the group of translations.

9.2. Representations and energy levels

Crucial to what follows is the notion of an operator which relates two representations to one another.

Definition 9.2.1. Let U and W be two representations of the same group G on spaces $\mathcal{H}$ and $\mathcal{K}$ respectively. An *intertwining operator* for U and W is a linear transformation, T, from $\mathcal{H}$ to $\mathcal{K}$ which satisfies

$$TU(x) = W(x)T$$

for all x in G. If there exists an invertible intertwining operator then U and W are said to be *equivalent.*

Remark 9.2.1. One also says that T *intertwines* U and W. Clearly the sum of two intertwining operators is an intertwining operator and so is a scalar multiple of an intertwining operator.

Example 9.2.1. Let G be the rotation group and let U be the representation on normalizable wave functions defined in Example 9.1.2. Let H be the Hamiltonian operator for a particle moving under the influence of a central force:

$$H = -\frac{\hbar^2}{2m}\nabla^2 + V(r). \tag{9.8}$$

Intuitively we know that since it describes a central force H is rotationally invariant, and the mathematical expression of this fact is that H is an intertwining operator. To see this consider

$$(U(A)H\psi)(\mathbf{r}) = (H\psi)(A^{-1}\mathbf{r}). \tag{9.9}$$

Introducing $\mathbf{s} = A^{-1}\mathbf{r}$ we may write this as

$$(U(A)H\psi)(\mathbf{r}) = (H\psi)(\mathbf{s}) = -\frac{\hbar^2}{2m}(\nabla_s^2\psi)(\mathbf{s}) + V(s)\psi(\mathbf{s}), \tag{9.10}$$

where we have written ∇_s^2 to make clear which variable we are differentiating. However, since $\mathbf{s}$ is related to $\mathbf{r}$ by a rotation and since ∇^2 is the same in any set of Cartesian coordinates, we have $\nabla_s^2 = \nabla_r^2$. Since we clearly have $V(s) = V(|A^{-1}\mathbf{r}|) = V(r)$ this means that

$$\begin{aligned}(U(A)H\psi)(\mathbf{r}) &= \left(\frac{\hbar^2}{2m}\nabla_r^2 + V(r)\right)\psi(A^{-1}\mathbf{r}) \\ &= (HU(A)\psi)(\mathbf{r}). \end{aligned} \tag{9.11}$$

Since this is true for any wave function ψ we deduce that

$$U(A)H = U(A)H; \tag{9.12}$$

in other words, that H is an intertwining operator. Similarly, if V is an even function then the Hamiltonian operator intertwines the action of the group generated by the parity operator. □

Symmetries of the quantum system often leave subspaces of $\mathcal{H}$ invariant, and so help us to split the problem into smaller, simpler pieces.

Definition 9.2.2. Let U be a representation of G on $\mathcal{H}$. A subspace $\mathcal{K}$ of $\mathcal{H}$ is said to be *invariant* under U if $U(x)\mathcal{K} \subseteq \mathcal{K}$ for all x in G.

Lemma 9.2.1. If T intertwines U and W then $\ker T$ is invariant under U and $\operatorname{im} T$ is invariant under W.

Proof. If $T\psi = 0$ then $TU(x)\psi = W(x)T\psi = 0$, so that $U(x)\psi$ is in $\ker T$ for all x in G. This means that $U(x)\ker T \subseteq \ker T$ so that $\ker T$ is invariant under U. Similarly, for any ψ in $\mathcal{H}$ we have

$$W(x)(T\psi) = T(U(x)\psi), \tag{9.13}$$

which shows that $W(x)$ maps the range of T to itself. □

Corollary 9.2.2. If T intertwines U with itself then every eigenspace of T is invariant under U.

Proof. The λ-eigenspace of T is $\ker(T - \lambda 1)$. Now the intertwining condition means that $U(x)T = TU(x)$ for all x in G, from which we readily deduce that

$$(T - \lambda 1)U(x) = U(x)(T - \lambda 1). \tag{9.14}$$

This means that $T - \lambda 1$ is also an intertwining operator, from which it follows that the λ-eigenspace is invariant under U. □

Applied to the case in which the Hamiltonian is an intertwining operator for some representation of a group G, we see that each of the energy levels must form an invariant subspace. In particular, if the potential V is an even function then the eigenspaces for H are invariant under the action of the parity group. This means that the eigenspaces for H will themselves split into eigenspaces for $\mathcal{P}$, that is we may as well take our energy eigenstates to be odd or even functions. This shows up very clearly with the harmonic oscillator where the eigenfunctions were either odd or even.

Definition 9.2.3. If $\mathcal{K}$ is a subspace of $\mathcal{H}$ which is invariant under U, then the restriction of $U(x)$ to $\mathcal{K}$ is called a *subrepresentation* of U.

This means that each energy level carries a representation of any symmetry group G for which the Hamiltonian is an intertwining operator, so by studying the possible representations of G we can obtain information about the energy levels. For example, if we can show that the representation is more than one dimensional then we know that the energy level must be degenerate. Usually, though, we are able to obtain more detailed information than that.

Before leaving the topic we note that invariant subspaces give some information about the rest of the space as well as about themselves.

Theorem 9.2.3. If $\mathcal{K}$ is an invariant subspace under the unitary representation U then so is $\mathcal{K}^\perp$, and U is the direct sum of the two subrepresentations obtained by restricting to the subspaces $\mathcal{K}$ and $\mathcal{K}^\perp$.

Proof. Let ξ be an element of $\mathcal{K}^\perp$, and ψ an element of $\mathcal{K}$. Then we see that

$$\langle U(x)\xi|\psi\rangle = \langle \xi|U(x)^*\psi\rangle = \langle \xi|U(x^{-1})\psi\rangle = 0, \tag{9.15}$$

since $U(x^{-1})\psi$ is in $\mathcal{K}$ by invariance, and ξ is in $\mathcal{K}^\perp$. Thus $U(x)\xi$ is in $\mathcal{K}^\perp$, showing that $\mathcal{K}^\perp$ is also invariant under U. The fact that U is a direct sum of its subrepresentations follows directly from the fact that $\mathcal{H} = \mathcal{K} \oplus \mathcal{K}^\perp$. □

This means that we can regard U as the direct sum of its restrictions to $\mathcal{K}$ and to $\mathcal{K}^\perp$, in the sense of the following definition.

Definition 9.2.4. If U_1 and U_2 are unitary representations of G on $\mathcal{H}_1$ and $\mathcal{H}_2$ respectively, then there is a representation $U = U_1 \oplus U_2$ on $\mathcal{H} = \mathcal{H}_1 \oplus \mathcal{H}_2$ called the *direct sum* defined by

$$U(x)\,(\psi_1 \oplus \psi_2) = U_1(x)\psi_1 \oplus U_2(x)\psi_2$$

for x in G, ψ_1 in $\mathcal{H}_1$, and ψ_2 in $\mathcal{H}_2$.

9.3. Irreducible representations

We have just seen that a representation that leaves a subspace invariant can be decomposed into two smaller pieces. This fact suggests a special role for those spaces that cannot be split any further.

Definition 9.3.1. A unitary representation U on a space $\mathcal{H}$ for which the only invariant subspaces are $\{0\}$ and $\mathcal{H}$ is said to be *irreducible.*

Remark 9.3.1. Any one-dimensional representation is automatically irreducible since its only subspaces are $\{0\}$ and $\mathcal{H}$. In one dimension every linear operator is multiplication by some scalar, and the operator is unitary if that scalar has modulus 1.

Theorem 9.3.1. Let U be a unitary representation on a finite-dimensional space. Then U is a direct sum of irreducible subrepresentations.

Proof. We proceed by induction on the dimension of the space $\mathcal{H}$. If $\dim \mathcal{H} = 1$ then U is itself irreducible and the result is obvious. If $\mathcal{H}$ is not irreducible then it contains a non-trivial invariant subspace $\mathcal{K}$, so that $\dim \mathcal{K}$ and $\dim \mathcal{K}^{\perp}$ are both less than $\dim \mathcal{H}$. By the inductive hypothesis both $\mathcal{K}$ and $\mathcal{K}^{\perp}$ are direct sums of irreducible subrepresentations, so that the result follows from Theorem 9.2.3. □

Remark 9.3.2. Most of the representations of groups on wave functions described in Section 9.1 are infinite dimensional, so that this result does not apply directly; however, we have seen that individual energy levels also carry representations and since there is usually only a finite degeneracy these are finite dimensional.

In infinite dimensions the situation is more complicated than this theorem might suggest, and some representations cannot be so decomposed into a direct sum of irreducibles even if one allows an infinite number of terms in the sum. However, for some groups such as the rotation group it can be shown that every representation does decompose in this way.

The group of rotations of the plane provides a very striking example of this when we consider its natural representation on periodic wave functions given by

$$(U(\alpha)\psi)(\theta) = \psi(\theta - \alpha) \tag{9.16}$$

for θ and α in $[0, 2\pi]$. The function ψ can be expanded uniquely into a Fourier series

$$\psi(\theta) = \sum_{n=-\infty}^{\infty} c_n e^{in\theta}. \tag{9.17}$$

The functions $e_n(\theta) = \exp(in\theta)$ thus form a basis for the space. Moreover,

$$(U(\alpha)e_n)(\theta) = e^{in(\theta-\alpha)} = e^{-in\alpha}e_n(\theta), \tag{9.18}$$

so that each basis vector spans an invariant subspace. Being only one dimensional these invariant subspaces are irreducible, so the Fourier series provides a way of decomposing the representation U into a direct sum of irreducibles. We can therefore regard the decomposition of a representation into irreducibles as a generalization of Fourier analysis. Bearing in mind the importance of Fourier series in problem solving, this suggests that the decomposition of a representation into irreducibles is likely to be an extremely valuable tool.

9.4. Abelian groups

It is no coincidence that the irreducible representations of the planar rotation group appearing in the Fourier series are one dimensional. We shall now show that this is always the case for abelian groups.

Theorem 9.4.1. (Schur's lemma) If T intertwines two irreducible representations U and W then either $T = 0$ or T is invertible.

Proof. By Lemma 9.2.1 $\ker T$ is a U-invariant subspace, and so by the irreducibility of U it must be either $\{0\}$ or $\mathcal{H}$. If $T \neq 0$ then we must have $\ker T = \{0\}$, which means that T is one–one. Similarly $\operatorname{im} T$ is W invariant, and if $T \neq 0$ then its image must be the whole representation space of W, so that T is onto. Thus if T is non-zero it is both one–one and onto, and therefore invertible. □

Corollary 9.4.2. If T intertwines an irreducible representation U with itself then T is a multiple of the identity operator.

Proof. The crucial point here is a theorem that for some complex number λ the operator $T - \lambda 1$ is not invertible. In finite dimensions this just says that T has an eigenvalue, which is a well-known consequence of the fact that the characteristic polynomial has at least one complex root. An analogous theorem holds more generally although we shall not prove it.

We have already observed in the course of proving Corollary 9.2.2 that when T is an intertwining operator so is $T - \lambda 1$. Since, by hypothesis, $T - \lambda 1$ is not invertible, Schur's lemma tells us that it must vanish, that is $T = \lambda 1$. □

Corollary 9.4.3. Every irreducible representation of an abelian group G is one dimensional.

Proof. Let U be a unitary representation of G. Since G is abelian we have

$$U(x)U(y) = U(xy) = U(yx) = U(y)U(x) \tag{9.19}$$

for all x and y in G. This means that for every x in G the operator $U(x)$ is an intertwining operator. If U is irreducible then the preceding corollary tells us that for each x, $U(x)$ is just multiplication by some complex scalar (which will depend on x). This means that any non-zero vector ψ spans an invariant one dimensional subspace, and by irreducibility this must be the whole of $\mathcal{H}$. □

This result opens the way to classifying all irreducible representations of abelian groups. It is convenient to introduce the notation $\mathbf{T}$ for the multiplicative group of complex numbers of modulus 1:

$$\mathbf{T} = \{z \in \mathbf{C} : |z| = 1\}. \tag{9.20}$$

Example 9.4.1. Take $G = \mathbf{Z}$. We have already noted that one-dimensional representations are just multiplications by scalars of modulus 1. We therefore seek a homomorphism

$$U : \mathbf{Z} \to \mathbf{T}. \tag{9.21}$$

Let us write $U(1) = \exp(i\theta)$. By the homomorphism property we then have

$$U(n) = U(1)^n = e^{in\theta}, \tag{9.22}$$

so that U is completely determined by the real number θ. On the other hand it is easy to see that this formula does define a representation of $\mathbf{Z}$ for any real value of θ, though values of θ that differ by 2π give the same representation.

Example 9.4.2. Let $G = SO(2)$, the rotation group of the plane. Writing α and β for angles of rotation, we want a $\mathbf{T}$-valued function U such that

$$U(\alpha)U(\beta) = U(\alpha + \beta). \tag{9.23}$$

Since the arguments of U are angles, the function must be periodic with period 2π. There are obvious solutions of the equation when $U(\alpha) = \exp(in\alpha)$ for some integer n, and we shall now show that there are no others, so that the irreducible representations of $SO(2)$ are parametrized by integers. Assuming that U is integrable we may multiply the defining identity by $\exp(-in\beta)$ and integrate to get information about its Fourier coefficients. When supplemented with a change of variable, this gives

$$\begin{aligned} U(\alpha) \int_0^{2\pi} U(\beta) e^{-in\beta}\, d\beta &= \int_0^{2\pi} U(\alpha + \beta) e^{-in\beta}\, d\beta \\ &= e^{in\alpha} \int_0^{2\pi} U(\phi) e^{-in\phi}\, d\phi, \end{aligned} \tag{9.24}$$

where the last step exploits the periodicity of U to change the limits of the integral after substituting $\phi = \alpha + \beta$. Since U is non-zero at least one of its Fourier coefficients, the n-th say, must be non-zero, and so can be divided out to obtain

$$U(\alpha) = e^{in\alpha}. \tag{9.25}$$

Using the isomorphism $\alpha \mapsto \exp(i\alpha)$ between $SO(2)$ and $\mathbf{T}$ we can deduce that the irreducible representations of $\mathbf{T}$ all have the form

$$U(z) = z^n. \tag{9.26}$$

Example 9.4.3. Let $G = \mathbf{R}$. Intuition suggests that the obvious homomorphisms from $\mathbf{R}$ to $\mathbf{T}$ given by $U(x) = \exp(ikx)$ for real k are the only ones, and we shall now prove this. For any homomorphism U from $\mathbf{R}$ to $\mathbf{T}$ we may find a real number κ such that $U(2\pi) = \exp(2\pi i\kappa)$. Then $W(x) = U(x)\exp(-i\kappa x)$ is not only an homomorphism but is also periodic of period 2π. By the preceding example we know that there is an integer n such that $W(x) = \exp(inx)$, so that

$$U(x) = W(x)e^{i\kappa x} = e^{i(n+\kappa)x}, \tag{9.27}$$

and putting $k = n + \kappa$ we obtain the expected result. (A similar argument may be used to obtain the irreducible representations of any abelian group G for which we know the irreducibles of both a subgroup Γ and the quotient G/Γ.)

Example 9.4.4. Let G be the direct product of abelian groups $G_1 \times G_2 \times \ldots \times G_n$. Any element x in G can be written as $(x_1, x_2, \ldots, x_n)$, so we may expand $U(x)$ as the product

$$U(x) = U_1(x_1)U_2(x_2)\ldots U_n(x_n) \tag{9.28}$$

where each U_j is a one-dimensional representation of G_j. For example, the irreducible representations of $\mathbf{R}^3$ have the form

$$U(\mathbf{x}) = e^{ik_1x_1}e^{ik_2x_2}e^{ik_3x_3} = e^{i\mathbf{k}.\mathbf{x}}. \tag{9.29}$$

This shows that plane waves can be interpreted as irreducible representations of $\mathbf{R}^3$. Similarly, the irreducible representations of $\mathbf{T}^n$ have the form

$$U(z_1, z_2, \ldots, z_n) = z_1^{k_1} z_2^{k_2} \ldots z_n^{k_n}, \tag{9.30}$$

for some integers $k_1, k_2, \ldots, k_n$.

9.5. Time evolution

We have now determined all the irreducible representations of a number of useful groups including $\mathbf{R}$, but reducible representations are also important. We saw in Section 6.4 that, for Hamiltonians which do not depend on time, Schrödinger's equation can be integrated to give

$$\psi_t = U(t)\psi_0, \tag{9.31}$$

where $U(t) = \exp(-iHt/\hbar)$ is a unitary operator satisfying

$$U(s+t) = U(s)U(t) \tag{9.32}$$

for all s and t in $\mathbf{R}$. These are precisely the conditions for U to be a representation of $\mathbf{R}$.

> **Proposition 9.5.1.** The operators $U(t)$ define a unitary representation of $\mathbf{R}$, that is for all $s, t \in \mathbf{R}$
>
> $$U(s)U(t) = U(s+t).$$

Proof. We have already seen that the operators $U(t)$ are unitary. Assuming (as may be proved) that exponentials behave in the usual way, we easily verify by a formal calculation that

$$U(s+t) = U(s)U(t), \tag{9.33}$$

as required. (When $s = -t$ this just tells us that $U(t)$ and $U(-t)$ are inverse, as needed for the unitarity.) □

Although the above arguments can all be made rigorous there is some advantage in reversing the argument and starting from the postulate that the time evolution for a conservative system is given by a unitary representation. (Unitarity ensures that probabilities and transition functions do not change with time.) The following theorem then ensures the existence of a Hamiltonian operator H, and Schrödinger's equation follows on differentiation.

Stone's theorem 9.5.2. Let $U(t)$ be a unitary representation of $\mathbf{R}$ on $\mathcal{H}$, such that $\langle\psi|U(t)\psi\rangle$ is a continuous function of t for all $\psi \in \mathcal{H}$. Then there is a unique self-adjoint operator H, called the *infinitesimal generator* of U, such that

$$U(t) = e^{-itH/\hbar}$$

for all $t \in \mathcal{H}$. The operator H is defined on the set of vectors, ψ, for which $(i\hbar/t)[U(t)-1]\psi$ converges to a limit as $t \to 0$; the limit is then $H\psi$.

Stone's theorem, whose proof lies beyond the scope of these notes, describes all the representations of $\mathbf{R}$ and not just the irreducibles. Its scope is far wider than might at first appear because many other groups contain subgroups related to $\mathbf{R}$ as subgroups. To be more precise we need some more terminology.

Let Γ and G be groups with a homomorphism ϕ from Γ to G. If U is a representation of G then the composition $U \circ \phi$ is a representation of Γ on the same space, since the composition of homomorphisms is a homomorphism. If Γ is a subgroup of G, then there is an inclusion homomorphism ϕ which sends an element of Γ to itself regarded as an element of G. Then $U \circ \phi$ is called the restriction of U to Γ. This is more simply expressed as follows:

Definition 9.5.1. If Γ is a subgroup of G, and U a representation of G, then the *restriction* of U to Γ takes y in Γ to $U(y)$.

There are many homomorphisms from $\mathbf{R}$ to the rotation group $SO(3)$, for given an axis in the direction of the unit vector $\mathbf{n} \in \mathbf{R}^3$ we may define $R_{\mathbf{n}}(t)$ to be the rotation in a positive sense through an angle t about the axis $\mathbf{n}$. It is easy to check that the map taking t to $R_{\mathbf{n}}(t)$ is a homomorphism, so given any representation U of $SO(3)$ we obtain a representation $U_{\mathbf{n}} = U \circ R_{\mathbf{n}}$ of $\mathbf{R}$. By Stone's theorem there exists a self-adjoint operator $J_{\mathbf{n}}$ such that

$$U_{\mathbf{n}}(t) = \exp\left(-\frac{it}{\hbar}J_{\mathbf{n}}\right) \tag{9.34}$$

and which is given by

$$J_{\mathbf{n}} = i\hbar\frac{d}{dt}U_{\mathbf{n}}(t)\bigg|_{t=0}. \tag{9.35}$$

The significance of $J_{\mathbf{n}}$ is easily seen by reference to Example 9.1.2. There we have

$$(U(R_{\mathbf{n}}(t))\,\psi)(\mathbf{r}) = \psi\left(R_{\mathbf{n}}(t)^{-1}\mathbf{r}\right) = \psi\left(R_{\mathbf{n}}(-t)\mathbf{r}\right), \tag{9.36}$$

so that

$$\begin{aligned}(J_{\mathbf{n}}\psi)(\mathbf{r}) &= i\hbar\frac{d}{dt}\psi\left(R_{\mathbf{n}}(-t)\mathbf{r}\right)\Big|_{t=0} \\ &= i\hbar\left[\left(\frac{d}{dt}\left(R_{\mathbf{n}}(-t)\mathbf{r}\right).\nabla\psi\right)\left(R_{\mathbf{n}}(-t)\mathbf{r}\right)\right]\Big|_{t=0}.\end{aligned} \tag{9.37}$$

Now $R_{\mathbf{n}}(0) = 1$ and $d\left(R_{\mathbf{n}}(-t)\mathbf{r}\right)/dt$, being the derivative of a steady rotation about $\mathbf{n}$, with unit speed, is given by the classical mechanical formula $-\mathbf{n}\times\mathbf{r}$. Alternatively one can differentiate the explicit formula

$$R_{\mathbf{n}}(-t)\mathbf{r} = (1-\cos t)(\mathbf{r}.\mathbf{n})\mathbf{n} + \cos t\mathbf{r} - \sin t(\mathbf{n}\times\mathbf{r}). \tag{9.38}$$

Thence

$$(J_{\mathbf{n}}\psi)(\mathbf{r}) = -i\hbar(\mathbf{n}\times\mathbf{r}).\nabla\psi = -i\hbar\mathbf{n}.(\mathbf{r}\times\nabla)\psi, \tag{9.39}$$

which is the formula for $\mathbf{n}.\mathbf{J}$ given in Section 8.1. As the notation suggested we can therefore interpret the infinitesimal generator $J_{\mathbf{n}}$ as angular momentum.

This is no accident. If we take the subgroup $\{t\mathbf{a} : t\in\mathbf{R}\}\subseteq\mathbf{R}^3$ and consider the representation of Example 15.1.4 we have the infinitesimal generator

$$\begin{aligned}(P_{\mathbf{a}}\psi)(\mathbf{r}) &= i\hbar\frac{d}{dt}\left(U(t\mathbf{a})\psi\right)(\mathbf{r})\Big|_{t=0} \\ &= i\hbar\frac{d}{dt}\left(\psi(\mathbf{r}-t\mathbf{a})\right)\Big|_{t=0} \\ &= i\hbar(-\mathbf{a}.(\nabla)\,\psi)(\mathbf{r}),\end{aligned} \tag{9.40}$$

so that $P_{\mathbf{a}}$ is just the momentum in the $\mathbf{a}$ direction.

Definition 9.5.2. The infinitesimal generator of a representation arising in this way is called a *generalized momentum.*

We have seen above that both the linear momentum and the angular momentum are special cases of this definition. In the case of the rotation group one can show directly by choosing subgroups of rotations about different axes that

$$[J_{\mathbf{m}}, J_{\mathbf{n}}] = i\hbar J_{\mathbf{m}\times\mathbf{n}} \tag{9.41}$$

so that the commutation relations for the angular momentum are a consequence of the fact that U is a representation of $SO(3)$. This accords with the heuristic observation that angular momentum is most important where there is rotational symmetry.

9.6. The irreducible representations of the rotation group

Given the connection between the rotation group and angular momentum described in the last section, it is easy to find all the irreducible representations of $SO(3)$. We already found in Theorem 8.3.1 that the only ways to satisfy the angular momentum commutation relations required a $(2j+1)$-dimensional space with a basis $\{\psi_m : -j \leq m \leq j\}$ of J_3-eigenvectors satisfying

$$J_3\psi_m = m\hbar\psi_m. \tag{9.42}$$

This relation exponentiates to tell us that for rotations $R_3(t)$ about the third coordinate axis

$$U(R_3(t))\psi_m = \exp\left(-\frac{it}{\hbar}J_3\right)\psi_m = e^{-imt}\psi_m. \tag{9.43}$$

If we require that $U(R_3(2\pi)) = 1$ then m must be integral. This in turn forces j to be integral and recovers precisely those representations that could be realized on wave functions. If one takes the explicit realization of the angular momentum operators on spherical harmonics described in Section 8.4 then one can show that the rotation group acts on the wave functions as in Example 9.1.2. We shall denote the $(2j+1)$-dimensional representation of the rotation group by D^j.

9.7. Characters

When one recalls the amount of work that went into finding the solution of the angular momentum commutation relations, one realizes how much more complicated multi-dimensional representations are than the one-dimensional representations of abelian groups. In this section we shall show that it is nonetheless possible to find a single complex-valued function that encapsulates all the information for any finite-dimensional representation.

Definition 9.7.1. Let U be a finite-dimensional unitary representation of G. The *character* χ_U of U is the complex-valued function on G defined by

$$\chi_U(x) = \mathrm{tr}\,(U(x)).$$

Remark 9.7.1. Clearly if U is one dimensional then it is just multiplication by χ_U, so that one-dimensional representations are sometimes called characters. In general the dimension of U is given by $\chi_U(1)$ where 1 denotes the identity in G.

The main properties of characters are summarized below.

Proposition 9.7.1. Let TUT^{-1} denote the representation that sends x in G to $TU(x)T^{-1}$. Then for any x and y in G and any finite-dimensional representations U and W we have
(i) $\chi_{TUT^{-1}}(x) = \chi_U(x)$;
(ii) $\chi_U(yxy^{-1}) = \chi_U(x)$;
(iii) $\chi_{U \oplus W}(x) = \chi_U(x) + \chi_W(x)$.

Proof. (i) We have by definition and the standard properties of traces

$$\chi_{TUT^{-1}}(x) = \mathrm{tr}(TU(x)T^{-1}) = \mathrm{tr}(T^{-1}TU(x)) = \mathrm{tr}(U(x)) = \chi_U(x). \tag{9.44}$$

(ii) By definition we have

$$\chi_U(yxy^{-1}) = \mathrm{tr}(U(y)U(x)U(y)^{-1}), \tag{9.45}$$

so setting $T = U(y)$ in the previous part we obtain $\chi_U(yxy^{-1}) = \chi_U(x)$.

(iii) If $\{u_j\}$ and $\{w_\alpha\}$ are orthonormal bases for the representation spaces of U and W respectively, then their union, $\{u_j\} \cup \{w_\alpha\}$, is an orthonormal basis for their direct sum. Thence we calculate that

$$\chi_{U \oplus W}(x) = \sum_j \langle u_j | U(x) u_j \rangle + \sum_\alpha \langle w_\alpha | W(x) w_\alpha \rangle = \chi_U(x) + \chi_W(x). \tag{9.46}$$

Remark 9.7.2. Part (i) shows that equivalent representations have the same character. The converse is also true and representations that have the same character are equivalent. We shall prove that for the rotation group in the next section.

9.8. The characters of the rotation group

According to Proposition 9.7.1 the value of a character depends only on the conjugacy class of its argument. Now if S is a rotation and $R_{\mathbf{n}}(t)$ denotes the rotation through an angle t about the axis $\mathbf{n}$ it is easy to check that

$$SR_{\mathbf{n}}(t)S^{-1} = R_{S\mathbf{n}}(t), \tag{9.47}$$

which shows that the conjugacy class of a rotation depends only on the angle through which it rotates and not on the axis. Let us write $\Delta_j(\theta)$ for the character of a rotation through θ in the $(2j+1)$-dimensional irreducible representation D^j.

Theorem 9.8.1. The characters of the rotation group are given by

$$\Delta_j(\theta) = \frac{\sin\left[(j+\frac{1}{2})\theta\right]}{\sin\left(\frac{1}{2}\theta\right)}$$

for integral j.

Proof. We already observed that j must be integral. We may as well evaluate the trace for a rotation about the third coordinate axis, and since this has the basis elements $\{\psi_m\}$ as eigenvectors with eigenvalues $\{\exp(-im\theta)\}$ we have

$$\begin{aligned}\Delta_j(\theta) &= \sum_{m=-j}^{j} e^{-im\theta} \\ &= e^{ij\theta}\frac{e^{-i(2j+1)\theta}-1}{e^{-i\theta}-1} \\ &= \frac{e^{i(j+\frac{1}{2})\theta}-e^{-i(j+\frac{1}{2})\theta}}{e^{i\frac{1}{2}\theta}-e^{-i\frac{1}{2}\theta}} \\ &= \frac{\sin\left[(j+\frac{1}{2})\theta\right]}{\sin\left(\frac{1}{2}\theta\right)}.\end{aligned}$$

□

The key to many of the further properties of the characters is provided by the following orthogonality theorem.

Theorem 9.8.2. For any k and j in $\frac{1}{2}\mathbf{Z}$

$$\int_0^{2\pi} \overline{\Delta_k(\theta)}\Delta_j(\theta)\frac{1-\cos\theta}{2\pi}\,d\theta = \delta_{kj}.$$

Remark 9.8.1. The complex conjugation of Δ_k might seem superfluous since our explicit formula shows the characters of the rotation group to be real valued; however, similar orthogonality relations hold for other groups whose characters may be genuinely complex, so it is more convenient to include it.

Proof. This follows by a direct calculation since $1 - \cos\theta = 2\sin^2\theta$, so that

$$\int_0^{2\pi} \overline{\Delta_k(\theta)}\Delta_l(\theta)\frac{1-\cos\theta}{2\pi}\,d\theta = \int_0^{2\pi} \Delta_k(\theta)\sin\left(\tfrac{1}{2}\theta\right)\Delta_l(\theta)\sin\left(\tfrac{1}{2}\theta\right)\frac{d\theta}{\pi}$$
$$= \int_0^{2\pi} \sin\left(k+\tfrac{1}{2}\right)\theta\sin\left(l+\tfrac{1}{2}\right)\theta\frac{d\theta}{\pi}. \quad (9.48)$$

The usual Fourier orthogonality relations for sines now show that this is δ_{kl}. □

Corollary 9.8.3. Any character Δ of the rotation group can be expanded uniquely in the form

$$\sum_{j\in\frac{1}{2}\mathbf{Z}} n_j\Delta_j,$$

where

$$n_j = \int_0^{2\pi} \Delta(\theta)\Delta_j(\theta)\frac{1-\cos\theta}{2\pi}\,d\theta.$$

Proof. The axis of a rotation is determined only up to a sign and we know that $R_{-\mathbf{n}}(\theta) = R_{\mathbf{n}}(-\theta)$, so we must have $\Delta(-\theta) = \Delta(\theta)$. In other words the characters are even functions. This means that $\Delta(\theta)\sin\left(\frac{1}{2}\theta\right)$ is an odd function that can be expanded uniquely as a series of sines

$$\Delta(\theta)\sin\left(\tfrac{1}{2}\theta\right) = \sum_j n_j\sin\left[\left(j+\tfrac{1}{2}\right)\theta\right], \quad (9.49)$$

with n_j defined by the stated formula. From this we obtain the series

$$\Delta(\theta) = \sum n_j\Delta_j(\theta),$$

as asserted. □

9.9. The spin representation of the rotation group

We only obtain representations of the rotation group when j is integral, but it is possible using the spin representation of angular momentum described in Section 8.8 to obtain representations of a closely related group for half-integral j as well.

Definition 9.9.1. The group $\mathbf{SU}(2)$ is the group of 2×2 unitary matrices (that is, $U^*U = 1$) with determinant 1.

Theorem 9.9.1. For each $U \in \mathbf{SU}(2)$ there is a rotation $R(U)$ such that for all $\mathbf{a} \in \mathbf{R}^3$

$$U(\boldsymbol{\sigma}.\mathbf{a})U^* = \boldsymbol{\sigma}.R(U)\mathbf{a}.$$

Proof. According to Theorem 8.8.1 the matrix $\boldsymbol{\sigma}.\mathbf{a}$ is traceless and self-adjoint. It is therefore immediate that $U(\boldsymbol{\sigma}.\mathbf{a})U^*$ is also self-adjoint and that

$$\operatorname{tr}(U(\boldsymbol{\sigma}.\mathbf{a})U^*) = \operatorname{tr}(U^*U(\boldsymbol{\sigma}.\mathbf{a})) = \operatorname{tr}(\boldsymbol{\sigma}.\mathbf{a}) = 0. \tag{9.50}$$

Theorem 8.8.1 now tells us that $U(\boldsymbol{\sigma}.\mathbf{a})U^* = \boldsymbol{\sigma}.\mathbf{a}'$ for some $\mathbf{a}' \in \mathbf{R}^3$. Moreover, this relationship between $\mathbf{a}$ and $\mathbf{a}'$ is clearly linear. We also have the identity

$$\det(\boldsymbol{\sigma}.\mathbf{a}) = \det\begin{pmatrix} a_3 & a_1 - ia_2 \\ a_1 + ia_2 & -a_3 \end{pmatrix} = -(a_1^2 + a_2^2 + a_3^2). \tag{9.51}$$

Since

$$\det(\boldsymbol{\sigma}.\mathbf{a}') = \det(U(\boldsymbol{\sigma}.\mathbf{a})U^*) = \det(U)\det(\boldsymbol{\sigma}.\mathbf{a})\det(U^*) = \det(\boldsymbol{\sigma}.\mathbf{a}), \tag{9.52}$$

we deduce that

$$-|\mathbf{a}'|^2 = -|\mathbf{a}|^2, \tag{9.53}$$

so that $\mathbf{a}' = R(U)\mathbf{a}$ for some orthogonal transformation $R(U)$.

From the identity

$$[U(\boldsymbol{\sigma}.\mathbf{a})U^*, U(\boldsymbol{\sigma}.\mathbf{b})U^*] = 2iU(\boldsymbol{\sigma}.(\mathbf{a} \times \mathbf{b}))U^*, \tag{9.54}$$

with the help of Corollary 8.8.2, we deduce that

$$R(U)\mathbf{a} \times R(U)\mathbf{b} = R(U)(\mathbf{a} \times \mathbf{b}) \tag{9.55}$$

so that $R(U)$ respects the orientation of the vector product and must be a rotation. □

Theorem 9.9.2. The map $U \mapsto R(U)$ of the last theorem defines a homomorphism from $\mathbf{SU}(2)$ onto the rotation group. Its kernel is $\{\pm 1\}$, so that the rotation group is isomorphic to $\mathbf{SU}(2)/\{\pm 1\}$.

Proof. Since

$$\boldsymbol{\sigma}.(R(UV)\mathbf{a}) = UV(\boldsymbol{\sigma}.\mathbf{a})V^*U^* = U(\boldsymbol{\sigma}.(R(V)\mathbf{a}))U^* = \boldsymbol{\sigma}.(R(U)R(V)\mathbf{a}), \tag{9.56}$$

we see that $R(UV)\mathbf{a} = R(U)R(V)\mathbf{a}$ for all $\mathbf{a} \in \mathbf{R}^3$ so that $R(UV) = R(U)R(V)$ and R is a homomorphism.

Using the unitarity of U, the kernel of R consists of those $U \in \mathbf{SU}(2)$ for which $U(\boldsymbol{\sigma}.\mathbf{a})U^* = \boldsymbol{\sigma}.\mathbf{a}$, or equivalently,

$$U(\boldsymbol{\sigma}.\mathbf{a}) = (\boldsymbol{\sigma}.\mathbf{a})U, \tag{9.57}$$

for all $\mathbf{a} \in \mathbf{R}^3$. Since, in particular, U commutes with σ_3 it must be a diagonal matrix, and since it commutes with σ_1 its diagonal entries must be the same. That means that U is a multiple of the identity, $\lambda 1$, say. Now $1 = \det(U) = \lambda^2$ forces $\lambda = \pm 1$ and gives $\ker(R) = \{\pm 1\}$.

One can prove that every rotation $R_{\mathbf{n}}(t)$ is in the image of R by checking that it is the image of $U = \cos(\frac{1}{2}t) - i\sin(\frac{1}{2}t)\boldsymbol{\sigma}.\mathbf{n}$. (See also Exercise 8.11.) The first isomorphism theorem for groups now tells us that the rotation group is isomorphic to $\mathbf{SU}(2)/\ker(R) = \mathbf{SU}(2)/\{\pm 1\}$. □

The homomorphism R gives an immediate connection between the representation theory of $\mathbf{SU}(2)$ and the rotation group.

Corollary 9.9.3. Any representation W of the rotation group lifts to a representation $R^*W = W \circ R$ of $\mathbf{SU}(2)$.

In the proof of Theorem 9.9.2 we saw that the rotation through an angle t about the third coordinate axis is the image of the diagonal matrix

$$U = \cos(\tfrac{1}{2}t) - i\sin(\tfrac{1}{2}t)\sigma_3 = \begin{pmatrix} e^{-\frac{1}{2}it} & 0 \\ 0 & e^{\frac{1}{2}it} \end{pmatrix}. \tag{9.58}$$

We write $\alpha_1 = \exp(\frac{1}{2}it)$ and $\alpha_2 = \exp(-\frac{1}{2}it)$ for the two diagonal entries. Using Theorem 9.8.1, we can obtain the following expression for the character:

$$R^*\Delta^l(\alpha) = \Delta^l(t) = \frac{\sin[(l+\frac{1}{2})t]}{\sin(\frac{1}{2}t)} = \frac{\alpha_1^{2l+1} - \alpha_2^{2l+1}}{\alpha_1 - \alpha_2}. \tag{9.59}$$

This last expression shows that $R^*\Delta^l$ makes perfectly good sense as a character of $\mathbf{SU}(2)$ when l is half-integral too. □

9.10. The hidden symmetries of hydrogen

The components of the Runge–Lenz vector, $\mathbf{A}$, in Section 8.7 provide additional constants of the motion for the hydrogen atom, suggesting that there might be additional symmetries. We shall now show how the spectrum of hydrogen can be obtained by group theory.

We start by Fourier transforming Schrödinger's equation. Writing $\widehat{\psi} = \mathcal{F}\psi$ and remembering that the Fourier transform of the product $V\psi$ is a convolution, we obtain

$$\frac{p^2}{2m}\widehat{\psi}(\mathbf{p}) + \frac{1}{(2\pi\hbar)^{\frac{3}{2}}}\int \widehat{V}(\mathbf{p}-\mathbf{q})\widehat{\psi}(\mathbf{q})d^3\mathbf{q} = E\widehat{\psi}(\mathbf{p}). \tag{9.60}$$

It is straightforward to show that the Fourier transform of $Ke^{-\kappa r}/r$ is $K\sqrt{2\hbar/\pi}/(p^2+\kappa^2\hbar^2)$, so that, letting $\kappa \to 0$, and substituting $E = -p_0^2/2m$, we have

$$\frac{p_0^2+p^2}{2m}\widehat{\psi}(\mathbf{p}) + \frac{K}{2\pi^2\hbar}\int |\mathbf{p}-\mathbf{q}|^{-2}\widehat{\psi}(\mathbf{q})d^3\mathbf{q} = 0. \tag{9.61}$$

So far the calculation has been quite conventional, but we now introduce the matrix variable,

$$Z = Z_0 + i\mathbf{Z}.\boldsymbol{\sigma} = (p_0 + i\mathbf{p}.\boldsymbol{\sigma})^2/(p_0^2+p^2), \tag{9.62}$$

which satisfies

$$Z^*Z = (p_0 - i\mathbf{p}.\boldsymbol{\sigma})^2(p_0 + i\mathbf{p}.\boldsymbol{\sigma})^2/(p_0^2+p^2)^2 = 1, \tag{9.63}$$

so that Z is unitary. Similarly $\det(Z) = 1$, so that Z is actually in the group $\mathbf{SU}(2)$. Letting $W = (p_0 + i\mathbf{q}.\boldsymbol{\sigma})^2/(p_0^2 + q^2)$, we may calculate that

$$|\mathbf{p} - \mathbf{q}|^2 = 4p_0^2 \det(Z - W)/\det(1 + Z)\det(1 + W), \tag{9.64}$$

so that writing $\Psi(Z) = \det(1 + Z)^{-2}\widehat{\psi}(\mathbf{p})$ and $dW = dW_1 dW_2 dW_3/\pi^2 W_0$, we have

$$\Psi(Z) = \frac{mK}{\hbar p_0}\int \det(Z - W)^{-1}\Psi(W)dW. \tag{9.65}$$

This shows that Schrödinger's equation for the hydrogen atom is equivalent to an integral equation on the group $\mathbf{SU}(2)$, which we shall now solve using its characters described in the previous section.

Lemma 9.10.1. Let $\chi_j = R^*\Delta^{j/2}$ denote the character of $\mathbf{SU}(2)$ which takes the value $(\alpha_1^{j+1} - \alpha_2^{-j-1})/(\alpha_1 - \alpha_2)$ for a group element whose eigenvalues are α_1 and α_2. Then

$$\det(Z - W)^{-1} = \sum_j \chi_j(WZ^{-1}).$$

Proof. We first observe that, if the eigenvalues of W are α_1 and α_2, then

$$\begin{aligned}\det(1 - W)^{-1} &= (1 - \alpha_1)^{-1}(1 - \alpha_2)^{-1} \\ &= \frac{1}{\alpha_1 - \alpha_2}\left(\frac{1}{1 - \alpha_1} - \frac{1}{1 - \alpha_2}\right) \\ &= \sum_j \frac{\alpha_1^{j+1} - \alpha_2^{-j-1}}{\alpha_1 - \alpha_2} \\ &= \sum_j \chi_j(W).\end{aligned} \tag{9.66}$$

In general we have

$$\det(Z - W) = \det(Z)\det(1 - Z^{-1}W) = \det(1 - Z^{-1}W), \tag{9.67}$$

from which the result now follows. □

From this we deduce that Ψ satisfies the equation

$$\Psi(Z) = \frac{mK}{\pi^2\hbar p_0}\sum_j \int \chi_j(Z^{-1}W)\Psi(W)dW. \tag{9.68}$$

Lemma 9.10.2. For any operator ρ on the d_k-dimensional space of the irreducible representation $C^k = R^* D^{k/2}$ of $\mathbf{SU}(2)$ one has the identity

$$d_k \int \frac{\operatorname{tr}(\rho C^k(W))}{\det(Z-W)}\, dW = \operatorname{tr}(\rho C^k(Z).$$

Proof. The representation C^k has character χ_k and $d_k = k+1$. We start by considering

$$\int \frac{\operatorname{tr}(\rho C^k(UWU^{-1}))}{\det(1-W)}\, dW, \tag{9.69}$$

and make the change of variable, $Y = UWU^{-1}$. According to Theorem 9.9.1 conjugation by U just rotates $\mathbf{W}$, so that $dY = dW$ and so the integral becomes

$$\int \frac{\operatorname{tr}(\rho C^k(Y))}{\det(1-U^{-1}YU)}\, dY. \tag{9.70}$$

Since $\det(1-U^{-1}YU) = \det(U^{-1}(1-Y)U) = \det(1-Y)$, this gives the identity

$$\int \frac{\operatorname{tr}(\rho C^k(UWU^{-1}))}{\det(1-W)}\, dW = \int \frac{\operatorname{tr}(\rho C^k(W))}{\det(1-W)}\, dW. \tag{9.71}$$

As this is true for all ρ we deduce that

$$C^k(U) \int \frac{C^k(W)}{\det(1-W)}\, dW C^k(U)^{-1} = \int \frac{C^k(W)}{\det(1-W)}\, dW, \tag{9.72}$$

showing the integral on the right to be an intertwining operator for the irreducible, C^k. By Schur's lemma 9.4.1, this means that it is a multiple of the identity:

$$\int \frac{C^k(W)}{\det(1-W)}\, dW = \kappa 1. \tag{9.73}$$

Taking traces, we have

$$\int \frac{\chi_k(W)}{\det(1-W)}\, dW = \kappa d_k. \tag{9.74}$$

On the other hand, using the expression (9.67) for $\det(1-W)^{-1}$ as a sum of characters, and the orthogonality relations analogous to Theorem 9.8.2, we see that the left-hand side is 1, so that $\kappa = d_k^{-1}$. This gives us

$$d_k \int \frac{C^k(W)}{\det(1-W)}\, dW = 1, \tag{9.75}$$

and, multiplying by $\rho C^k(Z)$ and taking the trace, we obtain

$$\operatorname{tr}(\rho C^k(Z)) = d_k \int \frac{C^k(ZW)}{\det(1-W)}\, dW. \tag{9.76}$$

We finish the proof with another change of variables, replacing W by $Z^{-1}W$, and using the identity (9.68), to arrive at the result. □

This lemma provides a ready source of solutions to the Schrödinger equation, since we may take $mK/\hbar p_0 = d_k$ and $\Psi(Z) = \operatorname{tr}(\rho C^k(Z))$ for any ρ. The corresponding energy is

$$E = -\frac{p_0^2}{2m} = -\frac{mK^2}{2\hbar^2 d_k^2}, \tag{9.77}$$

and there are d_k^2 independent choices of ρ, in agreement with the usual formulae for the energy and degeneracy.

To complete this approach we shall sketch why these are the only solutions. To do this we multiply equation (9.69) by $\chi_k(ZX^{-1})$ and integrate, to obtain

$$\int \chi_k(ZX^{-1})\Psi(Z) = \frac{mK}{\hbar p_0} \int \frac{\chi_k(X^{-1}Z)}{\det(Z-W)} \Psi(W)\, dW dZ. \tag{9.78}$$

Now, setting $\rho = C^k(X^{-1})$ and interchanging W and Z in the lemma we have

$$d_k \int \frac{\chi_k(X^{-1}Z)}{\det(W-Z)}\, dZ = \chi_k(X^{-1}W). \tag{9.79}$$

Since $\det(W-Z) = \det(Z-W)$, this may be substituted into equation (9.79) to give

$$d_k \int \chi_k(ZX^{-1})\Psi(Z) = \frac{mK}{\hbar p_0} \int \chi_k(X^{-1}W)\Psi(W)\, dW. \tag{9.80}$$

According to equation (9.69) the integral cannot vanish for all k unless $\Psi = 0$. For any k giving a non-zero integral we then have $mK/\hbar p_0 = d_k$, confirming our earlier result.

9.11. Wigner's theorem

Symmetries of a quantum system are not always realized by unitary representations. Physically all that matters is the action on states (which are unchanged when the vector is multiplied by a scalar) and that the transition probabilities, $|\langle\phi|\psi\rangle|^2/\|\phi\|^2\|\psi\|^2$, are unchanged by the action of the symmetry group. This could also be ensured by taking *antiunitary operators* $U(g)$, that is operators such that

$$\langle U(g)\phi|U(g)\psi\rangle = \langle\psi|\phi\rangle, \tag{9.81}$$

since this is just the conjugate of $\langle\phi|\psi\rangle$. In fact, we have the following result:

Wigner's theorem 9.11.1. Let G be a group that acts on the states of a space $\mathcal{H}$ in such a way as to preserve transition probabilities,

$$|\langle\phi|\psi\rangle|^2/\|\phi\|^2\|\psi\|^2.$$

Then up to scalar multiples the action of $g \in G$ can be written in the form $\psi \to U(g)\psi$, where each $U(g)$ is either unitary or antiunitary, and

$$U(xy) = \sigma(x,y)U(x)U(y)$$

for some complex number $\sigma(x,y)$ of modulus 1.

There are physically important examples for which the operators are antiunitary rather than unitary. Many physical systems are unchanged by time reversal and this provides one of the simplest examples of the occurrence of antiunitary operators. Let us define the action of the *time reversal* operator $\mathcal{T}$ on a wave function to be

$$(\mathcal{T}\psi)(t,\mathbf{r}) = \overline{\psi(-t,\mathbf{r})}. \tag{9.82}$$

If we just change t to $-t$ in Schrödinger's equation with a time-independent potential then it becomes

$$-i\hbar\frac{\partial}{\partial t}\psi(-t,\mathbf{r}) = -\frac{\hbar^2}{2m}\nabla^2\psi + V\psi \tag{9.83}$$

so that complex conjugation turns it back to the usual Schrödinger equation. The combination of reversing the sign of t and conjugating, which occur in $\mathcal{T}$, therefore gives a symmetry of the system. Another common example of an antiunitary operator is provided by the charge conjugation $\mathcal{C}$ which reverses the charge of each particle and conjugates its wave function. We shall discuss this in more detail in Section 18.3.

Nonetheless, for many systems only unitary operators can occur.

Weyl's lemma 9.11.2. Under the assumptions of the previous theorem, $U(g^2)$ is unitary for all $g \in G$.

Proof. We know that $U(g^2) = \sigma(g,g)U(g)^2$. If $U(g)$ is unitary then so also is its square, but even if $U(g)$ is antiunitary its square satisfies

$$\langle U(g)^2\phi|U(g)^2\psi\rangle = \langle U(g)\psi|U(g)\phi\rangle = \langle\phi|\psi\rangle, \tag{9.84}$$

and so is unitary. Since $\sigma(g,g)$ just multiplies by a scalar $U(g^2)$ is unitary in either case. □

In the rotation group every rotation through θ is the square of a rotation through $\frac{1}{2}\theta$ about the same axis, and so must be represented by a unitary operator. The same applies to most of the other groups that we have considered. It is not so easy to get rid of the scalar *multiplier* $\sigma(x,y)$. In fact, we have already encountered one example where it is needed, in the case of the representations of the rotation group when j is a half-integer but not integral. Then, because of the sign ambiguities, the most that can be said is that $U(xy) = \pm U(x)U(y)$, so that we are dealing with a multiplier that only takes the values ± 1. In general, when σ is needed, one says that U is a *projective representation* with multiplier σ. Fortunately, although some important physical examples are described by projective representations, in many cases it is possible to get rid of σ and take U to be a unitary representation.

Exercises

9.1 Show that the cyclic group $\mathbf{Z}_N$ of order N, with generator z, has representations given by $U(z^k) = \exp(2rk\pi i/N)$, for $r = 0, \ldots, N-1$, and that every irreducible representation is of this form.

9.2 Let a be the cyclic permutation (123) in the symmetric group S_3, and let b be the transposition (12). Show that b and a are of order 2 and 3, respectively, and that $ba = a^2b$. Show that there is a representation, U, of S_3 with

$$U(a) = \begin{pmatrix} \omega & 0 \\ 0 & \omega^{-1} \end{pmatrix}, \qquad U(b) = \begin{pmatrix} 0 & 1 \\ 1 & 0 \end{pmatrix},$$

provided that $\omega^3 = 1$. Show that if $\omega \neq 1$ then it is irreducible.

9.3 Let $\mathcal{H}$ be the space of complex-valued functions on a finite group, G, equipped with the inner product

$$\langle\phi|\psi\rangle = \sum_{g\in G} \overline{\phi(g)}\psi(g),$$

for any ϕ and ψ in $\mathcal{H}$. Show that

$$(U(x)\psi)(g) = \psi(x^{-1}g)$$

defines a unitary representation of G.

9.4° The vectors $\mathbf{e}_1$, $\mathbf{e}_2$, and $\mathbf{e}_3$ form a right-handed orthonormal triad in $\mathbf{R}^3$. A group multiplication is put on $\mathbf{R}^3$ by defining the product of two vectors to be

$$\mathbf{x} \circ \mathbf{y} = \mathbf{x} + \mathbf{y} + \tfrac{1}{2}[\mathbf{e}_3, \mathbf{x}, \mathbf{y}]\mathbf{e}_3,$$

where the square brackets denote the triple scalar product. If U is a representation of this group show that

$$U(\mathbf{x})^{-1}U(\mathbf{y})U(\mathbf{x}) = U(\mathbf{y} - [\mathbf{e}_3, \mathbf{x}, \mathbf{y}]\mathbf{e}_3).$$

Show also that, for $j = 1, 2$, or 3, the map taking the real number t to $U(t\mathbf{e}_j)$ defines a representation of $\mathbf{R}$. If A_j is the infinitesimal generator corresponding to $U(t\mathbf{e}_j)$ show that

$$U(\mathbf{x})^{-1}A_2U(\mathbf{x}) = A_2 - (\mathbf{x}.\mathbf{e}_1)A_3$$

and

$$U(\mathbf{x})^{-1}A_3U(\mathbf{x}) = A_3.$$

Deduce that

$$[A_1, A_2] = i\hbar A_3$$

and thence that, if U is irreducible, then either $A_3 = 0$ or $X = A_3^{-1}A_1$ and $P = A_2$ satisfy the canonical commutation relations.

9.5° Let U be an irreducible unitary representation of the rotation group $\mathbf{SO}(3)$ on a finite-dimensional inner product space V. Suppose that there is a non-zero vector Ω that is fixed by $U(h)$ whenever h is a rotation about the axis $\mathbf{k}$ in $\mathbf{R}^3$. Show that it is possible to define a one–one linear map T from V to the functions on the sphere $S^2 = \{\mathbf{u} \in \mathbf{R}^3 : |\mathbf{u}| = 1\}$ by

$$(T\psi)(g\mathbf{k}) = \langle U(g)\Omega|\psi\rangle,$$

for $g \in \mathbf{SO}(3)$ and $\psi \in V$. Show that

$$(TU(x)\psi)(\mathbf{u}) = (T\psi)(x^{-1}\mathbf{u})$$

for all $x \in \mathbf{SO}(3)$ and $\mathbf{u} \in S^2$. Show that there is a unique representation, U, on $V = \mathbf{C}^3$ whose action on real vectors is the usual rotational action on $\mathbf{R}^3$. By taking $\Omega = \mathbf{k}$, show that the image of T then consists of the restriction to S^2 of functions linear in the components of $\mathbf{u}$.

10 Measurements and paradoxes

> Quantum mechanics is very impressive, but hardly brings us any closer to the secrets of the Old One. I am at any rate convinced that *He* does not play dice.
>
> ALBERT EINSTEIN, letter to Max Born, 4 December 1926

10.1. The quantum Zeno paradox

We noted in Section 6.5 that quantum measurements change the system which is measured. The abrupt change of state caused by a measurement is quite different from the steady evolution described by Schrödinger's equation, and this shows up particularly clearly if we imagine carrying out frequent repeated measurements on an evolving system.

Let U be any operator on $\mathcal{H}$ and P_ϕ the projection onto a normalized vector ϕ, that is $P_\phi\psi = \langle\phi|\psi\rangle\phi$. Then $P_\phi U P_\phi \psi = \langle\phi|U\phi\rangle P_\phi\psi$, and by induction we see that

$$P_\phi U P_\phi U P_\phi \dots U P_\phi = \langle\phi|U\phi\rangle^n P_\phi, \tag{10.1}$$

where n is the number of Us appearing on the left. Now taking for U the unitary time evolution operator defined by the Hamiltonian H, and using Lemma 7.3.1, we have

$$\begin{aligned}\langle\phi|U_{t/n}\phi\rangle^n &= \exp\left(-n\frac{it}{n\hbar}\mathsf{E}_\phi(H) - n\frac{t^2}{2n^2\hbar^2}\right)\left[1 + o\left(\frac{t^2}{n^2}\right)\right] \\ &= \exp\left(-\frac{it}{\hbar}\mathsf{E}_\phi(H) - \frac{t^2}{2n\hbar^2}\right)\left[1 + o\left(\frac{t^2}{n^2}\right)\right].\end{aligned} \tag{10.2}$$

As $n \to \infty$ for fixed t this tends to $\exp(-it\mathsf{E}_\phi(H)/\hbar)$ so that

$$P_\phi U_{t/n} P_\phi U_{t/n} P_\phi \dots U_{t/n} P_\phi \to \exp\left(-\frac{it}{\hbar}\mathsf{E}_\phi(H)\right) P_\phi. \tag{10.3}$$

Thus a very large number of measurements regularly repeated at intervals t/n causes the system to evolve almost as though it were in an eigenstate with energy $\mathsf{E}_\phi(H)$. Since multiples do not affect the physical state the system never evolves away from ϕ. When the expectation value vanishes then we even have

$$P_\phi U_{t/n} P_\phi U_{t/n} \dots P_\phi U_{t/n} P_\phi \to P_\phi. \tag{10.4}$$

This effect is known as the *quantum watched pot* effect or the *quantum Zeno paradox* (by comparison with the classical Zeno paradoxes which seemed to show that motion is impossible).

10.2. Bell's inequality

Even more alarming than the abruptness of the projection which accompanies a measurement is the fact that the change in the wave function is not just local but instantaneously pervades the whole of space, in a way which seems to defy our classical notions of space and time. The moment that the energy of the harmonic oscillator is measured to be $\frac{1}{2}\hbar\omega$ the wave function is transmuted to a multiple of ψ_0 throughout the entire universe. This offends common sense and seems to contradict the assertion of relativity theory that information cannot be transmitted faster than the speed of light. This is at the nub of most of the famous paradoxes of quantum mechanics, and it led Einstein and Schrödinger, amongst others, to reject quantum theory as a complete theory of physics.

To appreciate the differences between classical and quantum measurements it is useful to examine some simple features of classical probability theories first. As long as one is dealing with only one observable at a time there is essentially no difference between classical and quantum probability. However, this changes as soon as one starts to consider questions involving two or more observables at the same time, such as joint distributions or conditional probabilities: interference effects can arise which make the quantum probabilities quite different from their classical analogues.

Let us first consider the classical probability, $\mathsf{P}(R \setminus S)$, of the event that R occurs but S does not.

Lemma 10.2.1. For any events Q, R, and S, we have

$$\mathsf{P}(Q \setminus R) + \mathsf{P}(R \setminus S) \geq \mathsf{P}(Q \setminus S).$$

Proof. This can easily be seen from the Venn diagram in Figure 10.1, or by the following argument. Any point $q \in Q \setminus S$ is either in R or not in R. In the first case it is in $R \setminus S$ and otherwise it is in $Q \setminus R$. This shows that $(Q \setminus S) \subseteq (Q \setminus R) \cup (R \setminus S)$, from which we deduce the inequality for the probabilities. □

Remark 10.2.1. It can actually be shown (see Exercise 10.3) that

$$(Q \setminus R) \cup (R \setminus S) = (Q \setminus S) \cup ((Q \cap S) \setminus R) \cup (R \setminus (Q \cap S)), \quad (10.5)$$

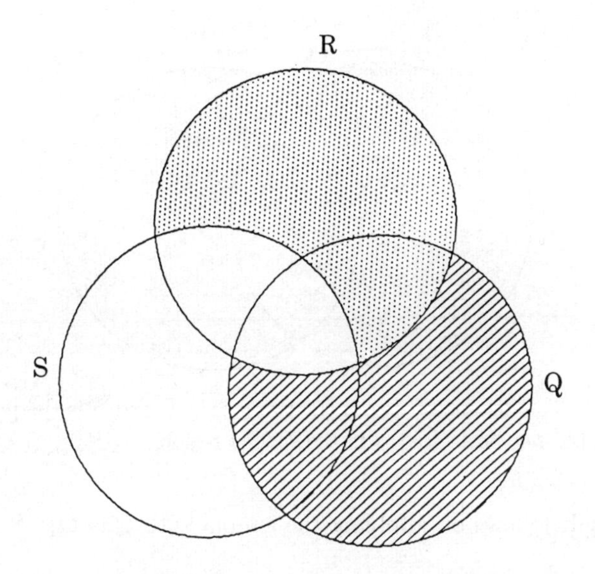

FIGURE 10.1. From the Venn diagram one may see that $\mathsf{P}(Q \setminus R) + \mathsf{P}(R \setminus S) \geq \mathsf{P}(Q \setminus S)$.

so that there is equality if and only if

$$Q \cap S \subseteq R \subseteq Q \cup S. \tag{10.6}$$

This readily extends to larger numbers of events.

Corollary 10.2.2. For $n \geq 2$ events $Q_1, Q_2, \ldots, Q_n$, we have the inequality

$$\sum_{j=1}^{n-1} \mathsf{P}(Q_j \setminus Q_{j+1}) \geq \mathsf{P}(Q_1 \setminus Q_n).$$

Proof. This can be proved by induction on n, starting with the case of $n = 2$, when the two sides are identical and there is equality. For $n + 1$ events the inductive hypothesis gives

$$\sum_{j=1}^{n} \mathsf{P}(Q_j \setminus Q_{j+1}) \geq \mathsf{P}(Q_1 \setminus Q_n) + \mathsf{P}(Q_n \setminus Q_{n+1}), \tag{10.7}$$

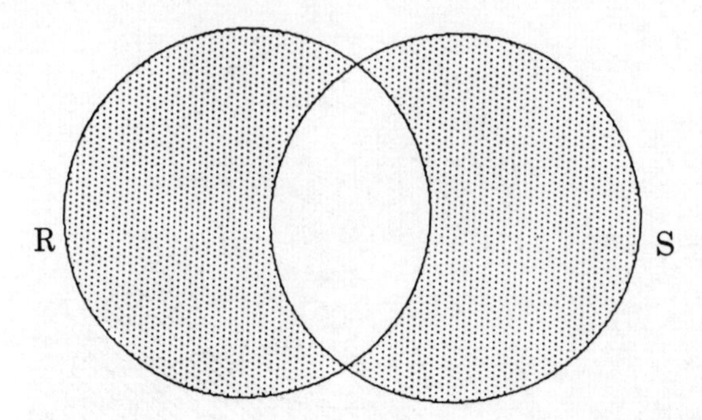

FIGURE 10.2. The distance between the two events R and S is the probability of being in one but not both, that is in the shaded region.

and the result follows on applying the lemma with $Q = Q_1$, $R = Q_n$, and $S = Q_{n+1}$. □

We can also deduce from the lemma a useful result about the probability that just one of two events R and S occurs, which is the sum of the probabilities $\mathsf{P}(R \setminus S)$ and $\mathsf{P}(S \setminus R)$ (see Figure 10.2).

Proposition 10.2.3. For any events Q, R, and S, $\mathsf{D}(R, S) = \mathsf{P}(R \cup S) - \mathsf{P}(R \cap S)$ satisfies:
(i) $\mathsf{D}(R, R) = 0$;
(ii) symmetry, $\mathsf{D}(R, S) = \mathsf{D}(S, R)$;
(iii) the triangle inequality, $\mathsf{D}(Q, S) \leq \mathsf{D}(Q, R) + \mathsf{D}(R, S)$.

Proof. The first property is obvious from the fact that $R \setminus R$ is empty, and the second from the fact that $\mathsf{D}(R, S)$ is defined symmetrically in R and S. The third follows by adding the result of the lemma for the sets Q, R, and S to that when their order is reversed. □

Remark 10.2.2. In topology a function D satisfying these three conditions and which only vanishes when $R = S$, is called a distance function or metric. In fact our $\mathsf{D}(R, S)$ vanishes if and only if R and S always occur together

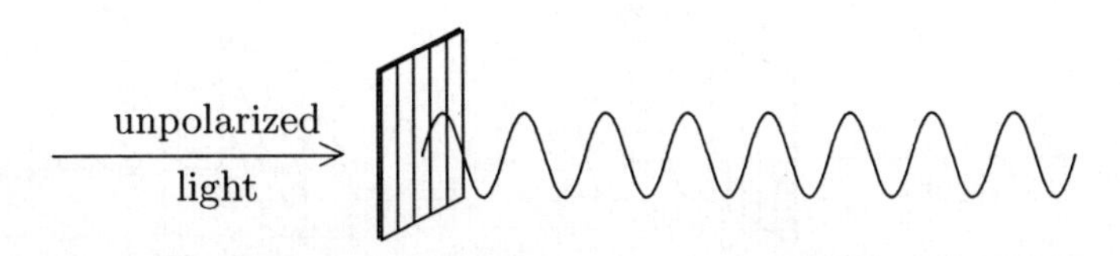

FIGURE 10.3. Unpolarized light incident on a vertical polarizing filter from the left emerges vertically polarized.

(never one without the other), so if we identify two events which can only occur together then D makes the set of events into a metric topological space. This result has been known to mathematicians since the 1920s, but its relevance to quantum theory was first appreciated by John Bell, who independently rediscovered an equivalent form of the triangle inequality for expectation values in 1964 (see Exercise 10.2). In 1969 Clauser, Horne, Shimony, and Holt extended the idea to four events (the $n = 4$ case of Corollary 10.2.2).

10.3. Polarization

Although at first sight it is rather disconcerting that a measurement can change the state of a quantum system, there is an everyday example of the same sort of phenomenon. Light can be polarized horizontally, vertically, or at any intermediate angle by passing it through an appropriate polarizing filter. When confused by the paradoxes of quantum measurements it is often helpful to take some sheets of polaroid and see how the light behaves in the corresponding situation. Polarization is also important because it has provided some of the most sensitive checks of quantum theory so far.

The effect of a vertically polarizing filter on a light wave can be visualized as in Figure 10.3 above. One can also measure the polarization of a light beam by passing it through a filter, and comparing the intensity of the light which emerges with that incident on the filter. Since the filter only transmits light polarized at the appropriate angle, the measurement has, just as in quantum theory, affected what is measured.

Mathematically the polarization states of light can be described by vectors in a two-dimensional inner product space $\mathcal{H}$ spanned by orthogonal vectors ξ and η, which represent vertical and horizontal polarizations re-

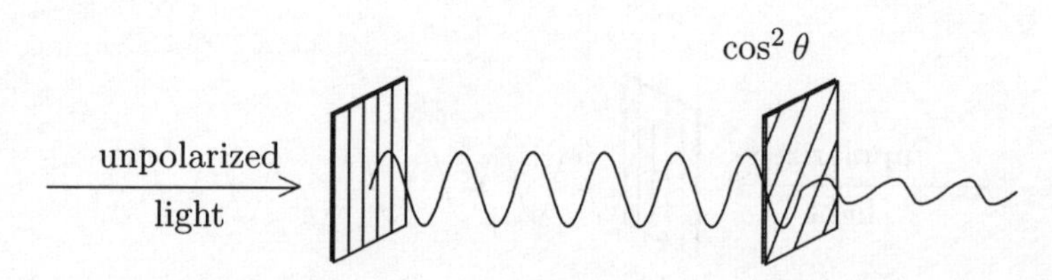

FIGURE 10.4. Malus' law tells us that a proportion $\cos^2\theta$ of the light transmitted by the first filter will also pass through a second filter polarized at an angle θ to the first.

spectively. If the polarization of the incoming beam was described by ψ then after passing through a vertically polarized filter it becomes $\langle\xi|\psi\rangle\xi$. In other words the filter simply projects out the component of ψ in the direction of ξ. This is exactly the same as the effect of quantum measurement. Light polarized at an angle θ to the vertical is described by the vector

$$\psi_\theta = \xi\cos\theta + \eta\sin\theta. \tag{10.8}$$

When the initial beam is polarized at an angle θ, so that $\psi = \psi_\theta$, this gives $\langle\xi|\psi_\theta\rangle\xi = \xi\cos\theta$. This means that when we think of light as consisting as photons rather than waves, the probability of a photon passing through the filter is $\cos^2\theta$, and the intensity of light emerging from the filter has the same angular dependence. This result, which can easily be verified, was discovered by Etienne Malus in 1809 and is known as Malus' law. (See Figure 10.4.) When $\theta = \pi/2$ no light can pass through both filters.

When thinking of light as a stream of photons one must guard against the temptation to regard these as classical particles. The dangers become apparent when one considers the effect of inserting a third filter between the first two. A naïve view of a filter is that it blocks photons which are not polarized in the appropriate direction, which suggests that the insertion of a third filter must result in even more photons being stopped, and so in a diminution of the light emerging from the system.

On the other hand we can easily calculate the intensity of the transmitted light directly. Let us suppose that the middle filter is polarized at an angle ϕ to the vertical (Figure 10.5). Since the middle filter makes an angle $\theta-\phi$ with the first a proportion $\cos^2(\theta-\phi)$ of the light passes through. At the third filter the probability of getting through is $\cos^2\phi$, giving an overall probability of transmission $\cos^2(\theta-\phi)\cos^2\phi$. This must be compared with

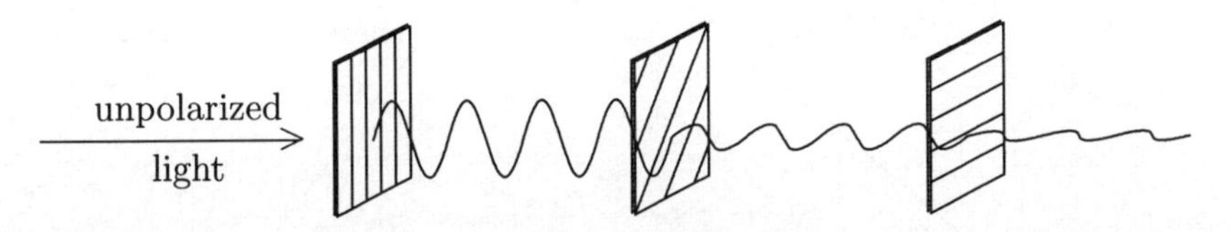

FIGURE 10.5. A third filter inserted at an angle ϕ to the vertical between the two filters allows a proportion $\cos^2\phi$ to pass and then a proportion $\cos^2(\theta-\phi)$ can pass the second filter as well. Overall a proportion $\cos^2\phi\cos^2(\theta-\phi)$ passes all three filters.

the probability $\cos^2\theta$ before the middle filter was inserted. One need only consider the case when the outer filters are at right angles ($\theta=\pi/2$), to see that the naïve picture is quite wrong. Without the middle filter the probability of transmission is $\cos^2\pi/2=0$, whilst with the middle filter it is

$$\cos^2\left(\frac{\pi}{2}-\phi\right)\cos^2\phi=\sin^2\phi\cos^2\phi=\frac{1}{4}\sin^2 2\phi, \tag{10.9}$$

which is positive for most angles. In other words, the middle filter, far from hindering itinerant photons, actually enhances their chances of passing through the horizontal filter. The predicted intensity of $\frac{1}{4}\sin^2 2\phi=\frac{1}{8}[1-\cos(4\phi)]$ has periodicity $\pi/2$, with maxima and minima separated by $\pi/4$, which matches what is seen in practice.

In fact, this argument identifies a subtle and dangerous probabilistic trap which can also be expressed more mathematically. Let r be a polaroid filter and let R denote the event that a given photon in an incident beam is transmitted by r. This can be calculated using Einstein's interpretation of the photoelectric effect, whereby the number of photons passing through r is proportional to the intensity of the transmitted beam. The probability, $\mathsf{P}(R)$, of a photon passing through r is therefore given by the ratio of the intensity of the transmitted beam to that of the incident beam. Reinterpreting our previous discussion in these terms we see that the probability that a photon which has been transmitted by r will also be transmitted by a second filter s at an angle ϕ is $\cos^2\phi$. The probability that a photon transmitted by r will not be transmitted by s is therefore $1-\cos^2\phi=\sin^2\phi$, and the probability $\mathsf{P}(R\setminus S)$ that a photon is transmitted by r but not by s is $\mathsf{P}(R)\sin^2\phi$. Considering a third filter q at an angle θ to s and $\theta-\phi$ to

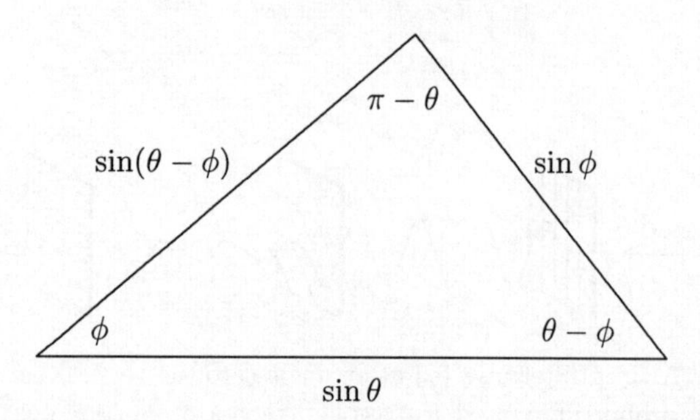

FIGURE 10.6. By the sine rule the sides of the triangle are proportional to $\sin \phi$, $\sin(\theta - \phi)$, and $\sin \theta$. Bell's inequality amounts to the condition that the side $\sin \theta$ should be less than the hypotenuse of a right-angled triangle with sides $\sin \phi$ and $\sin(\theta - \phi)$. This is clearly false if $\pi - \theta > \pi/2$, that is if $\pi/2 > \theta$.

r, and letting Q be the event that a photon passes q, we similarly calculate that $\mathsf{P}(Q \setminus R) = \mathsf{P}(Q) \sin^2(\theta - \phi)$ and $\mathsf{P}(Q \setminus S) = \mathsf{P}(Q) \sin^2(\theta)$.

The problem is that these need not satisfy the triangle-type inequality of Lemma 10.2.1. We shall see later that it is easy to create a beam for which $\mathsf{P}(Q) = \mathsf{P}(R)$. Then the triangle inequality is violated whenever $0 < \phi < \theta < \pi/2$. To see this, consider a triangle whose edges are parallel to the polarization directions of the three filters. (See Figure 10.6.) By the sine rule the sides have lengths proportional to $\sin \theta$, $\sin(\theta - \phi)$, and $\sin \phi$. By the cosine rule we have

$$\sin^2 \theta = \sin^2 \phi + \sin^2(\theta - \phi) - 2 \sin \phi \sin(\theta - \phi) \cos(\pi - \theta), \tag{10.10}$$

which can be rewritten as

$$\sin^2 \phi + \sin^2(\theta - \phi) - \sin^2 \theta = -2 \sin \phi \sin(\theta - \phi) \cos \theta. \tag{10.11}$$

Multiplying by $\mathsf{P}(Q) = \mathsf{P}(R)$ and taking $0 < \phi < \theta < \pi/2$, we have

$$\mathsf{P}(Q \setminus R) + \mathsf{P}(R \setminus S) - \mathsf{P}(Q \setminus S) < 0. \tag{10.12}$$

It is obvious that this apparent violation of our earlier result stems from a careless use of probability. It is clear, for example, that the event S that a photon passes filter s does not have the same meaning in $\mathsf{P}(Q \setminus S)$ as it has in $\mathsf{P}(R \setminus S)$, because the photon is changed during transmission by q or r.

When $\theta = 2\phi$, so that the angles between successive filters are ϕ, the violation of the triangle inequality is proportional to

$$2\sin^2\phi - \sin^2 2\phi = -2\sin^2\phi\cos 2\phi, \tag{10.13}$$

which is negative for all small angles ϕ, rather than positive as the triangle inequality would demand. One could do even better by adding another filter t at an angle ϕ to s. For light which is equally likely to be transmitted by any one of the filters, one obtains

$$\mathsf{P}(Q\backslash R)+\mathsf{P}(R\backslash S)+\mathsf{P}(S\backslash T)-\mathsf{P}(Q\backslash T) = \mathsf{P}(Q)\left(3\sin^2\phi - \sin^2 3\phi\right), \tag{10.14}$$

giving an even larger violation of Corollary 10.2.2.

10.4. The Einstein–Podolsky–Rosen paradox

Useful though the example of the polarization filters is, it does not disturb us unduly because we can easily comprehend how successive filtering of light is different from the conjunction of events in classical probability theory. However, there is a simple modification of the idea of filtering which does seem truly paradoxical. The idea goes back to Einstein who was never happy with the primacy of statistical laws in quantum theory. In 1935, in collaboration with Boris Podolsky and Nathan Rosen, he discovered one of the most puzzling paradoxes of the theory, which exhibits very clearly the grounds for his unease.

The paradox relies on the fact that conservation laws often provide information about one part of a system in terms of another. For example, if a stationary atom spontaneously decays into two fragments, their momenta must be equal and opposite. Measuring the momentum of fragment A tells us the momentum of the fragment B as well. This suggests that we might be able to beat the uncertainty principle by measuring the position of B and the momentum of A. Combining the information would give both the position and momentum of B.

Similar considerations apply in the case of an atom which emits two identically polarized photons. By conservation of momentum these travel in opposite directions, and we could arrange for each photon to encounter a polarizing filter at some distance from the atom. Since the photons have identical polarization their behaviour at the filters must be correlated. Thus a measurement of the polarization of either photon can be interpreted as a measurement of the polarization of the other (Figure 10.7). If the two measurements are made at sites so widely separated that even a light signal communicating the result of one measurement would arrive only

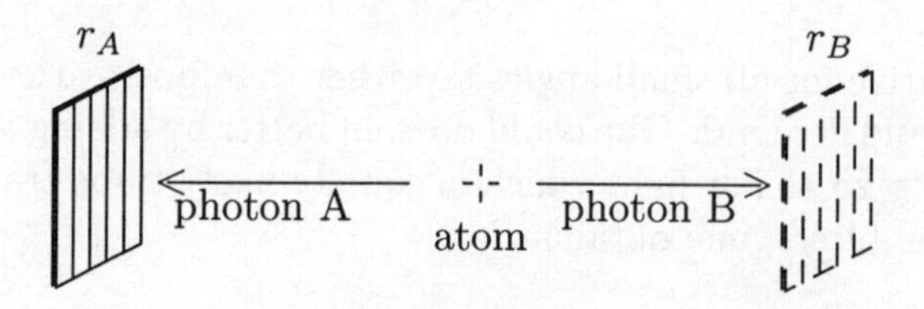

FIGURE 10.7. When an atom emits two identically polarized photons, A and B, the effect of passing A through r_A is the same as if B passed through r_B.

after the other measurement had occurred, it seems implausible that either measurement could be influenced by the other, for no known forces travel faster than light. More precisely, in these circumstances we expect that the transmission of photon B by a filter r_B really does depend only on the photon and the filter and not on anything which has happened to A, so that it constitutes a well-defined event R_B. We shall call this the *locality* assumption. It means that probabilities such as $\mathsf{P}(Q_A \setminus R_B)$, the probability that photon A is transmitted by filter q_A but photon B is not transmitted by r_B, may be unambiguously defined and should satisfy the normal probabilistic rules.

Now, in quantum theory there is a single state vector for the two photons

$$\psi(\mathbf{r}_A, \mathbf{r}_B) = \frac{1}{\sqrt{2}} \left[\xi(\mathbf{r}_A)\xi(\mathbf{r}_B) + \eta(\mathbf{r}_A)\eta(\mathbf{r}_B)\right], \tag{10.15}$$

which must immediately change whenever one of them passes through a filter. An elementary calculation shows that for any angle θ one can also write

$$\psi(\mathbf{r}_A, \mathbf{r}_B) = \frac{1}{\sqrt{2}} \left[\psi_\theta(\mathbf{r}_A)\psi_\theta(\mathbf{r}_B) + \psi_{\pi/2-\theta}(\mathbf{r}_A)\psi_{\pi/2-\theta}(\mathbf{r}_B)\right], \tag{10.16}$$

where $\psi_\theta = \xi\cos\theta + \eta\sin\theta$ and, similarly, $\psi_{\pi/2-\theta} = \eta\cos\theta - \xi\sin\theta$. This means that, whatever its polarization angle, θ, there is a probability of a half that a filter will transmit photon A. Then the wave function is projected onto $\psi_\theta(\mathbf{r}_1)\psi_\theta(\mathbf{r}_2)/\sqrt{2}$, and the second photon is also polarized at the same angle. The net effect would be the same had the second photon passed through a filter at that angle instead. Every change in the polarization state of one photon must therefore be matched by its distant companion no matter how far dispersed they may be.

This is the nub of the paradox: whilst it is easy to see that a photon might itself be changed by passing through a filter, it offends our understanding of local forces that another should immediately suffer the same fate, and we rather expect that distant photon to remain unchanged, at least for a time, so that measurements on it obey the classical probabilistic rules given earlier. Quantum mechanics, however, allows no respite: its effects are immediate and all-pervading. The effect on the wave function when photon A passing through a filter r_A is indistinguishable from that when photon B passes through a filter r_B at the same angle. If A encounters r_A and B encounters a filter s_B, the effect should therefore be the same as if B had encountered both r_B and s_B. This means that $\mathsf{P}(R_A \setminus S_B) = \mathsf{P}(R_B \setminus S_B)$. But this presents us with a problem (see Figure 10.8).

Proposition 10.4.1. Let Q_A and S_A be the events that a photon A is transmitted by the polaroid filters q_A or s_A, respectively, and similarly R_B and T_B the events that photon B is transmitted by filters r_B or t_B. Then for some filter angles the quantum mechanical probabilities satisfy

$$\mathsf{P}(Q_A \setminus R_B) + \mathsf{P}(R_B \setminus S_A) + \mathsf{P}(S_A \setminus T_B) < \mathsf{P}(Q_A \setminus T_B).$$

Proof. We know that the probability of transmission by a filter at any angle is $\frac{1}{2}$. We have also seen that

$$\begin{aligned}\mathsf{P}(Q_A \setminus R_B) + \mathsf{P}(R_B \setminus S_A) + \mathsf{P}(S_A \setminus T_B) - \mathsf{P}(Q_A \setminus T_B) \\ = \mathsf{P}(Q_B \setminus R_B) + \mathsf{P}(R_B \setminus S_B) + \mathsf{P}(S_B \setminus T_B) - \mathsf{P}(Q_B \setminus T_B),\end{aligned} \tag{10.17}$$

and we saw at the end of the last section that this can be negative. In particular, if the angles between the successive filters are all ϕ, one expects

$$\begin{aligned}\mathsf{P}(Q_A \setminus R_B) + \mathsf{P}(R_B \setminus S_A) + \mathsf{P}(S_A \setminus T_B) - \mathsf{P}(Q_A \setminus T_B) \\ = \mathsf{P}(Q)\left(3\sin^2\phi - \sin^2 3\phi\right),\end{aligned} \tag{10.18}$$

as in equation (10.14). □

This directly contradicts the result in Corollary 10.2.2, and this time we cannot take refuge in the argument that the events are not well defined, because the locality assumption says that they should be. Bell was

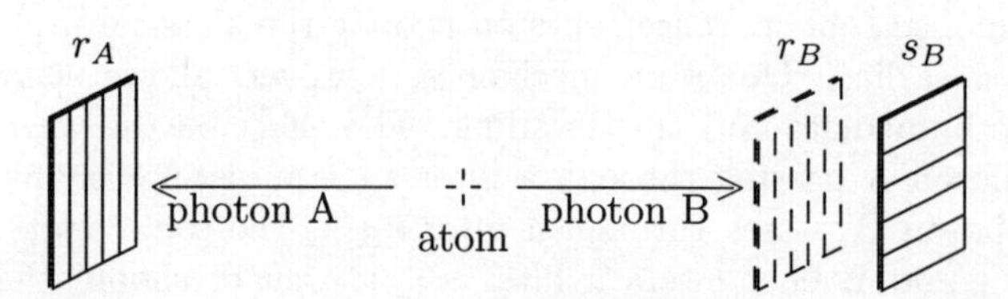

FIGURE 10.8. By carrying out measurements on both photons it is possible to determine quantities such as $\mathsf{P}(Q_A \setminus S_B)$, and so to check Bell's inequalities experimentally.

the first to realize that quantum mechanics gives predictions which are inconsistent with the inequalities for classical probability derived in Lemma 10.2.1. Many have believed that the statistical features of quantum theory simply arise out of our ignorance of what is going on at very small length scales, much as statistical mechanics overcomes our ignorance of the detailed motion of all the molecules in a gas. Quantum theory could then arise as an average over classical 'hidden variables', whose small size has so far precluded their detection. Many mathematicians and physicists investigated this idea and showed that the most obvious hidden variable theories could not reproduce some of the predictions of quantum mechanics, but Bell's observation shows that there are always differences between the predictions of any local hidden variable theory and quantum mechanics. It has formed the basis for many subsequent experiments to test the predictions of quantum measurement theory.

Clauser, Horne, Shimony, and Holt, following an idea of Bohm, suggested that this could provide a practical test of quantum theory against classical ideas. An atom of calcium is irradiated by two lasers which excite it to a state with higher energy. It subsequently decays back to its original state emitting two photons of wavelengths 551.3 and 422.7 nanometres, respectively. Since the angular momentum of the initial and final states is the same one can show that the two photons emitted in opposite directions are identically polarized, so that we are in the situation described above. By putting polarizing filters in the paths of the two photons one can find the correlations between them and check whether the inequality of Corollary 10.2.2 holds. More precisely one takes each filter at an angle ϕ to the preceding one, so that, using (10.18) and recalling that $\mathsf{P}(Q) = \frac{1}{2}$, quantum

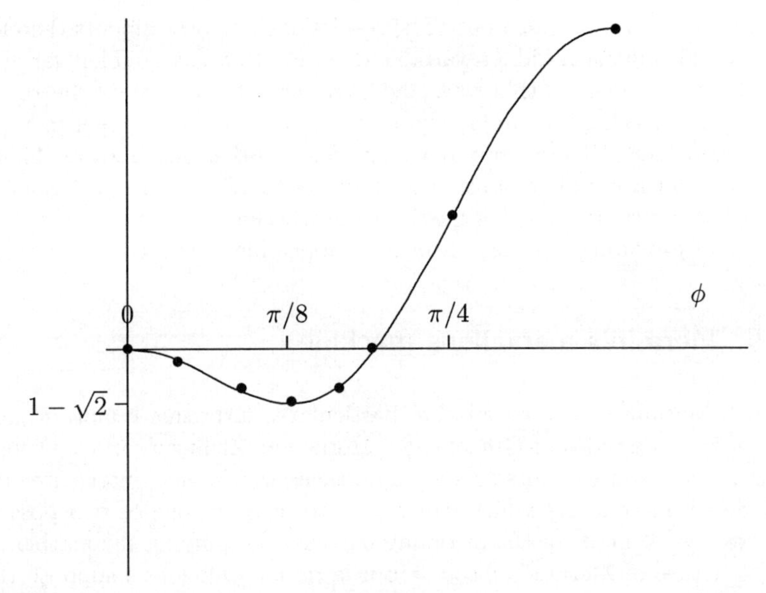

FIGURE 10.9. Schematic diagram of Aspect's measurements of the violation of the triangle inequality. The experimental points lie close to the theoretical curve $3\sin^2\phi - \sin^2 3\phi$, and well below 0.

mechanics predicts

$$\mathsf{P}(Q_A\backslash R_B)+\mathsf{P}(R_B\backslash S_A)+\mathsf{P}(S_A\backslash T_B)-\mathsf{P}(Q_A\backslash T_B) = \frac{1}{2}\left(3\sin^2\phi - \sin^2 3\phi\right), \tag{10.19}$$

whilst classical theory predicts the left-hand side to be non-negative. (In practice the experiments tend to use $\mathsf{D}(R_A, S_B) = \mathsf{P}(R_A\backslash S_B)+\mathsf{P}(S_B\backslash R_A)$, but, since the probabilities depend only on the angles, this just doubles everything and gives $3\sin^2\phi - \sin^2 3\phi$.)

Experiments of this kind were carried out by Freedman and Clauser and others in the early 1970s and in a more sensitive form by Aspect and his collaborators a decade later. Their results verified that the triangle inequality for probabilities is indeed violated in the way predicted by quantum theory (Figure 10.9).

In Aspect's experiments it was possible to change the polarization direction of the first filter very rapidly so that it could be chosen after the photons had left the atom. The filters were far enough apart that the second photon would pass through its polarizing filter before any signal could arrive (even at the speed of light) to reveal which polarization direction had been chosen for the first filter. In this way direct communication between the photons could be ruled out. The results of these experiments therefore seem to rule out local hidden variable theories. Non-local hidden variables cannot be excluded in this way, and some are able to mimic the results of quantum mechanics completely, so that they would be experimentally indistinguishable. These tend, however, to have other peculiarities (hidden variables which permit remote measurements to affect each other must be somewhat unusual), and it is questionable whether they achieve the original aim of providing a simple, intuitively appealing alternative to quantum theory.

10.5. Mermin's marvellous machine

In 1990 Mermin presented another particularly surprising quantum paradox, based on an idea of Greenberger, Horne and Zeilinger. In his thought experiment a source emits triples of particles which then enter three distant detectors, each of which can be set to measure one of two possible observables, X or Y, of the incoming particle. Simplifying the mathematical structure of Mermin's discussion in a minor way, let us suppose that each of X and Y can take only the values ± 1, so that $X^2 = 1 = Y^2$. Suppose now that the source is set up so that whenever two of the three detectors observe Y and the third X the product of the measured values is -1. Writing X_j and Y_j for the observation by the j-th detector of X or Y, respectively, we have $X_1Y_2Y_3 = -1$ and similarly $Y_1X_2Y_3 = -1$, $Y_1Y_2X_3 = -1$. From this we may deduce that

$$(X_1Y_2Y_3)(Y_1X_2Y_3)(Y_1Y_2X_3) = (-1)^3 = -1. \tag{10.20}$$

In the classical case, where all observables commute, this reduces to

$$(X_1Y_1^2)(X_2Y_2^2)(X_3Y_3^2) = -1, \tag{10.21}$$

and then, since $Y^2 = 1$, to

$$X_1X_2X_3 = -1. \tag{10.22}$$

If the detectors are distant enough that we do not expect them to influence each other, then the result of an observation of X by detector j should be independent of which observables the other two detectors are measuring, and $X_1X_2X_3$ should also be the product of the results when all three detectors are switched to observe X. The answer in this case must therefore be -1.

Let us now consider a quantum mechanical example of the same sort of measurement, in which the three particles each have spin $\frac{1}{2}$, and the observables are chosen to be $X = \sigma_1$ and $Y = \sigma_2$. The detector emits the three particles in the state $\psi = (\psi_{+++} + \psi_{---})/\sqrt{2}$, where $\psi_{\pm\pm\pm}$ denotes the state in which all three beams have σ_3 eigenvalue ± 1. For a single spin $\frac{1}{2}$ particle Definition 8.3.1 gives $\sigma_1 \psi_\pm = \psi_\mp$, and $\sigma_2 \psi_\pm = \pm i \psi_\mp$, so that

$$X_1 Y_2 Y_3 \psi = i^2 \psi = -\psi. \tag{10.23}$$

More generally, whenever two detectors measure Y they obtain a product -1, so that our hypothesis holds. Observables for different particles still commute, so that we again have

$$\begin{aligned}(X_1 Y_2 Y_3)(Y_1 X_2 Y_3)(Y_1 Y_2 X_3) &= (X_1 Y_1^2)(Y_2 X_2 Y_2)(Y_3^2 X_3) \\ &= X_1 (Y_2 X_2 Y_2) X_3.\end{aligned} \tag{10.24}$$

However, this time the anticommutation relations for Pauli spin matrices mean that $Y_2 X_2 = -X_2 Y_2$, so that the product reduces to $-X_1 X_2 X_3$, and we have

$$-X_1 X_2 X_3 = (X_1 Y_2 Y_3)(Y_1 X_2 Y_3)(Y_1 Y_2 X_3) = -1. \tag{10.25}$$

The result of measuring X with each detector must therefore give the product $+1$. In other words the quantum mechanical experiment would yield exactly the opposite answer to that which is expected classically.

10.6. Schrödinger's cat

Schrödinger shared some of Einstein's mistrust of quantum theory. (He had originally hoped to interpret $|\psi(x)|^2$ as the actual charge density of an electron in an atom and not just its probability density.) The Einstein–Podolsky–Rosen paper appeared as Schrödinger arrived in Oxford as an early refugee from the Nazi regime in Germany. Some physicists had dismissed the paradox as interesting but no cause for alarm, since one should never have expected the microscopic quantum world to be consistent with an intuition based on everyday experience. Schrödinger's rebuttal of this argument pointed out that microscopic events could have important consequences in the everyday world too (see Figure 10.10):

> One can also imagine quite burlesque cases. A cat is penned up in a steel chamber with the following fiendish contraption (which must be secured against direct interference by the cat): a Geiger counter containing a minute quantity of radioactive material, *so* small that in an hour *perhaps* one of the atoms may decay, but equally probably none will. If a decay occurs it is detected by the Geiger counter,

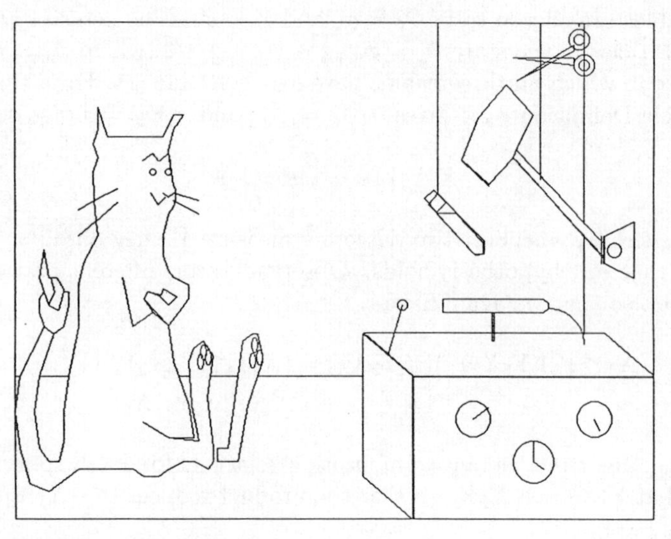

FIGURE 10.10. Schrödinger's cat.

which activates a small hammer through a relay and smashes a phial of prussic acid. If one leaves this entire system alone, at the end of one hour one can say that the cat is still alive provided that no atom has decayed. The first decay would have poisoned it. The ψ-function for the whole system would express this by containing the living and dead cat mixed or blended together in equal portions.

It is typical of such cases that an uncertainty which is originally confined to the atomic domain is transformed into a gross uncertainty which can be distinguished by direct observation. That presents an obstacle to naive acceptance of the 'blurred model' as a picture of reality. In itself there is nothing ambiguous or contradictory about it. It is the difference between a blurred photograph or one out of focus and a picture of clouds or fog-banks.

At first sight it might appear that one could equally well have replaced the radioactive atom and phial of poison by a robot which tosses a coin and then shoots the cat if the coin comes down tails. However, in that case at the end of the allotted time the cat would certainly be either alive or dead inside the box although we would not yet know which. The new ingredient introduced by quantum mechanics is that until we open the box it is apparently possible for the superposition of live and dead cats inside it to create interference effects. In fact, as Asher Peres pointed out in 1978, this paradox could have even more bizarre consequences. The two

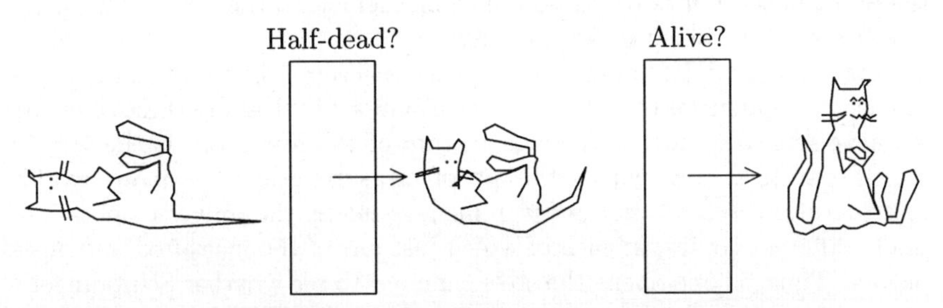

FIGURE 10.11. Peres' quantum mechanical device for resuscitating dead cats.

possible states of Schrödinger's cat (live and dead) can be described in the space $\mathbf{C}^2$ spanned by vectors ψ_{live} and ψ_{dead}. Mathematically this two-dimensional space is identical to that used for describing the polarization states of the photon, so we can try translating the polarizing filter ideas into the language of cats. Suppose that at the end of the hour we found the cat to be dead, that is in the state ψ_{dead}. We might then carry out an observation to see if it were in some half-dead state described by the vector $\cos\theta\psi_{\text{dead}} + \sin\theta\psi_{\text{live}}$. We could then look again to see if it was alive. The effect would be much the same as inserting a third polarizing filter between two crossed filters, and there would be a probability of $\frac{1}{4}\sin^2 2\theta$ that the dead cat had been resuscitated! (See Figure 10.11.)

In 1961 Wigner suggested another variant of the idea, which essentially replaces the release of poison by the flash of a light, and the unfortunate cat by Wigner's friend. One is then faced with the conundrum of whether the wave function describing the light and friend projects down as soon as the friend observes the light, or only when the door is opened.

10.7. Many worlds and one world

In the half century since these paradoxes were first proposed they have fascinated mathematicians, physicists, and philosophers alike. The attempts to resolve them are too numerous to recount, although some can now be ruled out in the light of Aspect's results. Some have tried to change quantum mechanics in some way, so that the collapse of the wave function during a measurement is governed by a modified Schrödinger equation, and so fits into the same dynamical framework as the usual unitary time evolution. Others accept the mathematical formulation of quantum theory but seek to interpret it in a way which accords more with our intuition derived from classical mechanics. We shall briefly mention just two of these, one chosen because it is often described in popular accounts of quantum theory, the other because it is based on a nice mathematical theorem. (Two more theories are mentioned in Section 15.1.)

Hugh Everett III's 'many worlds' picture accepts the mathematical formulation of quantum theory, but gives it a new physical interpretation. As we have seen the most perplexing feature of the formalism is the way in which a state is changed by the act of measurement. The many worlds interpretation is that, as a measurement is made, the universe splits into many different copies, in each of which just one of the measured outcomes occurs. Thus, as one opens the steel chamber to see whether Schrödinger's cat is alive or dead, the universe bifurcates into one universe containing a decayed atom, a broken phial, and a dead cat, and another in which the atom and the phial are in pristine condition and a tetchy cat emerges unscathed. With its science fiction flavour, this is perhaps the best-known attempt to resolve the paradoxes.

At the opposite extreme is an idea which stresses the unity of the universe. No system is ever truly isolated: it interacts with the outside world by gravitational, electromagnetic, or other forces, or else we could not observe it at all. The very act of measurement forces the system to interact with the measuring apparatus, and for an accurate measurement this interaction must be strong and swift.

The following theorem, whose proof is given in Appendix A2, describes one possible consequence of studying the system and environment together.

Theorem 10.7.1. Let V be a two-dimensional inner product space and Ω a vector in V. Then there exists an inner product space $\mathcal{H}$, a family of unitary operators U_t, a homomorphism $\phi : \mathcal{L}(V) \to \mathcal{L}(\mathcal{H})$ which respects adjoints, and for each vector $\psi \in V$ a vector $\Psi \in \mathcal{H}$ such that for all $A \in \mathcal{L}(V)$ we have

$$\langle \Psi | \phi(A) \Psi \rangle = \langle \psi | A \psi \rangle$$

$$\lim_{t \to \infty} \langle U_t \Psi | \phi(A) U_t \Psi \rangle = \langle \Omega | A \Omega \rangle,$$

where the inner products on the left are in $\mathcal{H}$ and those on the right are in V.

We interpret V as the space for a system with two independent states (such as the polarized photon, spinning electron, or Schrödinger's cat), and

$\mathcal{H}$ as that for the environment as well. (The restriction to two states is for simplicity only and is not essential.) The homomorphism ϕ tells us how the observables for the two-state system can be interpreted as observables for that system together with its environment. Whatever the initial state, for large times the expectation value of any observable for the two-state system tends to the value given by the vector Ω. Although the time evolution is given by perfectly ordinary unitary operators U_t, the effect on the system is just the same as the collapse to Ω during a measurement, except that it is only asymptotic and not immediate. However, the asymptotic state is approached exponentially like $\exp(-\eta t)$ where η is the strength of the coupling between the system and its environment. The time-scale for the collapse should therefore be about the same as for the measurement and, since on the subatomic scale things tend to happen fast, could easily be of order 10^{-15} seconds. It would not be easy to distinguish such a swift exponential decay from instantaneous collapse. The 'open system' interpretation suggests that this is what happens: the description of measurements in terms of projections is just a useful approximation to the effect on the system alone of ordinary time evolution during periods of rapid change.

One serious objection to this interpretation is the fact that the collapse is infinitely protracted: after any finite time it is still possible in principle to reverse the process. This might, however, be very difficult in practice if energy is radiated during the measurement, since within seconds this will be far out in space and there is no practical way of recovering it. Nonetheless, this might offer the possibility of testing such theories experimentally.

The reader should not be worried if neither of these interpretations seems wholly convincing. Whether or not there are many worlds, there are certainly many world views, and few explanations seem persuasive to more than a small band of enthusiasts.

Exercises

10.1 Show that, under suitable conditions, for any projection, P, and unitary evolution group, $U_t = \exp(-itH/\hbar)$,

$$PU_tP = (1 - itPHP/\hbar)P + o(t^2),$$

and deduce that

$$PU_{t/n}PU_{t/n}\dots PU_{t/n}P \to \exp\left(-\frac{it}{\hbar}PHP\right)P.$$

10.2 Let A_{RS} be a random variable whose value is λ if just one of the events R and S occurs and μ otherwise. Show that

$$\mathsf{E}(A_{RS}) = \lambda\mathsf{D}(R,S) + \mu[1 - \mathsf{D}(R,S)] = \mu + (\lambda - \mu)\mathsf{D}(R,S),$$

and hence that

$$\mathsf{E}(A_{RS}) - \mu \geq \mathsf{E}(A_{QS}) - \mathsf{E}(A_{QR}).$$

Deduce that

$$-\mu + \mathsf{E}(A_{RS}) \geq |\mathsf{E}(A_{QS}) - \mathsf{E}(A_{QR})|.$$

[This last inequality is Bell's inequality. When $\mu = 0$ it reduces to the triangle inequality.]

10.3 Show that for any classical events Q, R, and S,

$$\mathsf{P}(Q \setminus R) + \mathsf{P}(R \setminus S) - \mathsf{P}(Q \setminus S) = \mathsf{P}((Q \cap S) \setminus R) + \mathsf{P}(R \setminus (Q \cup S)).$$

10.4 Show that, for $n + 1 \geq 3$ filters $q_1, \ldots, q_{n+1}$, each at an angle ϕ to its predecessor, and Q_j the event that a photon is transmitted by q_j, one has

$$\sum_{j=1}^{n} \mathsf{P}(Q_j \setminus Q_{j+1}) - \mathsf{P}(Q_1 \setminus Q_{n+1}) = n \sin^2 \phi - \sin^2 n\phi.$$

10.5 Show that for $n > 1$, $f(\phi) = n \sin^2 \phi - \sin^2 n\phi$ has extrema where $\phi = (k + \frac{1}{2})\pi/(n+1)$ for $k \in \mathbf{Z}$ or ϕ is a multiple of $\pi/(n-1)$. When $\phi = \pi/2(n+1)$ show that f has a minimum value of

$$(n+1) \sin^2 \frac{\pi}{2(n+1)} - 1.$$

Deduce that this minimum decreases with increasing n and find its value for $n = 2, 3$.

11 Alternative formulations of quantum theory

Heisenberg's new work, which will very soon appear, looks very mystical, but it is certainly correct and deep.

MAX BORN, letter to Albert Einstein, 15 July 1925

11.1. Pictures of quantum mechanics

The mathematical description of quantum theory rests on the idea of states and observables. However, the dynamical description of these two described in Section 6.4 is quite different. When the Hamiltonian is time independent, the states evolve according to the equation

$$\psi_t = U_t \psi_0, \tag{11.1}$$

where U_t is the unitary operator $\exp(-iHt/\hbar)$. Observables, such as the position and momentum, which have no explicit time dependence, are constant. (Observables may have explicit time dependence, for example $X + tP$. In this section a typical Schrödinger observable at time t will be written as A_t^S.)

Definition 11.1.1. This description of the dynamics of states and observables is called the *Schrödinger picture.*

Heisenberg's description of quantum mechanics is quite different: the states are constant and the observables evolve.

Definition 11.1.2. In the *Heisenberg picture* of quantum mechanics the states remain constant $\psi_t = \psi_0$, and an observable, described in the Schrödinger picture by A_t^S, evolves according to the equation $A_t = U_t^* A_t^S U_t$.

In fact, both the Heisenberg and Schrödinger pictures of quantum mechanics are special cases of the interaction picture, which compares the actual evolution with another evolution chosen to serve as a reference.

Definition 11.1.3. Let $V_t = \exp(-iH_0 t/\hbar)$ be the reference evolution. In the *interaction* or *Dirac picture* the states and observables evolve according to the equations

$$\psi_t = V_t^* U_t \psi_0$$
$$A_t = V_t^* A_t^S V_t.$$

Remark 11.1.1. When $V_t \equiv 1$ we obtain the Schrödinger picture and when $V_t \equiv U_t$ the Heisenberg picture.

The quantities that are of physical interest, such as expectation values, depend on both the states and the observables, and it is the evolution of these which really matters.

Theorem 11.1.1. In the interaction picture the expectation of an observable A_t in a state ψ_t is independent of the choice of V_t.

Proof. In the interaction picture

$$\begin{aligned}\langle\psi_t|A_t\psi_t\rangle &= \langle V_t^* U_t\psi_0|V_t^* A_t^S V_t V_t^* U_t\psi_0\rangle \\ &= \langle U_t\psi_0|A_t^S U_t\psi_0\rangle,\end{aligned} \tag{11.2}$$

since V_t is unitary and its inverse is V_t^*. Similarly $\langle\psi_t|\psi_t\rangle = \langle U_t\psi_0|U_t\psi_0\rangle$, so that

$$\frac{\langle\psi_t|A_t\psi_t\rangle}{\|\psi_t\|^2} = \frac{\langle U_t\psi_0|A_t^S U_t\psi_0\rangle}{\|U_t\psi_0\|^2}, \tag{11.3}$$

independent of V_t. □

Corollary 11.1.2. The Heisenberg, Schrödinger, and interaction pictures all give the same expectation values.

11.2. Differential equations for the time evolution

In practice we have generally found the time evolution of wave functions using Schrödinger's equation, so it is useful to know its analogues in the other pictures.

Theorem 11.2.1. In the interaction picture the time evolution of states and observables is governed by the differential equations:

$$i\hbar \frac{d\psi_t}{dt} = H_t' \psi_t;$$

$$\frac{dA_t}{dt} = \frac{\partial A_t}{\partial t} - \frac{i}{\hbar} [A_t, H_0];$$

where $H_t' = V_t^* (H - H_0) V_t$, and $\partial A_t / \partial t = V_t^* \left(dA_t^S / dt \right) V_t$.

Proof. By definition

$$i\hbar \frac{dU_t}{dt} = HU_t = U_t H, \tag{11.4}$$

and

$$i\hbar \frac{dV_t^*}{dt} = i\hbar \frac{dV_{-t}}{dt} = V_{-t} (-H_0) = -V_t^* H_0. \tag{11.5}$$

So, differentiating $\psi_t = V_t^* U_t \psi_0$, we obtain

$$\begin{aligned} i\hbar \frac{d\psi_t}{dt} &= V_t^* (-H_0) U_t \psi_0 + V_t^* H U_t \psi_0 \\ &= V_t^* (H - H_0) V_t V_t^* U_t \psi_0 \\ &= H_t' \psi_t. \end{aligned} \tag{11.6}$$

Similarly

$$\begin{aligned} i\hbar \frac{dA_t}{dt} &= (-H_0) V_t^* A_t^S V_t + i\hbar V_t^* \frac{dA_t^S}{dt} V_t + V_t^* A_t^S V_t H_0 \\ &= i\hbar V_t^* \frac{dA_t^S}{dt} V_t - H_0 A_t + A_t H_0 \\ &= i\hbar \frac{\partial A_t}{\partial t} + [A_t, H_0], \end{aligned} \tag{11.7}$$

and the second formula follows immediately. □

Setting $V_t \equiv U_t$ we obtain the following result:

Corollary 11.2.2. In the Heisenberg picture where states are constant, observables evolve according to the equation

$$\frac{dA}{dt} = \frac{\partial A}{\partial t} - \frac{i}{\hbar}[A, H].$$

It follows immediately that a Hamiltonian which does not explicitly depend on time (so that $\partial H/\partial t = 0$) is actually constant. One can also deal with the case of time-dependent Hamiltonians by using the differential equation of the corollary to define the evolution of observables in the Heisenberg picture.

11.3. Time-dependent perturbation theory

The interaction picture enables us to compare two time evolutions. This is particularly useful for complicated Hamiltonians, where Schrödinger's equation cannot be integrated directly, but one can compare it with some simpler reference system which is supposed to be understood. The differential equation of Theorem 11.2.1 can be integrated to give

$$\psi_t = \psi_0 + \frac{1}{i\hbar}\int_0^t H'_s \psi_s \, ds. \tag{11.8}$$

If there were no perturbation, H', then V_t and U_t would be identical, and we should be in the Heisenberg picture, where ψ_t is constant. We therefore take this as the starting point for an iterative scheme with

$$\psi_t^{(0)} = \psi_0, \tag{11.9}$$

and

$$\psi_t^{(N)} = \psi_0 + \frac{1}{i\hbar}\int_0^t H'_{t_N} \psi_{t_N}^{(N-1)} \, dt_N, \tag{11.10}$$

as the N-th order approximation, for $N \geq 1$. This expression can also be written as

$$\psi_t^{(N)} = \psi_0 + \sum_{n=1}^{N} (i\hbar)^{-n} \int_0^t \cdots \int_0^{t_2} H'_{t_n} H'_{t_{n-1}} \cdots H'_{t_1} \psi_0 \, dt_1 dt_2 \ldots dt_n, \tag{11.11}$$

as is readily proved by induction. Unfortunately, in some of the most important applications the integrals are not well defined, and even when the integrals are well behaved the sequence $\psi_t^{(N)}$ often diverges. Nonetheless, this is an important theoretical tool.

When it does converge we may rewrite it as a formula for the time evolution,

$$V_t^* U_t = 1 + \sum_{n=1}^{\infty} (i\hbar)^{-n} \int_0^t \dots \int_0^{t_2} H'_{t_n} H'_{t_{n-1}} \dots H'_{t_1} \, d^n t. \tag{11.12}$$

Multiplying by V_t we arrive at the following result:

Proposition 11.3.1. (The Feynman–Dyson expansion) The time evolution operator U_t enjoys a formal expansion as

$$U_t = V_t + \sum_{n=1}^{\infty} (i\hbar)^{-n} \int_0^t \dots \int_0^{t_2} V_{t-t_n} H' V_{t_n - t_{n-1}} H' \dots H' V_{t_1} \, d^n t,$$

where $H' = H - H_0$.

Remark 11.3.1. The integrand in the n-th order term of the series

$$V_{t-t_n} H' V_{t_n - t_{n-1}} \dots V_{t_2 - t_1} H' V_{t_1} \tag{11.13}$$

can be interpreted as describing a system in which the perturbation $H' = (H - H_0)$ is turned on only at times $t_1, t_2, \dots, t_n$. In between the system evolves as though there were no perturbation at all, and the Hamiltonian were just H_0.

11.4. Fermi's golden rule

Let us suppose that we start with a system that is in an eigenstate ϕ_1 of energy ϵ_1 for the unperturbed Hamiltonian. We wish to estimate the transition probability that after a time t a measurement will show it to have evolved to an eigenstate ϕ_2 of different energy $\epsilon_2 = \epsilon_1 - \epsilon$. After time t the Schrödinger wave function is $U_t\phi_1$, so that the transition probability is

$$\begin{aligned} |\langle \phi_2 | U_t \phi_1 \rangle|^2 &= |\langle V_t^* \phi_2 | V_t^* U_t \phi_1 \rangle|^2 \\ &= |\langle e^{-i\epsilon_2 t/\hbar} \phi_2 | V_t^* U_t \phi_1 \rangle|^2 \\ &= |\langle \phi_2 | V_t^* U_t \phi_1 \rangle|^2. \end{aligned} \tag{11.14}$$

As the eigenvector corresponding to a distinct eigenvalue ϕ_2 is orthogonal to ϕ_1, so we have

$$\begin{aligned}\langle \phi_2 | V_t^* U_t \phi_1 \rangle &= \langle \phi_2 | \phi_1 \rangle + \frac{1}{i\hbar} \int_0^t \langle \phi_2 | H_s' V_s^* U_s \phi_1 \rangle \, ds \\ &= \frac{1}{i\hbar} \int_0^t \langle \phi_2 | H_s' V_s^* U_s \phi_1 \rangle \, ds. \qquad (11.15)\end{aligned}$$

From this we see that already the first-order approximation to $V_t^* U_t \phi_1$ will provide a second-order approximation to the transition probability. This gives as an approximate value

$$\begin{aligned}\langle \phi_2 | V_t^* U_t \phi_1 \rangle &\sim \frac{1}{i\hbar} \int_0^t \langle \phi_2 | H_s' \phi_1 \rangle \, ds \\ &= \frac{1}{i\hbar} \int_0^t \langle \phi_2 | V_s^* H_0' V_s \phi_1 \rangle \, ds \\ &= \frac{1}{i\hbar} \int_0^t \langle V_s \phi_2, H_0' V_s \phi_1 | \, \rangle ds \\ &= \frac{1}{i\hbar} \int_0^t \exp[-i\,(\epsilon_1 - \epsilon_2)\, s/\hbar]\, \langle \phi_2 | H_0' \phi_1 \rangle \, ds \\ &= \frac{e^{-i\epsilon t/\hbar} - 1}{\epsilon} \langle \phi_2 | H_0' \phi_1 \rangle \\ &= e^{-i\epsilon t/2\hbar} \frac{\sin(\epsilon t/2\hbar)}{i\epsilon/2} \langle \phi_2 | H_0' \phi_1 \rangle. \qquad (11.16)\end{aligned}$$

From this we now deduce the transition probability by squaring the modulus.

Proposition 11.4.1. (Fermi's golden rule) The second-order approximation to the transition probability after time t is

$$|\langle \phi_2 | U_t \phi_1 \rangle|^2 \sim \left(\frac{\sin\,(\epsilon t/2\hbar)}{\epsilon/2} \right)^2 |\langle \phi_2 | H' \phi_1 \rangle|^2,$$

where ϵ is the energy difference between the two levels.

Fermi's golden rule confirms in a very explicit way the remarks about time–energy uncertainty relations made in Section 7.3. It is instructive to

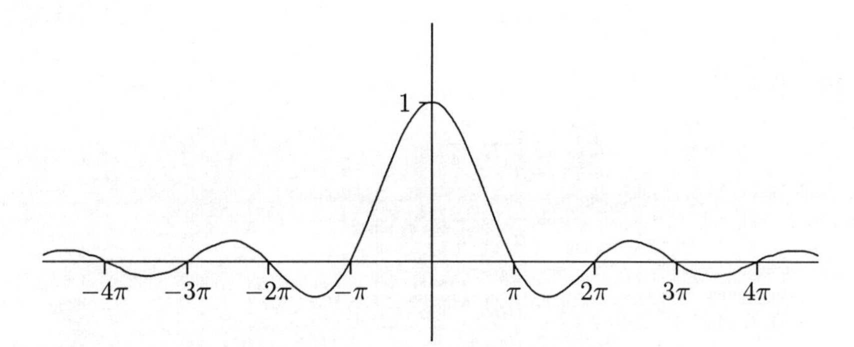

FIGURE 11.1. The graph of $\sin x/x$, whose square gives the approximate probability amplitude for a transition from one energy level to another.

plot this approximate transition probability against the energy difference, see Figure 11.1. At a given time t some transitions are much more probable than others. Viewed the other way round, any given transition is more probable at some times than others. One should not be unduly worried by the fact that the central peak has height $|\langle\phi_2|H_0'\phi_1\rangle t/\hbar|^2$ in defiance of the probabilistic interpretation for large t. This is, of course, just an indication that the approximation fails under those conditions. (It does, however, serve as a reminder that our arguments have been rather cavalier: a proper mathematical treatment of the Fermi rule is far from trivial.)

11.5. The harmonic oscillator in the Heisenberg picture

In the Heisenberg picture the differential equation of motion differs from that of classical Hamiltonian mechanics only in the replacement of the Poisson bracket by $i/\hbar$ times the commutator. This means that it is often possible to solve the quantum problem by more or less the same method as in the classical theory.

For the one-dimensional harmonic oscillator with Hamiltonian $H = P^2/2m + \frac{1}{2}m\omega^2X^2$ one has

$$\begin{aligned}\frac{dP}{dt} &= \frac{i}{\hbar}[H,P]\\ &= \frac{i}{\hbar}\left[\frac{1}{2}m\omega^2X^2, P\right]\end{aligned}$$

$$\begin{aligned} &= \frac{im\omega^2}{2\hbar}(X[X,P] + [X,P]X) \\ &= \frac{im\omega^2}{2\hbar}(2i\hbar X) \\ &= -m\omega^2 X, \end{aligned} \tag{11.17}$$

and similarly

$$\begin{aligned} \frac{dX}{dt} &= \frac{i}{\hbar}[H, X] \\ &= \frac{i}{\hbar}\left[\frac{1}{2m}P^2, X\right] \\ &= \frac{i}{2m\hbar}(P[P,X] + [P,X]P) \\ &= \frac{1}{m}P. \end{aligned} \tag{11.18}$$

These look exactly like the classical equations of motion and can be solved by the same observation (Exercise 7.18) that

$$\frac{d}{dt}(P \pm im\omega X) = -m\omega^2 X \pm i\omega P = \pm i\omega(P \pm im\omega X). \tag{11.19}$$

This can immediately be integrated to give

$$(P \pm im\omega X)_t = e^{\pm i\omega t}(P \pm im\omega X)_0. \tag{11.20}$$

The reader may wonder why we have concealed this extraordinary simplicity for so long. Unfortunately it is only for quadratic potentials that one finds such close agreement between classical and quantum mechanics; for anything more complicated the Heisenberg picture becomes considerably harder to handle. In any case more work is needed to obtain the bound state energy levels, and we postpone that to Example 11.10.

11.6*. Statistical mechanical states

In view of the strong role already played by probability in quantum theory it may come as a surprise to learn that we have not yet accounted for all the statistical aspects of physics. In practice physics is often concerned with large assemblages of particles of which we have only a very imperfect knowledge. We can scarcely hope to specify the states of all 10^{22}–10^{23} molecules in a litre of gas, let alone their constituent protons, neutrons, electrons, or even quarks. Even classical mechanics handles such systems by means of statistical averages and we should like to do the same in quantum theory. There is a clear difference between the probabilities that enter here as a means of covering our ignorance, and the probabilities encountered already in quantum theory. Consider, for example, a photon passing through a screen with two slits, and let ψ_1 and ψ_2 denote the wave functions corresponding to passage through the first slit or the second slit, respectively. When both slits are open the appropriate wave function is the superposition of these, $\psi = \psi_1 + \psi_2$, and this leads to interference effects. Suppose, on the other hand, that a laboratory technician closes one slit decided by the toss of a coin but forgets to record which one. Probabilities again enter, but we would no longer see any interference patterns, since the photon encountered only one open slit, even though we do not know which it was.

In order to handle this situation mathematically, we recall that vectors that are multiples of each other correspond to the same physical state. In other words, it is really the one-dimensional subspace spanned by ψ which is physically significant, and not ψ itself. This subspace is determined by the orthogonal projection Q_ψ that projects onto it:

$$Q_\psi \xi = \frac{\langle \psi | \xi \rangle}{\|\psi\|^2} \psi. \tag{11.21}$$

In fact, there are simple formulae for the physically significant quantities directly in terms of Q_ψ. In particular, if A is an observable then

$$\langle \xi | Q_\psi A \xi \rangle = \langle \xi | \frac{\langle \psi | A \xi \rangle}{\|\psi\|^2} \psi \rangle = \frac{\langle \xi | \psi \rangle \langle \psi | A \xi \rangle}{\|\psi\|^2}. \tag{11.22}$$

Summing over the vectors ξ in an orthonormal basis, we obtain the trace

$$\operatorname{tr}(Q_\psi A) = \frac{\langle \psi | A \psi \rangle}{\|\psi\|^2} = \mathsf{E}_\psi(A), \tag{11.23}$$

which shows explicitly how to recover the expectation of A from the projection.

Suppose now that we know only that there is a probability p_k that the state is described by the vector ψ_k in the orthonormal set $\{\psi_j : j = 1, 2, \ldots, n\}$. The expectation value should then be given by the weighted average

$$\sum_j p_j \mathsf{E}_{\psi_j}(A) = \sum_j p_j \operatorname{tr}(Q_{\psi_j} A) = \operatorname{tr}\left(\sum_j p_j Q_{\psi_j} A \right). \tag{11.24}$$

This suggests that such a statistical system is best described by the operator

$$\rho = \sum_j p_j Q_{\psi_j}. \tag{11.25}$$

This is a self-adjoint operator because the projections are self-adjoint and the probabilities real. It is positive because each $p_k \geq 0$, and also

$$\mathrm{tr}(\rho) = \sum_j p_j \mathrm{tr}(Q_{\psi_j}) = \sum_j p_j = 1. \tag{11.26}$$

Definition 11.6.1. A positive operator ρ that satisfies $\mathrm{tr}\rho = 1$ is called a *density operator*.

Definition 11.6.2. In a quantum statistical system whose state is described by a density operator ρ the expectation of the observable A is given by

$$\mathsf{E}_\rho(A) = \mathrm{tr}(\rho A).$$

In infinite dimensions it is already a strong constraint on an operator that it have a finite trace at all: the series that should give the trace is usually divergent. In fact, an operator that has a trace can be written in the form of a (possibly infinite) sum $\sum \lambda_j Q_j$, where the Q_j form a family of mutually orthogonal projections (that is, $Q_j Q_k = 0$ if $j \neq k$), and

$$\sum |\lambda_j| < \infty. \tag{11.27}$$

Thus the form $\sum p_j Q_{\psi_j}$ is essentially the most general possible. There are, as we shall see in Section 11.11, some infinite quantum systems whose states are better described by a slight generalization of a density operator, but for most finite systems density operators suffice.

If we know that the wave function is precisely ψ then we can take $\rho = Q_\psi$, and effectively return to ordinary quantum theory.

Definition 11.6.3. States described by one-dimensional projections

$$\rho = Q_\psi$$

are called *pure states.* States described by more general density operators are called *mixed states.*

Thus all the states that we have used in the earlier chapters have really been pure states. There is an easy way to recognize density operators and to see which of them describe pure states. For this we recall that if the difference $A - B$ between two operators A and B is positive then we say that $B \leq A$.

Proposition 11.6.1. The density operator satisfies the conditions

$$0 \leq \rho^2 \leq \rho \leq 1.$$

The state is pure if and only if $\rho^2 = \rho$.

Proof. It is clear that, as the square of a self-adjoint operator, ρ^2 is positive. For any orthonormal basis $\{\psi_j\}$ we have

$$\langle \psi_j | \psi_j \rangle = 1 = \mathrm{tr}(\rho) = \sum_j \langle \psi_j | \rho \psi_j \rangle \geq \langle \psi_j | \rho \psi_j \rangle. \tag{11.28}$$

So $1 - \rho \geq 0$, and since ρ commutes with $1 - \rho$ and is also positive, we have $\rho - \rho^2 = \rho(1 - \rho) \geq 0$. This has now established all the inequalities.

It is also clear that if $\rho = Q_\psi$ is pure then $\rho^2 = \rho$. Conversely, if $\rho^2 = \rho$ then ρ is a projection. By considering matrices with respect to suitable bases, the trace of a projection is easily seen to be the dimension of its range. Thus ρ has rank $\mathrm{tr}(\rho) = 1$, and is therefore a one-dimensional projection, and so takes the form Q_ψ, for some ψ. □

11.7*. Spin systems

To develop a feeling for density operators it is useful to consider the case of a two-dimensional inner product space $\mathcal{H}$, which could describe a spinning electron or the polarization states of a photon. We have already noted in Theorem 8.8.1 that any self-adjoint operator ρ can be written in the form

$$\rho = \tfrac{1}{2}\left(P_0 1 + \mathbf{P}.\boldsymbol{\sigma}\right), \tag{11.29}$$

where $P_0 = \mathrm{tr}(\rho)$ and $P_j = \mathrm{tr}(\rho\sigma_j)$ for $j = 1, 2, 3$.

For a density operator we require that $\mathrm{tr}(\rho) = 1$, so

$$\rho = \tfrac{1}{2}(1 + \mathbf{P}.\boldsymbol{\sigma}). \tag{11.30}$$

Definition 11.7.1. The vector $\mathbf{P} \in \mathbf{R}^3$ is called the *polarization vector.*

Proposition 11.7.1. The possible density operators for a spin $\frac{1}{2}$ system are characterized by a polarization vector $\mathbf{P}$ that lies in the unit ball in $\mathbf{R}^3$. The pure states are associated with polarization vectors lying on the surface, that is in the unit sphere.

Proof. Applying Theorem 8.7.1(ii) to $2\rho - 1 = \mathbf{P}.\boldsymbol{\sigma}$ we have

$$4\rho^2 - 4\rho + 1 = |\mathbf{P}|^2, \tag{11.31}$$

from which we deduce that

$$4(\rho - \rho^2) = (1 - |\mathbf{P}|^2). \tag{11.32}$$

It is therefore clear that $\rho^2 \leq \rho$ if and only if $|\mathbf{P}|^2 \leq 1$, and that ρ is pure (that is, $\rho^2 = \rho$) if and only if $|\mathbf{P}| = 1$. □

Any observable A is a self-adjoint operator and so can be written in the form

$$A = \tfrac{1}{2}\left(A_0 1 + \mathbf{A}.\boldsymbol{\sigma}\right). \tag{11.33}$$

We therefore have

$$\begin{aligned} \mathsf{E}_\rho(A) &= \mathrm{tr}(\rho A) \\ &= \mathrm{tr}\left(\tfrac{1}{2}\left(1 + \mathbf{P}.\boldsymbol{\sigma}\right)\tfrac{1}{2}\left(A_0 1 + \mathbf{A}.\boldsymbol{\sigma}\right)\right) \\ &= \tfrac{1}{4}\mathrm{tr}(A_0 + \mathbf{A}.\mathbf{P} + i(\mathbf{P} \times \mathbf{A}).\boldsymbol{\sigma}) \\ &= \tfrac{1}{2}\left(A_0 + \mathbf{A}.\mathbf{P}\right). \end{aligned} \tag{11.34}$$

Remark 11.7.1. The state $\rho = \frac{1}{2}1$ is the density operator for an unpolarized state in which $\mathbf{P} = (P_1, P_2, P_3) = 0$. Unpolarized light is well described by a state of this kind. Suppose that ψ is the state that can be transmitted by a particular polarized filter and let Q_ψ be the projection onto the space spanned by ψ. The probability that an unpolarized beam described by the density operator ρ will pass through the filter is

$$\operatorname{tr}(\rho Q_\psi) = \langle\psi|\rho\psi\rangle/\|\psi\|^2 = \tfrac{1}{2}. \tag{11.35}$$

Thus an unpolarized beam has an even chance of being passed by any filter.

11.8*. Gibbs' states

In the example of a box full of gas mentioned at the start of Section 11.6, the system would in practice settle into dynamical equilibrium in which the probabilities of different states depended on their energies. Josiah Willard Gibbs worked out the precise relationship for classical systems at the end of the last century, and his answer carries straight over to quantum theory.

Definition 11.8.1. If H, the Hamiltonian operator for a system, does not depend on time, and $\exp(-\beta H)$ has a finite trace, then the state described by the density operator

$$\rho = \frac{1}{\operatorname{tr}(e^{-\beta H})} e^{-\beta H}$$

is called the *Gibbs state* for inverse temperature β.

The inverse temperature corresponds to an actual temperature $1/k\beta$ K, where $k \sim 1.4 \times 10^{-23}$ joules per degree is known as the *Boltzmann constant.* The Gibbs state is that appropriate to a system which has come into thermal equilibrium at temperature $1/k\beta$ K.

Example 11.8.1. As an example consider the one-dimensional harmonic oscillator with Hamiltonian $H = P^2/2m + \frac{1}{2}m\omega^2X^2$. The eigenvalues of H are $\left(n+\frac{1}{2}\right)\hbar\omega$, so those of $\exp(-\beta H)$ are $\exp\left[-\left(n+\frac{1}{2}\right)\beta\hbar\omega\right]$. The trace, being the sum of the eigenvalues, is

$$\sum_{n=0}^{\infty} \exp\left[-\left(n+\tfrac{1}{2}\right)\beta\hbar\omega\right] = \frac{e^{-\frac{1}{2}\beta\hbar\omega}}{1-e^{-\beta\hbar\omega}} = \tfrac{1}{2}\operatorname{cosech}\left(\tfrac{1}{2}\beta\hbar\omega\right). \tag{11.36}$$

The probability of finding the energy to be $E_n = \left(n + \frac{1}{2}\right)\hbar\omega$ is therefore

$$\exp\left[-\left(n + \tfrac{1}{2}\right)\beta\hbar\omega\right] \times \left(1 - e^{-\beta\hbar\omega}\right) e^{\frac{1}{2}\beta\hbar\omega} = e^{-n\beta\hbar\omega}\left(1 - e^{-\beta\hbar\omega}\right). \quad (11.37)$$

We therefore have a geometric distribution of energies.

Rather than considering harmonic oscillators it might seem more natural to start with freely moving particles. Physically this means setting $\omega = 0$ to obtain the free particle Hamiltonian $H = P^2/2m$. However, as $\omega \to 0$ the expression for the trace diverges owing to the cosech term. When the probability distribution expressing our ignorance is continuous rather than discrete, as in the case of the free particle, the states of the system cannot be described by a density operator. This is not really surprising since unrestricted freely moving particles are physically unable to come into equilibrium. For free particles in a finite box the trace is finite and equilibrium is possible.

11.9*. The KMS condition

There is a useful generalization of Gibbs' states, which can be obtained by applying density operators in the Heisenberg picture. The connection comes from the relation between ρ and the time evolution operators:

$$e^{-\beta H} = e^{i(i\beta\hbar H)/\hbar} = U_{-i\beta\hbar}. \quad (11.38)$$

Theorem 11.9.1. (The KMS condition) Let E_β denote the expectation in the Gibbs state with density operator a multiple of $\exp(-\beta H)$, and suppose that the Heisenberg evolution of observables, $A_t = U_t^* A_0 U_t$, extends to complex values of t. Then

$$\mathsf{E}_\beta(A_t B) = \mathsf{E}_\beta(B A_{t+i\beta\hbar}).$$

Proof. We first note that, since $U_s^* = U_{-s}$, the Heisenberg equation of motion yields

$$U_{-s}A_t = U_{-s}A_t U_s U_{-s} = A_{t+s}U_{-s}. \quad (11.39)$$

Setting $s = i\hbar\beta$, multiplying by B, and taking the trace, we obtain

$$\mathrm{tr}(e^{-\beta H}A_t B) = \mathrm{tr}(U_{-i\beta\hbar}A_t B) = \mathrm{tr}(A_{t+i\beta\hbar}U_{-i\beta\hbar}B). \quad (11.40)$$

Since $\mathrm{tr}(CB) = \mathrm{tr}(BC)$ we deduce that

$$\mathrm{tr}(e^{-\beta H} A_t B) = \mathrm{tr}(U_{-i\beta\hbar} B A_{t+i\beta\hbar}) = \mathrm{tr}(e^{-\beta H} B A_{t+i\beta\hbar}), \qquad (11.41)$$

and dividing by $\mathrm{tr}(\exp(-\beta H))$ the result follows. □

Remark 11.9.1. This identity was first noted by Kubo, Martin, and Schwinger, whose initials now commemorate their contribution. In finite dimensions we can take $\beta = 0$, and then the KMS condition simply reduces to the identity $\mathrm{tr}(A_t B) = \mathrm{tr}(B A_t)$ for traces.

Gibbs' states are not the only ones to satisfy the KMS identity: there are others, which are also useful in physics. Moreover, the KMS condition on its own can supply a wealth of interesting information about a system, as we play off its reversal of the order of observables against commutation relations.

Example 11.9.1. Let us use the KMS condition to calculate the expectation value of the energy in a Gibbs state. For this we take $A = P + im\omega X$, $B = P - im\omega X$, and $t = 0$. Recalling equation (11.20) we see that

$$(P + im\omega X)_{i\beta\hbar} = e^{-\omega\beta\hbar}(P + im\omega X)_0. \qquad (11.42)$$

Dropping the subscript 0, the KMS condition therefore states that

$$\mathsf{E}_\beta((P + im\omega X)(P - im\omega X)) = e^{-\omega\beta\hbar}\mathsf{E}_\beta((P - im\omega X)(P + im\omega X)). \qquad (11.43)$$

Using Lemma 7.5.1 we can rewrite this identity as

$$\mathsf{E}_\beta(2m\left(H - \tfrac{1}{2}\hbar\omega\right)) = e^{-\omega\hbar\beta}\mathsf{E}_\beta(2m\left(H + \tfrac{1}{2}\hbar\omega\right)), \qquad (11.44)$$

which can be rearranged to give

$$(1 - e^{-\omega\beta\hbar})\mathsf{E}_\beta(H) = (1 + e^{-\omega\beta\hbar})\left(\tfrac{1}{2}\hbar\omega\right). \qquad (11.45)$$

The expectation value of the energy is therefore

$$\mathsf{E}_\beta(H) = \tfrac{1}{2}\hbar\omega(1 + e^{-\beta\hbar\omega})/(1 - e^{-\beta\hbar\omega}) = \tfrac{1}{2}\hbar\omega\coth\left(\tfrac{1}{2}\beta\hbar\omega\right). \qquad (11.46)$$

This can be checked using the exponential probability distribution of energies derived earlier.

This is the example that lay at the core of the problem that originally led Planck to introduce the quantum hypothesis. The energy difference between the expectation value of H and the ground state energy is

$$\mathsf{E}_\beta(H) - \tfrac{1}{2}\hbar\omega = \hbar\omega(e^{-\beta\hbar\omega})/(1 - e^{-\beta\hbar\omega}). \qquad (11.47)$$

In the limit as $\hbar \to 0$ and quantum effects are suppressed, this tends to $1/\beta$, which means that the average energy of each classical oscillator is independent of its frequency. The classical problem arose because it is easier to fit short wavelength (high frequency) waves into a given size of cavity, so if the energy per wave does not depend on its frequency, then most of the energy should be carried by the higher frequency waves. (The normal frequencies of waves in a cubic box of side a are given by $\omega = \pi\sqrt{j^2 + k^2 + l^2}/a$, for integer j, k, and l. The number of normal frequencies less than a given ω is proportional to ω^3.) Planck's law avoids this problem because the exponential damping factor, $\exp(-\beta\hbar\omega)$, in the formula more than compensates for the polynomial growth in the number of short wavelength waves that can be fitted into the cavity.

11.10*. Partition functions and the harmonic oscillator

The formula for the expectation value of the harmonic oscillator energy in a KMS state can be used to provide an independent derivation of the energy spectrum. According to Theorem 11.9.1 a Gibbs state satisfies the KMS condition, and so its mean energy must be given by

$$\frac{\mathrm{tr}\big(He^{-\beta H}\big)}{\mathrm{tr}(e^{-\beta H})} = \tfrac{1}{2}\hbar\omega\coth\big(\tfrac{1}{2}\beta\hbar\omega\big). \tag{11.48}$$

This has a convenient reformulation in terms of the partition function:

Definition 11.10.1. The *partition function*, $Z(\beta)$, is defined by

$$Z(\beta) = \mathrm{tr}\big(e^{-\beta H}\big),$$

when the right-hand side makes sense.

Remark 11.10.1. If the energy levels $\{E_1, E_2, \ldots\}$ of the Hamiltonian have degeneracies $\{d_1, d_2, \ldots\}$, then the partition function is

$$Z(\beta) = \sum_{n=1}^{\infty} d_n e^{-\beta E_n}, \tag{11.49}$$

provided that the sum converges.

Differentiating the definition and using our earlier expression for the mean energy, we have

$$\frac{Z'(\beta)}{Z(\beta)} = -\frac{\operatorname{tr}\big(He^{-\beta H}\big)}{\operatorname{tr}(e^{-\beta H})} = -\tfrac{1}{2}\hbar\omega \coth\big(\tfrac{1}{2}\beta\hbar\omega\big). \tag{11.50}$$

Introducing an integrating factor we obtain

$$\frac{d}{d\beta}\left[\sinh(\tfrac{1}{2}\beta\hbar\omega)Z(\beta)\right] = 0, \tag{11.51}$$

so that for some constant C

$$\begin{aligned} Z(\beta) &= C\operatorname{cosech}(\tfrac{1}{2}\beta\hbar\omega) \\ &= 2Ce^{-\beta\hbar\omega/2} / \left(1 - e^{-\beta\hbar\omega}\right) \\ &= 2C\sum_{n=1}^{\infty} e^{-(n+\frac{1}{2})\beta\hbar\omega}. \end{aligned} \tag{11.52}$$

Comparing this with the general formula in terms of energies and degeneracies shows that the energies can be expressed as $E_n = (n + \frac{1}{2})\hbar\omega$, and they all have the same degeneracy.

11.11*. Algebraic quantum theory

In Heisenberg's picture of quantum mechanics the observables play the dominant role, whilst the state vectors are immutable and, except for their vestigial role in the calculation of expectation values, superfluous. This fact is exploited in a simpler though more abstract version of Heisenberg's approach, which uses only the algebraic properties of the observables without insisting that they should be linear transformations, and replaces states by the expectation values that they served to define.

We have been using the following features of linear transformations on an inner product space:

(i) The linear transformations, $\mathcal{L}(\mathcal{H})$, themselves define a complex vector space, that is one can form linear combinations $\alpha A + \beta B$ of observables A and B with complex coefficients α and β.

(ii) One can form products of linear transformations and these distribute over sums so that $\mathcal{L}(\mathcal{H})$ is a ring in the sense of algebra. This ring has an identity, 1, and the product is related to the vector space structure by $(\alpha 1)A = \alpha A$.

(iii) There is an adjoint map $A \mapsto A^*$ which is conjugate linear,

$$(\alpha A + \beta B)^* = \overline{\alpha}A^* + \overline{\beta}B^*, \tag{11.53}$$

and satisfies $(AB)^* = B^*A^*$ and $A^{**} = A$.

Definition 11.11.1. Suppose that $\mathcal{A}$ has the structure both of a vector space and, with the same addition, of a ring with identity, 1, in such a way that, for all $A \in \mathcal{A}$ and $\alpha \in \mathbf{C}$, $(\alpha 1)A = \alpha A$. Then $\mathcal{A}$ is said to be an *algebra*. If in addition there is a map $* : \mathcal{A} \to \mathcal{A}$ such that $(\alpha A + \beta B)^* = \overline{\alpha} A^* + \overline{\beta} B^*$, $(AB)^* = B^*A^*$, and $A^{**} = A$, then $\mathcal{A}$ is said to be a $*$-*algebra*.

The algebraic formulation of quantum theory assumes that the observables are described by the self-adjoint elements of a $*$-algebra, that is elements A that satisfy $A^* = A$. For future reference we note that $1^* = 1^*1$ is self-adjoint and so $1^* = 1^{**} = 1$.

Definition 11.11.2. A *state* on a $*$-algebra $\mathcal{A}$ is a linear functional $E; \mathcal{A} \to \mathbf{C}$ that satisfies:
(i) $E(A^*A)$ is real and non-negative for all $A \in \mathcal{A}$,
(ii) $E(1) = 1$.

Remark 11.11.1. These assumptions are based on the properties of expectation values given in Proposition 6.3.1. The fact that E is a linear functional expresses the linearity of expectations, (iv), whilst the two assumptions correspond to (iii) and (i). The reality property (ii) need not be included since it follows from the other conditions.

Proposition 11.11.1. For any state, E, on a $*$-algebra $\mathcal{A}$, and for any $A, B \in \mathcal{A}$,

$$E(A^*B) = \overline{E(B^*A)}.$$

In particular $E(A^*) = \overline{E(A)}$, so that $E(A)$ is real for self-adjoint A. Furthermore, there is a Cauchy–Schwarz–Bunyakowski inequality:

$$E(A^*A)E(B^*B) \geq |E(B^*A)|^2.$$

Proof. For any $A, B \in \mathcal{A}$ and $\lambda \in \mathbf{C}$ we have

$$E((A+\lambda B)^*(A+\lambda B)) = E(A^*A) + |\lambda|^2 E(B^*B) + \lambda E(A^*B) + \overline{\lambda} E(B^*A). \tag{11.54}$$

The left-hand side of this identity, like the first two terms on the right-hand side, is positive and so real. This forces the sum of the two remaining terms to be real and gives

$$\lambda E(A^*B) + \overline{\lambda}E(B^*A) = \overline{\lambda}\overline{E(A^*B)} + \lambda\overline{E(B^*A)}. \tag{11.55}$$

Rearranging this we obtain

$$\lambda(E(A^*B) - \overline{E(B^*A)}) = \overline{\lambda}(\overline{E(A^*B)} - E(B^*A)), \tag{11.56}$$

which, since true for all complex λ, forces both sides to vanish and gives $E(A^*B) = \overline{E(B^*A)}$. Taking $B = 1$ gives $E(A^*) = \overline{E(A)}$ and so, in particular, when $A^* = A$ we deduce that $E(A)$ is real.

Finally, taking $\lambda = tE(B^*A) = t\overline{E(A^*B)}$ for real t we see that

$$0 \leq E(A^*A) + t^2|E(A^*B)|^2E(B^*B) + 2t|E(A^*B)|^2, \tag{11.57}$$

and the Cauchy–Schwarz–Bunyakowski inequality follows from the fact that the discriminant of this quadratic in t must be negative. □

Curiously this abstract version of quantum theory is much closer to the normal formulation given in Section 6.3 than might at first appear. In the following result the term homomorphism means that it is both a ring homomorphism and a linear transformation.

Theorem 11.11.2. (Gel'fand, Naimark, Segal) Let E be a state on a $*$-algebra $\mathcal{A}$. There exists an inner product space $\mathcal{H}_E$, a unit vector $\Omega_E \in \mathcal{H}_E$, and a homomorphism $\gamma : \mathcal{A} \to \mathcal{L}(\mathcal{H}_E)$, such that for all $A \in \mathcal{A}$

$$E(A) = \langle \Omega_E | \gamma(A)\Omega_E \rangle.$$

Proof. The space $\mathcal{H}_E$ will be a subspace of the dual space $\mathcal{A}'$ of linear functionals on $\mathcal{A}$. We can define a homomorphism $\gamma : \mathcal{A} \to \mathcal{L}(\mathcal{A}')$ by setting, for $f \in \mathcal{A}'$ and $X \in \mathcal{A}$,

$$(\gamma(A)f)(X) = f(XA). \tag{11.58}$$

Then

$$(\gamma(AB)f)(X) = f(XAB) = (\gamma(B)f)(XA) = (\gamma(A)\gamma(B)f)(X), \tag{11.59}$$

so that $\gamma(AB) = \gamma(A)\gamma(B)$, whilst $\gamma(\alpha A + \beta B) = \alpha\gamma(A) + \beta\gamma(B)$ follows from the linearity of f.

We have yet to define the inner product, and we now restrict attention to the subspace

$$\mathcal{H}_E = \{\gamma(A)E : A \in \mathcal{A}\}, \tag{11.60}$$

which is clearly invariant under the action of any $\gamma(B)$. On this we define

$$\langle\gamma(A)E|\gamma(B)E\rangle = E(A^*B). \tag{11.61}$$

This is well defined, since it can also be written as

$$E(A^*B) = (\gamma(B)E)\,(A^*), \tag{11.62}$$

showing that it depends on B only through $\gamma(B)E$. The preceding proposition tells us immediately that

$$\langle\gamma(A)E|\gamma(B)E\rangle = E(A^*B) = \overline{E(B^*A)} = \overline{\langle\gamma(B)E|\gamma(A)E\rangle}, \tag{11.63}$$

showing that the inner product depends on A through $\gamma(A)E$, and also giving the conjugate symmetry. The linearity properties of the inner product follow from the linearity of E. It is also clear from the definition of E that

$$\langle\gamma(A)E|\gamma(A)E\rangle = E(A^*A) \geq 0. \tag{11.64}$$

Finally the Cauchy–Schwarz–Bunyakowski inequality tells us that

$$|(\gamma(A)E)(B^*)|^2 = |E(B^*A)|^2 \leq E(B^*B)E(A^*A), \tag{11.65}$$

so that $E(A^*A)$ can vanish only if $(\gamma(A)E)(B^*) = 0$ for all B, which forces $\gamma(A)E = 0$, and shows that the inner product is strictly positive as required.

Finally, we note that the space $\mathcal{H}_E$ contains the distinguished linear functional E itself, and that

$$\langle E|\gamma(A)E\rangle = E(1^*A) = E(A), \tag{11.66}$$

so that E is the obvious candidate for the vector Ω_E. □

Remark 11.11.2. Interpreted in one way this result shows that, despite appearances, the abstract formulation of quantum theory that we have just introduced is no more general than that which we have been using since Chapter 6. We can always find an inner product space, use operators on it as observables, and calculate expectations defined by vectors in the usual way. However, this is slightly misleading for two reasons. The first is that the space $\mathcal{H}_E$ is often much bigger than one would otherwise need and not every linear transformation is of the form $\gamma(A)$. For example, when we define $E(A) = \text{tr}(\rho A)$ for some density operator ρ the space $\mathcal{H}_E$ is much larger than the space on which ρ acts. This is inevitable, for in the large space $\mathcal{H}_E$ all the statistical uncertainty included in ρ has been banished and states are represented by vectors again. We should not find this at all surprising, for if we were dealing with the statistical behaviour of molecules in a box we know that there is a larger space describing the quantum mechanics of each and every molecule in detail where statistical uncertainties disappear. However, we chose to work with density operators precisely to avoid getting entangled in that kind of detail.

The second reason for caution in interpreting the significance of this theorem is that in many important cases different states of a given physical system may lead to inequivalent spaces $\mathcal{H}_E$. It is the algebra $\mathcal{A}$ which is associated with the physical system, not the inner product space. For example, a ferromagnetic material is described by different spaces, $\mathcal{H}_E$, according to whether it is magnetized or not. The algebra, however, is the same.

Even this algebraic formulation of quantum theory is not the most general. Some physical observables, such as times when a photon hits a counter, cannot easily be interpreted as elements of the algebra, and for them there are still more general approaches.

Exercises

11.1° A particle is free to move in the x direction so that its Hamiltonian H is given by

$$H = P^2/2m.$$

At time $t = 0$ the system is in a state with the wave function

$$\psi = Ne^{-x^2/4\sigma^2},$$

where N and σ are real constants. Find the initial expectation values $\mathsf{E}_\psi(X^2)$, $\mathsf{E}_\psi(P^2)$, and $\mathsf{E}_\psi(PX + XP)$. Using the Heisenberg picture or otherwise show that at time t the expectation of X^2 is

$$\frac{\hbar^2 t^2}{4m^2\sigma^2} + \sigma^2,$$

and find the expected values of P^2 and $XP + PX$.

11.2° A particle moving in one dimension has the Hamiltonian

$$\frac{P^2}{2m} + V(X).$$

Using the Heisenberg picture or otherwise show that under suitable assumptions the expectation values of position and momentum satisfy

$$\frac{d}{dt}\mathsf{E}(X(t)) = \frac{\mathsf{E}(P(t))}{m},$$
$$\frac{d}{dt}\mathsf{E}(P(t)) = -\mathsf{E}(V'(X(t))).$$

11.3° The Hamiltonian for the one-dimensional harmonic oscillator is given by

$$H = \frac{1}{2m}\left(P^2 + m^2\omega^2 X^2\right).$$

Show that

$$\mathsf{E}\left(X^2_{t+\pi/2\omega}\right) + \mathsf{E}\left(X^2_t\right)$$

is constant.

11.4° The Hamiltonian for a one-dimensional system is given by

$$H = \frac{P^2}{2m} - mkX,$$

where k is a constant. Show that the expectation of the position observable at time t is given by

$$\mathsf{E}(X(t)) = \frac{1}{2}kt^2 + \mathsf{E}\left(\frac{P(0)t}{m} + X(0)\right).$$

Show further that

$$\lim_{t\to\infty}\frac{\Delta(X(t))}{t} = \frac{\Delta(P(0))}{m}.$$

11.5 The Hamiltonian for a spinning body is L_3. By considering the expectation value of L_+L_- in a KMS state show that the partition function $Z(\beta)$ satisfies the differential equation

$$Z'' + \hbar\coth\left(\tfrac{1}{2}\beta\hbar\right)Z' - \lambda\hbar^2 Z = 0,$$

where $\lambda\hbar^2$ is the eigenvalue of L^2. By substituting $f = \sinh(\frac{1}{2}\beta\hbar)Z$, or otherwise, deduce that

$$Z = A \sinh\left(\sqrt{(\lambda + \tfrac{1}{4})}\beta\hbar\right) / \sinh\left(\tfrac{1}{2}\beta\hbar\right),$$

where A is a constant. By considering the case of $\beta = -it$ and using periodicity in t, deduce that Z is a multiple of

$$\frac{\sinh[(l + \frac{1}{2})\beta\hbar]}{\sinh(\frac{1}{2}\beta\hbar)},$$

for some non-negative integer l.

11.6° Starting from the Schrödinger picture define the Heisenberg picture for a system in which the Hamiltonian, H, does not depend explicitly on time. Obtain the equation of motion satisfied by an operator that is not explicitly time dependent, and deduce that H is constant.

A particle of mass m moves along the x-axis under the influence of a uniform electric field with potential Fx. By using the Heisenberg picture, or otherwise, show that the dispersion of P is independent of time and find an expression for the dispersion of X.

Show also that

$$\frac{\mathsf{E}(P)^2}{2m} + F\mathsf{E}(X)$$

is constant during the motion, where $\mathsf{E}(A)$ denotes the expectation value of the observable A.

11.7° The operators J_1, J_2, and J_3 representing the components of angular momentum are hermitian and satisfy the commutation relations

$$[J_2, J_3] = i\hbar J_1, \qquad [J_3, J_1] = i\hbar J_2, \qquad [J_1, J_2] = i\hbar J_3.$$

Let $S = J_1 - iJ_2$ and let Φ_m be an eigenvector of J_3 with eigenvalue $m\hbar$, where $m > 1$. Show that

$$J_3 S\Phi_m = (m-1)\hbar S\Phi_m.$$

The Hamiltonian for a quantum mechanical system is given by

$$H = \frac{J_1^2}{A} + \frac{J_2^2}{A} + \frac{J_3^2}{C},$$

where A and C are positive constants. Obtain the equations of motion for J_3 and S in the Heisenberg picture. By making the substitution

$$S(t) = \exp\left[i\left(\frac{1}{A}-\frac{1}{C}\right)\frac{J_3(0)t}{\hbar}\right] T(t) \exp\left[i\left(\frac{1}{A}-\frac{1}{C}\right)\frac{J_3(0)t}{\hbar}\right],$$

or otherwise, find $S(t)$ in the Heisenberg picture in terms of $J_3(0)$ and $S(0)$. Initially the system is in an eigenstate of J_3. Show that the expectation value of J_1^2 is constant.

11.8° A particle with charge e forming a linear harmonic oscillator with unperturbed Hamiltonian

$$H_0 = \frac{P^2}{2m} + \frac{1}{2}m\omega^2 X^2$$

is placed in a weak electric field along the x-axis given by

$$F(t) = \frac{A}{t\sqrt{\pi}} e^{-t^2/\tau^2},$$

where A and τ are constants. At $t = -\infty$ the oscillator is in its ground state. Find to a first approximation the probability that it will be in its first excited state at $t = +\infty$.

11.9 Show that any operator of the form $\rho = \sum p_j Q_{\psi_j}$, with $p_j \in [0,1]$, is positive, that is

$$\langle\psi|\rho\psi\rangle \geq 0,$$

for any vector ψ.

11.10 For any inner product space $\mathcal{H}$, let $\mathcal{S}$ denote the subspace of operators, A, for which $\mathrm{tr}(A^*A)$ is finite, with the inner product

$$\langle A|B\rangle = \mathrm{tr}(A^*B).$$

Show that $\Omega = \sum \sqrt{p_j} Q_{\psi_j}$ is in $\mathcal{S}$, and that it is related to $\rho = \sum p_j Q_{\psi_j}$ by

$$\mathrm{tr}(\rho A) = \langle\Omega|A\Omega\rangle / \|\Omega\|^2.$$

12 Stationary perturbation theory

A beautiful Christmas and a good New Year filled with hydrogen transition probabilities, theory of helium etc.

WERNER HEISENBERG, letter to Wolfgang Pauli, 24 December 1925

12.1. Rayleigh–Schrödinger perturbation theory

For most physically interesting systems it is not possible to find simple closed formulae for the energy levels and wave functions. Generally the best that one can do is to find numerical approximations and iterative schemes. Since Schrödinger's equation is a differential equation there are many standard numerical methods that can supply approximate solutions, but there are also various special techniques tailored to this particular situation, which we shall describe over the next chapters.

The most obvious technique is to try to compare solutions of the equation

$$H\psi = E\psi \tag{12.1}$$

with the solutions for a more tractable Hamiltonian H_0. One natural way to link the behaviour of two Hamiltonians H_0 and H is to consider the family

$$H_u = (1-u)H_0 + uH = H_0 + uH', \tag{12.2}$$

where $H' = H - H_0$ is the perturbation, and the real parameter u just controls the strength of the perturbation. As the notation suggests, $H_u = H_0$ when $u = 0$, whilst $H_1 = H$. We would hope that each Hamiltonian has associated energy levels E_u and eigenstates ψ_u so that

$$H_u\psi_u = E_u\psi_u. \tag{12.3}$$

We shall assume, for convenience, that ψ_0 is normalized, and that

$$\langle\psi_0|\psi_u\rangle = 1, \tag{12.4}$$

for all u in the interval under consideration. (If ψ_u is to be a good approximation to ψ_0 then we expect $\langle\psi_0|\psi_u\rangle \neq 0$; indeed, since it is 1 when $u = 0$, there is an interval around 0 in which it does not vanish. Replacing ψ_u by $\psi_u/\langle\psi_0|\psi_u\rangle$ there, we may as well assume that $\langle\psi_0|\psi_u\rangle = 1$.)

Rayleigh–Schrödinger perturbation theory proceeds by supposing that both E_u and ψ_u have power series expansions,

$$\begin{aligned} E_u &= E_0 + uE' + u^2E'' + \dots, \\ \psi_u &= \psi_0 + u\psi' + u^2\psi'' + \dots. \end{aligned} \tag{12.5}$$

This is a very strong assumption, which can, nonetheless, sometimes be justified. When it is valid, one can simply compare coefficients of u in Schrödinger's equation,

$$(H_0 + uH')(\psi_0 + u\psi' + \dots) = (E_0 + uE' + \dots)(\psi_0 + u\psi' + \dots), \tag{12.6}$$

to obtain a sequence of equations for the various terms. The constant term gives back the unperturbed equation

$$H_0\psi_0 = E_0\psi_0, \tag{12.7}$$

showing that ψ_0 and E_0 are an eigenstate and energy level of the unperturbed problem. The coefficient of u gives

$$H_0\psi' + H'\psi_0 = E_0\psi' + E'\psi_0, \tag{12.8}$$

for the first-order corrections E', ψ' to the energy and eigenstate.

Theorem 12.1.1. Suppose that $\phi_1, \dots, \phi_D$ form an orthonormal basis for the E_0-eigenstates of H_0 with respect to which ψ_0 can be written as $\psi_0 = \sum c_r\phi_r$. Then the column vector of coefficients $(c_1, \dots, c_D)$ is an eigenvector of the self-adjoint matrix with entries $\langle\phi_r|H'\phi_s\rangle$ with eigenvalue E', so that

$$\det(\langle\phi_r|H'\phi_s\rangle - E'\delta_{rs}) = 0.$$

In particular, when E_0 is non-degenerate, one may take $\psi_0 = \phi_1$ and $E' = \langle\psi_0|H'\psi_0\rangle$. The corrections to the wave function, ψ', ψ'', ..., can be chosen to be orthogonal to ψ_0, and the first-order correction is given by a solution ψ' of

$$(E_0 - H_0)\psi' = (H' - E')\psi_0,$$

which is orthogonal to ψ_0.

Proof. Taking the inner product of equation (12.8) with ϕ_r we obtain

$$\langle\phi_r|H_0\psi'\rangle + \langle\phi_r|H'\psi_0\rangle = E_0\langle\phi_r|\psi'\rangle + E'\langle\phi_r|\psi_0\rangle. \tag{12.9}$$

Exploiting the self-adjointness of H_0 and simplifying we see that

$$\langle\phi_r|H_0\psi'\rangle = \langle H_0\phi_r|\psi'\rangle = E_0\langle\phi_r|\psi'\rangle, \tag{12.10}$$

so that the first terms on each side cancel leaving

$$\langle\phi_r|H'\psi_0\rangle = E'\langle\phi_r|\psi_0\rangle. \tag{12.11}$$

Substituting the expansion of ψ_0 in terms of the basis $\{\phi_r\}$, and using orthonormality, now shows that

$$\sum_{s=1}^{D}\langle\phi_r|H'\phi_s\rangle c_s = E'c_r, \tag{12.12}$$

giving the stated eigenvector property. The determinant equation for E' is just the condition for this equation to have a non-trivial solution. The fact that the matrix is self-adjoint follows from the identity

$$\langle\phi_r|H'\phi_s\rangle = \langle H'\phi_r|\phi_s\rangle = \overline{\langle\phi_s|H'\phi_r\rangle}. \tag{12.13}$$

When E_0 is non-degenerate, so that $D = 1$, $\psi_0 = c_1\phi_1$, and we may choose $c_1 = 1$. The condition on E' also reduces to

$$E' = \langle\phi_1|H'\phi_1\rangle = \langle\psi_0|H'\psi_0\rangle, \tag{12.14}$$

as asserted.

Substituting the series into equation (12.4) gives

$$\sum u^n\langle\psi_0|\psi^{(n)}\rangle = 1, \tag{12.15}$$

and taking the coefficient of u^n for $n > 0$, we see that $\psi^{(n)}$ is then orthogonal to ψ_0. □

Remark 12.1.1. There are more hidden assumptions lurking in the background of this chapter than of most others. Examples show that the domains of the operators H and H_0 may intersect trivially, so that $H - H_0$ is only defined on the zero vector. Even when its domain is a larger and more interesting set there can be problems in establishing some of the later properties. Nonetheless, perturbation theory is an invaluable theoretical tool, and there are many genuine examples for which its use can be justified. We shall therefore proceed formally assuming that the difference, $H' = H - H_0$, makes good sense, and that it is, in some sense, small.

Suppose that whenever $H_0\psi$ is defined so is $H'\psi$, and for some constants a and b independent of ψ one has the inequality

$$\|H'\psi\| \leq a\|H_0\psi\| + b\|\psi\|; \tag{12.16}$$

then there is a positive R such that E_u and ψ_u are analytic functions of u within the region where $|u| < R$. The helium atom (Section 12.3) falls into the category of examples covered by this criterion, so that in that case the Rayleigh–Schrödinger method can be justified.

12.2. Examples

To demonstrate the method we shall first consider an example that can be solved exactly, so that we may check our results. We take the Hamiltonian H for a two-dimensional oscillator in a potential $V = \frac{1}{2}m\omega^2(x^2+y^2)+uxy$. Diagonalizing the potential energy, as in Section 3.2, gives

$$\det\begin{pmatrix} m\omega^2-\lambda & u \\ u & m\omega^2-\lambda \end{pmatrix} = 0, \tag{12.17}$$

so that $\lambda = m\omega^2 \pm u$, and the frequencies, $\sqrt{\lambda/m}$, are given by

$$\omega_\pm = \omega\left(1 \pm \frac{u}{m\omega^2}\right)^{\frac{1}{2}} = \omega\left(1 \pm \frac{u}{2m\omega^2} + \ldots\right). \tag{12.18}$$

The true energies are therefore of the form

$$\left(n_+ + \tfrac{1}{2}\right)\hbar\omega_+ + \left(n_- + \tfrac{1}{2}\right)\hbar\omega_- = (n_+ + n_- + 1)\hbar\omega + (n_+ - n_-)\frac{u\hbar}{2m\omega} + \ldots. \tag{12.19}$$

We shall now compare the two lowest energy states with those of the isotropic oscillator whose potential is $\frac{1}{2}m\omega^2(x^2+y^2)$, taking uxy as the perturbation. The unperturbed Hamiltonian H_0 has energy levels $(n_1+\frac{1}{2})\hbar\omega + (n_2+\frac{1}{2})\hbar\omega = (n_1+n_2+1)\hbar\omega$, with eigenfunctions $\varphi_{n_1n_2} = \phi_{n_1}(x)\phi_{n_2}(y)$, where ϕ_n is the wave function for a one-dimensional oscillator. For the non-degenerate ground state φ_{00}, we have immediately

$$\begin{aligned} E' &= \langle\varphi_{00}|H'\varphi_{00}\rangle \\ &= \int_{\mathbf{R}^2} \overline{\phi_0(x)\phi_0(y)}xy\phi_0(x)\phi_0(y)\,dxdy \\ &= \int_{\mathbf{R}} x|\phi_0(x)|^2\,dx \int_{\mathbf{R}} y|\phi_0(y)|^2\,dy. \end{aligned} \tag{12.20}$$

Now $x|\phi_0(x)|^2$ is an odd function so its integral must vanish, leaving $E' = 0$. This agrees with the exact solution above, which has no first-order term in u when $n_+ = 0 = n_-$.

The first excited state is doubly degenerate with the orthonormal basis φ_{10} and φ_{01}, so that we must consider the matrix with entries $\langle \varphi_{kl} | H' \varphi_{rs} \rangle$ where $k + l = 1 = r + s$. Arguing as for the ground state, each diagonal element is of the form

$$\langle \varphi_{rs} | H' \varphi_{rs} \rangle = \int_{\mathbf{R}} x |\phi_r(x)|^2 \, dx \int_{\mathbf{R}} y |\phi_s(y)|^2 \, dy, \tag{12.21}$$

and since the integrands are again odd, this vanishes. By Theorem 12.1.1, we know that the matrix whose eigenvalues and eigenvectors are sought is self-adjoint and so has the form

$$\begin{pmatrix} 0 & a \\ \overline{a} & 0 \end{pmatrix}. \tag{12.22}$$

The eigenvalues are easily found to be $E' = \pm |a|$, and eigenvectors $(1, \pm 1)$, so that, on normalizing, $\psi_0 = (\varphi_{01} \pm \varphi_{10})/\sqrt{2}$. To find the energy correction we note that

$$\begin{aligned} a &= \langle \varphi_{01} | H' \varphi_{10} \rangle \\ &= \int_{\mathbf{R}} x \overline{\phi_0(x)} \phi_1(x) \, dx \int_{\mathbf{R}} y \overline{\phi_1(y)} \phi_0(y) \, dy \\ &= \left| \int_{\mathbf{R}} x \overline{\phi_0(x)} \phi_1(x) \, dx \right|^2 . \end{aligned} \tag{12.23}$$

It follows from Theorem 7.7.3 and equation (7.50) that the first excited state is related to the ground state by $\phi_1(x) = (2m\omega/\hbar)^{\frac{1}{2}} x \phi_0(x)$, so that

$$\int_{\mathbf{R}} x \overline{\phi_0(x)} \phi_1(x) \, dx = \sqrt{\frac{\hbar}{2m\omega}} \int_{\mathbf{R}} |\phi_1(x)|^2 \, dx = \sqrt{\frac{\hbar}{2m\omega}}, \tag{12.24}$$

and $E' = \pm |a| = \pm \hbar / 2m\omega$. This agrees with our precise formula when $n_+ = 1$ and $n_- = 0$ or vice versa.

12.3. The ground state of the helium atom

We shall now apply perturbation theory to a more interesting example. The helium atom has two electrons orbiting a nucleus of charge 2. By changing coordinates we can separate out the centre of mass and obtain the Hamiltonian

$$H = \left(-\frac{\hbar^2}{2m}\nabla_1^2 - \frac{\hbar^2}{2m}\nabla_2^2 - \frac{2e^2}{4\pi\epsilon_0 r_1} - \frac{2e^2}{4\pi\epsilon_0 r_2} + \frac{e^2}{4\pi\epsilon_0|\mathbf{r}_1 - \mathbf{r}_2|}\right), \tag{12.25}$$

where $\mathbf{r}_1$ and $\mathbf{r}_2$ are position vectors for the two electrons. Apart from the final term, which describes the repulsion between the two electrons, this looks just like the sum of two hydrogen-like Hamiltonians with nuclear charge $Z = 2$, so we take

$$H_0 = \left(-\frac{\hbar^2}{2m}\nabla_1^2 - \frac{Ze^2}{4\pi\epsilon_0 r_1}\right) + \left(-\frac{\hbar^2}{2m}\nabla_2^2 - \frac{Ze^2}{4\pi\epsilon_0 r_2}\right). \tag{12.26}$$

It will be useful to work with a general charge Z, as we shall need the results of the calculation again in Chapter 14.

The lowest energy state of H_0 is non-degenerate and according to 4.3.1 is described by the product of two hydrogenic ground states:

$$\begin{aligned}\psi_0(\mathbf{r}_1, \mathbf{r}_2) &= \left(\frac{Z^3}{\pi a^3}\right)^{\frac{1}{2}} e^{-Zr_1/a} \left(\frac{Z^3}{\pi a^3}\right)^{\frac{1}{2}} e^{-Zr_2/a} \\ &= \frac{Z^3}{\pi a^3} e^{-Z(r_1+r_2)/a}.\end{aligned} \tag{12.27}$$

The obvious approximation to the ground state of helium therefore starts with ψ_0 and energy $E_0 = -Z^2e^2/4\pi\epsilon_0 a$, which is the sum of two hydrogen-like ground state energies.

The perturbation is

$$H' = \frac{e^2}{4\pi\epsilon_0|\mathbf{r}_1 - \mathbf{r}_2|}. \tag{12.28}$$

To find the first-order correction to the energy we need to evaluate

$$\langle\psi_0|H'\psi_0\rangle = \left(\frac{Z^3}{\pi a^3}\right)^2 \int_{\mathbf{R}^3\times\mathbf{R}^3} \frac{e^2}{4\pi\epsilon_0|\mathbf{r}_1 - \mathbf{r}_2|} e^{-2Z(r_1+r_2)/a}\, d^3\mathbf{r}_1 d^3\mathbf{r}_2. \tag{12.29}$$

The only angular dependence occurs in the term

$$|\mathbf{r}_1 - \mathbf{r}_2|^{-1} = \left(r_1^2 + r_2^2 - 2r_1 r_2 \cos\theta\right)^{-\frac{1}{2}}, \tag{12.30}$$

where θ is the angle between r_1 and r_2. Considering just the angular part obtained by integrating this term over $\mathbf{r}_2$ we have

$$\int_0^\pi \int_0^{2\pi} \left(r_1^2 + r_2^2 - 2r_1 r_2 \cos\theta\right)^{-\frac{1}{2}} \sin\theta\, d\theta d\phi$$

$$= 2\pi \left[\frac{\left(r_1^2 + r_2^2 - 2r_1 r_2 \cos\theta\right)^{\frac{1}{2}}}{r_1 r_2} \right]_0^{\pi}$$

$$= \frac{2\pi}{r_1 r_2} \left[(r_1 + r_2) - |r_1 - r_2| \right]$$

$$= \begin{cases} 4\pi/r_2 & \text{if } r_2 > r_1 \\ 4\pi/r_1 & \text{if } r_1 \geq r_2, \end{cases} \tag{12.31}$$

a result well known in potential theory.

Since this no longer depends on the angles, the integration over the possible directions of $\mathbf{r}_1$ just multiplies by 4π. The integral for $\langle \psi_0 | H' \psi_0 \rangle$ therefore reduces to the sum of

$$\left(\frac{Z^3}{\pi a^3} \right)^2 \frac{e^2}{4\pi\epsilon_0} (4\pi)^2 \left(\int_0^\infty \int_{r_1}^\infty \frac{1}{r_2} e^{-2Z(r_1+r_2)/a} r_2^2 \, dr_2 r_1^2 dr_1 \right) \tag{12.32}$$

and a similar term with r_1 and r_2 interchanged, giving all together

$$32 \left(\frac{Z}{a} \right)^6 \frac{e^2}{4\pi\epsilon_0} \left(\int_0^\infty \int_{r_1}^\infty e^{-2Zr_2/a} r_2 \, dr_2 e^{-2Zr_1/a} r_1^2 dr_1 \right). \tag{12.33}$$

A simple integration by parts argument (or repeated differentiation with respect to k of the $n = 0$ case) shows that

$$\int_R^\infty e^{-kr} r^n \, dr = n! k^{-(n+1)} \left(e^{-kR} \sum_{j=0}^n \frac{1}{j!} (kR)^j \right), \tag{12.34}$$

so that, on doing the first integration, the previous integral reduces to

$$32 \left(\frac{Z}{a} \right)^6 \frac{e^2}{4\pi\epsilon_0} \left(\frac{a}{2Z} \right)^2 \left[\int_0^\infty \left(1 + \frac{2Zr_1}{a} \right) e^{-4Zr_1/a} r_1^2 dr_1 \right]. \tag{12.35}$$

Using (12.34) with $R = 0$ to do the final integration, we arrive at

$$8 \left(\frac{Z}{a} \right)^4 \frac{e^2}{4\pi\epsilon_0} \left[2! \left(\frac{a}{4Z} \right)^3 + 3! \left(\frac{2Z}{a} \right) \left(\frac{a}{4Z} \right)^4 \right] = \frac{5}{8} \frac{Ze^2}{4\pi\epsilon_0 a}. \tag{12.36}$$

Recalling that in this case $Z = 2$, we see that the first-order estimate of the energy is therefore

$$E_1 = -\frac{e^2}{\pi\epsilon_0 a} \left(1 - \frac{5}{16} \right) = -0.6875 \frac{e^2}{\pi\epsilon_0 a}. \tag{12.37}$$

Compared with the experimental value of $-0.73 e^2/\pi\epsilon_0 a$ this is about 5% too high.

12.4. Higher order Rayleigh–Schrödinger theory

For simplicity we shall consider higher order corrections only when the energy level E_0 is non-degenerate, so that the E_0-eigenvector, ψ_0, is uniquely determined up to multiples.

The coefficient of u^k in the expansion of the Schrödinger equation gives

$$H_0\psi^{(k)} + H'\psi^{(k-1)} = E_0\psi^{(k)} + E'\psi^{(k-1)} + \ldots + E^{(k)}\psi_0, \qquad (12.38)$$

and, taking the inner product with ψ_0 and using the orthogonality of ψ_0 and $\psi^{(j)}$, we obtain

$$\langle\psi_0|H_0\psi^{(k)}\rangle + \langle\psi_0|H'\psi^{(k-1)}\rangle = E_0\langle\psi_0|\psi^{(k)}\rangle + E^{(k)}\langle\psi_0|\psi_0\rangle. \qquad (12.39)$$

As in the first-order case the first two terms cancel (both vanish), leaving

$$\langle\psi_0|H'\psi^{(k-1)}\rangle = E^{(k)}\|\psi_0\|^2. \qquad (12.40)$$

Actually this is only one of many formulae for $E^{(k)}$ that can be obtained from the following useful result.

Lemma 12.4.1. For any $u \neq v$ we have

$$\langle\psi_u|H'\psi_v\rangle = \left(\frac{E_u - E_v}{u - v}\right)\langle\psi_u|\psi_v\rangle.$$

Proof. We first note that

$$\langle\psi_u|(H_u - H_v)\psi_v\rangle = \langle H_u\psi_u|\psi_v\rangle - \langle\psi_u|H_v\psi_v\rangle = (E_u - E_v)\langle\psi_u|\psi_v\rangle, \quad (12.41)$$

from which the result follows on dividing by $u - v$ and noting that $H_u - H_v = (u - v)H'$. □

By comparing coefficients of v^{k-1} we obtain our previous formula for E_k, but there are many other possibilities.

Corollary 12.4.2. The energy correction $E^{(2k+1)}$ depends only on the lower order energy corrections and the first k corrections to the wave function.

Proof. We first note that

$$\left(\frac{E_u - E_v}{u - v}\right) = \sum_{n=1}^{\infty} E^{(n)} \frac{u^n - v^n}{u - v} = \sum_{n=1}^{\infty} E^{(n)} \left(\sum_{j=0}^{n-1} u^j v^{n-1-j}\right). \quad (12.42)$$

Comparing coefficients of $u^k v^k$ in the lemma, we see that there is a term $E^{(2k+1)}\|\psi_0\|^2$, and all the other terms involve lower order energy corrections and the wave functions up to $\psi^{(k)}$. □

The uv coefficient, which after some simplification yields

$$\langle \psi' | H' \psi' \rangle = E''' \|\psi_0\|^2 + E' \|\psi'\|^2, \quad (12.43)$$

shows, in particular, that E''' depends only on ψ', and not on ψ'', as the earlier formula would have suggested.

Nonetheless, even with this improvement, we need ψ' in order to get beyond the first-order energy correction.

Theorem 12.4.3. Suppose that $\{\psi_1, \psi_2, \ldots\}$ is an orthonormal basis of eigenvectors for H_0, with corresponding eigenvalues $E_\alpha \neq E_0$, for $\alpha \neq 0$. Then we have

$$\psi' = \sum_{\alpha \neq 0} \frac{\langle \psi_\alpha | H' \psi_0 \rangle}{E_0 - E_\alpha} \psi_\alpha,$$

$$E'' = \sum_{\alpha \neq 0} \frac{|\langle \psi_\alpha | H' \psi_0 \rangle|^2}{E_0 - E_\alpha}.$$

Proof. The vector ψ' is orthogonal to ψ_0 and so its expansion with respect to the orthonormal basis takes the form

$$\psi' = \sum_{\alpha \neq 0} \langle \psi_\alpha | \psi' \rangle \psi_\alpha. \quad (12.44)$$

Now, the inner product of the first-order equation with ψ_α gives

$$\langle \psi_\alpha | H_0 \psi' \rangle - E_0 \langle \psi_\alpha | \psi' \rangle = E' \langle \psi_\alpha | \psi_0 \rangle - \langle \psi_\alpha | H' \psi_0 \rangle. \quad (12.45)$$

The first term on the right vanishes, because ψ_α and ψ_0 correspond to distinct eigenvalues and so are orthogonal. By the usual argument we also have

$$\langle\psi_\alpha|H_0\psi'\rangle = \langle H_0\psi_\alpha|\psi'\rangle = E_\alpha\langle\psi_\alpha|\psi'\rangle, \tag{12.46}$$

so that

$$(E_0 - E_\alpha)\langle\psi_\alpha|\psi'\rangle = \langle\psi_\alpha|H'\psi_0\rangle. \tag{12.47}$$

Substituting this into the earlier expansion we obtain

$$\psi' = \sum_{\alpha\neq 0}\frac{\langle\psi_\alpha|H'\psi_0\rangle}{E_0 - E_\alpha}\psi_\alpha, \tag{12.48}$$

as asserted, and substituting this into $E'' = \langle\psi_0|H'\psi'\rangle$ gives the formula for the second-order energy correction. □

12.5. The Berry phase

Rayleigh–Schrödinger perturbation theory is concerned with the behaviour of a parametrized family of Hamiltonians, $\{H_u\}$, for small changes in the parameter, u. Recently it has been realized that subtle changes can occur even when the Hamiltonian returns to its original form after a series of changes.

We need not restrict ourselves to a real parameter u as we did in perturbation theory, but rather take u to lie in a subset $\mathcal{U}$ of $\mathbf{R}^n$ for some $n \geq 1$. It will be useful to introduce a different normalization rule to that of perturbation theory, one more appropriate to large parameter changes.

Definition 12.5.1. Vectors will be normalized so that $\|\psi_0\| = 1$, and

$$\langle\psi_u|\frac{\partial\psi_u}{\partial u_j}\rangle = 0,$$

for $j = 1, 2, \ldots, n$. We shall often abbreviate this to

$$\langle\psi_u|\mathrm{grad}(\psi_u)\rangle = 0.$$

This normalization rule arises naturally for a system whose Hamiltonian is changing, but slowly enough that the state can adjust and always be in an eigenstate. (This is the adiabatic approximation.) This rule specifies that infinitesimal changes to ψ_u are orthogonal to ψ_u, whereas the previous rule ensured that all changes were orthogonal to ψ_0. We also have

$$\frac{\partial}{\partial u_j}\langle\psi_u|\psi_u\rangle = \langle\psi_u|\frac{\partial\psi_u}{\partial u_j}\rangle + \langle\frac{\partial\psi_u}{\partial u_j}|\psi_u\rangle = 0, \tag{12.49}$$

so that $\|\psi_u\|^2$ is a constant, and ψ_u can remain normalized for all u. However, the rule does more than just ensuring normalization: it also fixes the phase of ψ_u as the following result shows.

Proposition 12.5.1. Suppose that, for each $u \in \mathcal{U}$, E_u is non-degenerate, and let ϕ_u be a normalized E_u-eigenvector of H_u. Then ϕ_u can be expressed as $\lambda_u^{-1}\psi_u$ where

$$\lambda_u = \lambda_0 \exp\left(-\int_C \langle\phi_u|d\phi_u\rangle\right),$$

C is any curve in $\mathcal{U}$ that joins 0 to u, and

$$\langle\phi_u|d\phi_u\rangle = \sum_{j=1}^{n}\langle\phi_u|\frac{\partial\phi}{\partial u_j}\rangle\, du_j.$$

Proof. Since E_u is non-degenerate, ϕ_u must take the form $\lambda_u^{-1}\psi_u$, for some $\lambda_u \in \mathbf{C}\setminus\{0\}$. By the normalization rule

$$\begin{aligned} 0 &= \langle\lambda_u\phi_u|\frac{\partial\lambda}{\partial u_j}\phi_u + \lambda\frac{\partial\phi_u}{\partial u_j}\rangle \\ &= \overline{\lambda_u}\left(\frac{\partial\lambda}{\partial u_j}\|\phi_u\|^2 + \lambda\langle\phi_u|\frac{\partial\phi_u}{\partial u_j}\rangle\right). \end{aligned} \tag{12.50}$$

Since $\lambda \neq 0$ and $\|\phi_u\| = 1$ we have

$$\frac{\partial\lambda}{\partial u_j}/\lambda = -\langle\phi_u|\frac{\partial\phi_u}{\partial u_j}\rangle. \tag{12.51}$$

Integrating

$$\begin{aligned} \ln(\lambda_u) - \ln(\lambda_0) &= \int_C \sum_j \left(\frac{\partial\lambda}{\partial u_j}/\lambda\right) du_j \\ &= -\int_C \sum_j \langle\phi_u|\frac{\partial\phi_u}{\partial u_j}\rangle\, du_j \\ &= -\int_C \langle\phi_u|d\phi_u\rangle, \end{aligned} \tag{12.52}$$

whence the result follows by exponentiation. □

The case when we return to 0 along a closed curve C is especially interesting.

Corollary 12.5.2. If u traverses a closed curve C then the wave function is multiplied by the phase factor

$$\exp\left(-\int_C \langle\phi_u|d\phi_u\rangle\right).$$

Definition 12.5.2. The factor $\exp(-\int_C\langle\phi_u|d\phi_u\rangle)$ is called the *Berry phase factor* associated to the curve C.

If the closed curve C can be spanned by a surface $S \subseteq \mathcal{U}$, then Stokes' theorem allows us to express the factor as

$$\exp\left(-\int_S \langle\frac{\partial\phi}{\partial u_j}|\frac{\partial\phi}{\partial u_k}\rangle du_j du_k\right). \tag{12.53}$$

The general existence of this large scale phase factor for adiabatic systems was noted by Michael Berry in 1984. A phase factor of this kind occurs for many systems, and in particular for polarized photons in an optical fibre, where it was demonstrated experimentally a few years later by Tomita and Chiao.

12.6. The Bloch–Floquet theorems

Consider a particle moving on the x-axis under the influence of a potential, V, that is periodic of period a. (This might provide a simple model of a crystal in which the atoms are regularly spaced.) Let us introduce the family of Hamiltonians

$$H_u = -\frac{\hbar^2}{2m}\frac{d^2}{dx^2} + V(x - ua), \tag{12.54}$$

for $u \in [0,1]$.

Since V is periodic $H_1 = H_0$, and we have a closed curve in the space of Hamiltonians. If ϕ is an eigenfunction of H_0 with energy E then clearly $\phi_u(x) = \phi(x - ua)$ will be an eigenfunction of H_u with the same energy, since we have only changed variable.

By the chain rule

$$\begin{aligned}\frac{d}{du}\phi_u(x) &= -a\phi'(x - ua) \\ &= -\frac{ia}{\hbar}P\phi(x - au) \\ &= -\frac{ia}{\hbar}P\phi_u,\end{aligned} \tag{12.55}$$

so

$$\langle \phi_u | d\phi_u \rangle = -\frac{ia}{\hbar} du \langle \phi_u | P\phi_u \rangle = -\frac{ia}{\hbar}\mathsf{E}_{\phi_u}(P)du. \tag{12.56}$$

By changing variable the inner products with ϕ and ϕ_u are the same, so the expectation values for the two states are the same. On integrating from $u = 0$ to $u = 1$ we obtain the following result:

Theorem 12.6.1. (Floquet's theorem) The Berry phase factor for a particle moving in one dimension in the presence of a potential that is periodic of period a is

$$\exp\left(-\int_0^1 \langle \phi_u | d\phi_u \rangle\right) = \exp\left(\frac{ia}{\hbar}\mathsf{E}_\phi(P)\right).$$

This property of ordinary differential equations with periodic coefficients was discovered by Floquet, and the factor is known as the monodromy. One important consequence of the formula for the monodromy is that it is the same if $\mathsf{E}_\phi(P)$ is increased by an integer multiple of $2\pi\hbar/a$. This gives rise to a periodicity in momentum space.

Bloch discovered the corresponding result in quantum theory, and generalized it to three dimensions, where one has a crystal lattice spanned by three vectors $\mathbf{a}_1$, $\mathbf{a}_2$, and $\mathbf{a}_3$. We shall use square brackets denote triple scalar products, so that $[\mathbf{a}_1, \mathbf{a}_2, \mathbf{a}_3] = \mathbf{a}_1 . (\mathbf{a}_2 \times \mathbf{a}_3)$.

Theorem 12.6.2. (Bloch's theorem) Suppose that a particle moves in $\mathbf{R}^3$ in the presence of a potential, V, that satisfies

$$V(\mathbf{x}+\mathbf{a}_j) = V(\mathbf{x})$$

for $j = 1, 2, 3$. Translation through $\mathbf{n} = n_1\mathbf{a}_1 + n_2\mathbf{a}_2 + n_3\mathbf{a}_3$ multiplies the wave function by a factor

$$\exp\left(\frac{i}{\hbar}\mathsf{E}_\phi(\mathbf{n}.\mathbf{P})\right).$$

This time the factors are the same (for all $\mathbf{n}$) if the expectation of $\mathbf{P}$ is increased by a vector in the reciprocal lattice spanned by $\mathbf{a}_2 \times \mathbf{a}_3/[\mathbf{a}_1, \mathbf{a}_2, \mathbf{a}_3]$, $\mathbf{a}_3 \times \mathbf{a}_1/[\mathbf{a}_1, \mathbf{a}_2, \mathbf{a}_3]$, and $\mathbf{a}_1 \times \mathbf{a}_2/[\mathbf{a}_1, \mathbf{a}_2, \mathbf{a}_3]$.

The Bloch theorem is fundamental to much solid state physics, since many materials are crystalline, and so have periodic potentials. The periodicity in momentum space when the momentum is increased by a vector in the reciprocal lattice gives rise to a band structure in the permissible energies. When the bands are separated by wide gaps, considerable energy has to be expended to raise an electron from one band to another. If a band is full, this can make it difficult to accelerate its electrons to give a current, and the material behaves as an insulator. If a band is only partly full the electrons can easily be raised to higher energy states within the band, and a current is easily produced, so that one has a good conductor. Between these two extremes are some semi-conductors, which have bands which are very narrow or overlap.

12.7. Historical notes

Schrödinger was able to use perturbation theory, an idea borrowed from classical mechanics, to explain some of the features of Johannes Stark's observations of the spectrum of atoms in an electric field. Schrödinger added the potential $H' = \mathbf{F}.\mathbf{r}$ for a uniform electric field $\mathbf{F}$ to the hydrogen atom Hamiltonian H_0, and was able to show that the degenerate excited energy levels split. The change in the energy levels reflects the distortion of the atom caused by the field. It is this distortion which is responsible for the fact that the dielectric constant, ϵ, within matter differs from its value in empty space.

Mathematically, this example has to be treated with great caution. Actually, E.C. Titchmarsh showed that the Hamiltonian H has no eigenvectors, so that we can hardly expect to find them by perturbation theory. In fact, the series $E_0 + uE' + \ldots$ can be proved to be divergent. One should really expect this physically. Quantum mechanically the electron can tunnel through the potential barrier, and then the electric field will accelerate it away from the nucleus. This is just saying that the field can ionize the atom by snatching its electron. Truly bound states are thus impossible and so there are no energies to compute by perturbation theory or any other means. However, a detailed calculation shows that for realistic fields **F** this dissociation process takes a very long time. On ordinary timescales the wave functions are very nearly time independent, and behave as though they described bound states. It is these 'metastable states' whose approximate energy the perturbation theory evaluates, and the coefficients of the perturbation series encode important physical information.

Exercises

12.1 Prove formula (12.43) for E'''.

12.2° A quantum mechanical system has Hamiltonian

$$H = \frac{P^2}{2m} + \frac{1}{2}m\omega^2 X^2 + \lambda X^4,$$

with λ a small real parameter. Show that to first order in λ the energy eigenvalues calculated using perturbation theory are

$$\left(n + \tfrac{1}{2}\right)\hbar\omega + \left(2n^2 + 2n + 1\right)\frac{3\hbar^2\lambda}{4m^2\omega^2}.$$

[*Hint*: You may find it useful to use the creation and annihilation operators and the relation $[a_-, a_+^n] = na_+^{n-1}$, for $n = 1, \ldots, \infty$.]

12.3° A quantum mechanical system has Hamiltonian

$$H_0 = \frac{P^2}{2m} + \frac{1}{2}m\omega^2 X^2.$$

Calculate the second-order perturbation theoretic corrections to the energy levels for each of the following perturbations:

(i) $H' = \epsilon X$;
(ii) $H' = \epsilon P$;
(iii) $H' = \epsilon(PX + XP)$.

Where possible compare your results with the exact answers.

12.4° Find the eigenstates and energies for a particle of mass m which is confined to a two-dimensional square box $0 \le x \le a$, $0 \le y \le a$.

Comment on the degeneracy of the ground state and first excited state. If the system is now subjected to a small perturbation with potential energy $V = \epsilon xy$, find the energy change of the ground state and the first excited state to first order in ϵ. Construct the corresponding zero-order wave functions for the perturbed system for the first excited state.

12.5° In a model of the hydrogen atom that allows for the finite size of the nucleus, the electron moves in the potential

$$V(r) = \begin{cases} -Ze^2/4\pi\epsilon_0 r & r \geq R \\ -Ze^2/4\pi\epsilon_0 R & 0 \leq r \leq R. \end{cases}$$

If H_0 is chosen to be the Hamiltonian for the electron in a Coulomb field due to central charge Ze, find the first-order correction to the ground state energy, and show that for $R \ll a$ it is quadratic in R.

12.6° The Hamiltonian of a rigid rotator in a magnetic field perpendicular to the y-axis is of the form

$$AL^2 + BL_3 + CL_1.$$

If A and B are very much larger than C find the energy eigenstates correct to first order in C and the energy eigenvalues to second order. Compare your results with the exact answer.

12.7° The energy levels of a hydrogen atom with Hamiltonian H_0 are known to be E_n for each eigenstate ψ_{nlm}, where n is the principal quantum number and the labels l and m are associated with the total angular momentum and L_3. In the presence of a weak magnetic field the Hamiltonian becomes

$$H = H_0 - \frac{eB}{2\mu} L_3.$$

Describe the splitting of the first excited energy level, giving the degeneracies.

12.8 A particle of mass m moves in two dimensions under the influence of a potential $\frac{1}{2}m\omega^2((1+\epsilon)x^2+(1-\epsilon)y^2)$. What are the possible energy levels if $\epsilon = 0$? When ϵ does not vanish, use degenerate perturbation theory to calculate the corrections to the energy of the first excited state to order ϵ and the corresponding unperturbed states. Calculate the energy levels of the system directly and compare the exact answer with that given by perturbation theory.

[The normalized wave functions for the two lowest states of the harmonic oscillator whose Hamiltonian is $P^2/2m + \frac{1}{2}m\omega^2X^2$ are $\psi_0 = (\pi a^2)^{-\frac{1}{4}}\exp(-x^2/2a^2)$ and $\sqrt{2}(x/a)\psi_0(x)$ with energies $\frac{1}{2}\hbar\omega$ and $\frac{3}{2}\hbar\omega$ respectively, where $a^2 = \hbar/m\omega$.]

12.9° A particle of mass M moves in the xy-plane under the influence of a potential ϵU, where

$$U(x,y) = \begin{cases} xy & \text{for } 0 < x < \pi,\ 0 < y < \pi \\ +\infty & \text{otherwise} \end{cases}$$

and $\epsilon \ll \hbar^2/M$. Show that when $\epsilon = 0$ the lowest two eigenvalues are $\hbar^2/M$ and $5\hbar^2/2M$ and find the corresponding eigenvectors. What are the degeneracies in each case?

Use time-independent perturbation theory to show that, correct to the first order in ϵ, the lowest three energy levels are

$$\frac{\hbar^2}{M} + \epsilon a^2, \qquad \frac{5\hbar^2}{2M} + \epsilon(a^2 - b^2), \qquad \frac{5\hbar^2}{2M} + \epsilon(a^2 + b^2),$$

where $a = \pi/2$ and $b = 16/9\pi$.

12.10° For H_0 and H' self-adjoint operators and $\lambda \in \mathbf{R}$ a family of Hamiltonians is defined by

$$H_\lambda = H_0 + \lambda H'.$$

Show that, with ψ_0 normalized, the second-order correction can be written as

$$E'' = \langle (H' - E')\psi_0 | \psi' \rangle.$$

A particle of mass m moves in the interval $[0, a]$ under the influence of a potential $V(x) = \lambda\cos(N\pi x/a)$ with N a positive integer. Given that the energy is approximately $\hbar^2k^2\pi^2/2ma^2$, find the correction to the energy to first order in λ and show that it vanishes for all but one value of N. Find the second-order correction to the energy level in those cases where the first-order correction vanishes.
[It may be assumed that the normalized wave functions for a particle moving freely in the interval $[0, a]$ are given for $k = 1, 2, \ldots$ by

$$\phi_k(x) = \sqrt{\frac{2}{a}}\sin\frac{k\pi x}{a}.]$$

12.11° A small additional term H' is added to the Hamiltonian H_0. Find an expression for the first-order correction to the wave function when

$\phi = H'\psi$ is an eigenfunction of H_0 with energy $\mathcal{E} \neq E_0$, and show that to second order the energy is

$$\mathcal{E} + \frac{\lambda^2 \|\phi\|^2}{E_0 - \mathcal{E}}.$$

By choosing $H_0 = P^2/2m + m\omega^2 X^2/2 + \kappa$ for suitable ω and κ, show that the ground state energy of

$$H = \frac{P^2}{2m} + \frac{1}{2}m\alpha^2 X^2 + \lambda X^4$$

is approximately

$$\frac{1}{2}\hbar\omega\left[1 - \frac{3}{2}\frac{\lambda\hbar}{m^2\omega^3} - \left(\frac{\lambda\hbar}{m^2\omega^3}\right)^2\right],$$

where $\omega^3 - \alpha^2\omega - 6\lambda\hbar/m^2 = 0$.
[The normalized wave function for the third excited state of $H_0 = P^2/2m + m\omega^2 X^2/2$ is

$$\frac{1}{\sqrt{24}}\left[4\left(\frac{m\omega x^2}{\hbar}\right)^2 - 12\left(\frac{m\omega x^2}{\hbar}\right) + 3\right]\psi_0,$$

where ψ_0 is the ground state wave function.]

13* Iterative perturbation theory

Even perturbation theory is no more complicated than the forced vibrations of a string.

ERWIN SCHRÖDINGER, letter to Willy Wien, 22 February 1926

13.1. The Brillouin–Wigner iteration

The Rayleigh–Schrödinger theory described in the preceding chapter is by no means the only approach to stationary perturbation theory, nor is it usually the most powerful. As one of many possible alternatives we shall now describe an iterative procedure for approximating stationary states, sometimes called Wigner–Brillouin perturbation theory, which resembles the Feynman–Dyson method of Section 11.3.

It is sensible to approximate ψ by the closest appropriate eigenvector of H_0, that is by its projection onto the E_0-eigenspace. We therefore suppose that $\psi_0 = P_0\psi$ where P_0 is the projection operator onto the space of all H_0-eigenvectors with eigenvalue E_0. (Clearly we require that $P_0\psi \neq 0$, otherwise we should have chosen a different energy E_0.) It will also be useful to introduce the complementary projection $Q = 1 - P_0$ onto $\ker(E_0 - H_0)^{\perp}$.

Lemma 13.1.1. If ψ is normalized so that $\psi_0 = P_0\psi$ is a unit vector then the eigenstates for H and H_0 are related by the equations

$$\psi = \psi_0 + (E_0 - H_0)^{-1}Q(H' + E_0 - E)\psi.$$

Proof. We know that

$$\begin{aligned}(H' + E_0 - E)\psi &= (H - H_0 + E_0 - E)\psi \\ &= (H - E)\psi + (E_0 - H_0)\psi \\ &= (E_0 - H_0)\psi. \qquad (13.1)\end{aligned}$$

By definition $(H_0 - E_0)P_0$ vanishes, and, taking its adjoint, so does $P_0(H_0 - E_0)$. We therefore see that $P_0(H' + E_0 - E)\psi = 0$, and consequently that

$$(H' + E_0 - E)\psi = Q(H' + E_0 - E)\psi. \qquad (13.2)$$

Since E_0 is not an eigenvalue of H_0 on the image of Q, the inverse of $(E_0 - H_0)$ is well defined there and we deduce that

$$Q\psi = (E_0 - H_0)^{-1} Q(H' + E_0 - E)\psi. \tag{13.3}$$

The formula for ψ now follows on substituting this into the identity

$$\psi = (P_0 + Q)\psi = \psi_0 + Q\psi. \qquad \square$$

Remark 13.1.1. When E_0 is non-degenerate, the formula for ψ can be given more explicitly by choosing an orthonormal basis $\{\psi_0, \phi_1, \phi_2, \phi_3, \ldots\}$ consisting of eigenvectors of H_0 satisfying $H_0\phi_\alpha = \epsilon_\alpha \phi_\alpha$ for $\alpha = 1, 2, \ldots$. Then, writing Q in terms of the orthonormal basis of eigenvectors we have

$$\begin{aligned}
\psi &= \psi_0 + (E - H_0)^{-1} QH'\psi \\
&= \psi_0 + (E - H_0)^{-1} \sum_{\alpha=1}^{\infty} \langle \phi_\alpha | H'\psi \rangle \phi_\alpha \\
&= \psi_0 + \sum_{\alpha=1}^{\infty} \langle \phi_\alpha | H'\psi \rangle \, (E - \epsilon_\alpha)^{-1} \, \phi_\alpha,
\end{aligned} \tag{13.4}$$

which is reminiscent of the formula (12.48) for the first-order Rayleigh–Schrödinger wave function.

The above lemma, coupled with the fact that the energy, E, is the expectation of the Hamiltonian, suggests the following iterative scheme for approximating eigenvalues and eigenvectors.

Definition 13.1.1. Let E_0 and ψ_0 be as above, and for $n = 0, 1, \ldots$ define the Brillouin–Wigner approximation

$$\begin{aligned}
E_{n+1} &= \langle \psi_n | H\psi_n \rangle / \|\psi_n\|^2 \\
\psi_{n+1} &= \psi_0 + (E_0 - H_0)^{-1} Q(H' + E_0 - E_{n+1})\psi_n.
\end{aligned}$$

It is reasonable to hope that if one has chosen H_0, E_0, and ψ_0 wisely then these sequences will converge to the true values E and ψ.

Lemma 13.1.2. The approximate wave functions satisfy the normalization condition

$$\langle \psi_0 | \psi_n \rangle = 1$$

for all positive integers n. The first-order approximation to the energy is given by

$$E_1 = E_0 + \langle \psi_0 | H' \psi_0 \rangle.$$

Proof. Since Q commutes with H_0 and ψ_0 is orthogonal to the range of Q, we have

$$\begin{aligned}\langle \psi_0 | \psi_{n+1} \rangle &= \langle \psi_0 | \psi_0 \rangle + \langle \psi_0 | Q(E_0 - H_0)^{-1} (H' + E_0 - E_{n+1}) \psi_n \rangle \\ &= \langle \psi_0 | \psi_0 \rangle \\ &= 1. \end{aligned} \tag{13.5}$$

(It is similarly possible to show that $P_0 \psi_n = \psi_0$ for all n.) The definition also gives

$$\begin{aligned} E_1 &= \langle \psi_0 | H \psi_0 \rangle \\ &= \langle \psi_0 | H_0 \psi_0 \rangle + \langle \psi_0 | H' \psi_0 \rangle \\ &= E_0 + \langle \psi_0 | H' \psi_0 \rangle, \end{aligned} \tag{13.6}$$

which completes the proof. □

It is worth remarking that

$$\|\psi_n - \psi_0\|^2 = \|\psi_n\|^2 - \langle \psi_n | \psi_0 \rangle - \langle \psi_0 | \psi_n \rangle + \|\psi_0\|^2 = \|\psi_n\|^2 - 1. \tag{13.7}$$

13.2. Convergence of the iteration scheme

When looking for higher order approximations it is often useful to use slightly different formulae.

Theorem 13.2.1. Let E_0 be a non-degenerate energy level, and set $\delta_n = \psi_n - \psi_{n-1}$. Then the approximate wave function ψ_n satisfies the equation

$$(E_0 - H_0)\delta_n = (H - E_n)\psi_{n-1} - \langle\psi_0|(H - E_n)\psi_{n-1}\rangle\psi_0$$

and, in particular, $(E_0 - H_0)\psi_1 = (H' + E_0 - E_1)\psi_0$. The energy difference between successive approximations is given by

$$(E_{n+1} - E_n)\,\|\psi_n\|^2 = \langle\delta_n|\,[(H - E_n) + 2\,(E_0 - H_0)]\,\delta_n\rangle.$$

Proof. Applying $E_0 - H_0$ to the recursive definition 13.1.1, and using the fact that $(E_0 - H_0)\psi_n$ vanishes, gives

$$(E_0 - H_0)\psi_n = Q(H' + E_0 - E_n)\psi_{n-1}. \tag{13.8}$$

Similarly, the first term in

$$(E_0 - H_0)\psi_{n-1} = P_0(E_0 - H_0)\psi_{n-1} + Q(E_0 - H_0)\psi_{n-1} \tag{13.9}$$

vanishes, so that

$$\begin{aligned}(E_0 - H_0)\delta_n &= Q(H' + E_0 - E_n)\psi_{n-1} - Q(E_0 - H_0)\psi_{n-1}\\ &= Q(H - E_n)\psi_{n-1}.\end{aligned} \tag{13.10}$$

Since Q is the complement of the projection onto ψ_0, we have, for any ϕ, $Q\phi = (1 - P_0)\phi = \phi - \langle\psi_0|\phi\rangle\psi_0$, from which the first formula follows. When $n = 1$ there is a simplification since, by definition, $\langle\psi_0|(H - E_1)\psi_0\rangle = 0$. This gives

$$(E_0 - H_0)\psi_1 = (E_0 - H_0)\psi_0 + (H - E_1)\psi_0, \tag{13.11}$$

which reduces to the stated formula.

To obtain the formula for the energy difference, we start with the expression

$$(E_{n+1} - E_n)\,\|\psi_n\|^2 = \langle\psi_n|(H - E_n)\psi_n\rangle, \tag{13.12}$$

and substitute $\psi_n = \psi_{n-1} + \delta_n$. Since $\langle\psi_{n-1}|(H - E_n)\psi_{n-1}\rangle$ vanishes by definition of E_n, this gives

$$\langle\delta_n|(H - E_n)\delta_n\rangle + \langle\psi_{n-1}|(H - E_n)\delta_n\rangle + \langle\delta_n|(H - E_n)\psi_{n-1}\rangle. \tag{13.13}$$

The middle terms may also be simplified since

$$\langle\delta_n|(H - E_n)\psi_{n-1}\rangle = \langle\delta_n|(E_0 - H_0)\delta_n\rangle + \langle\delta_n|\psi_0\rangle\langle\psi_0|(H - E)\psi_{n-1}\rangle, \tag{13.14}$$

and $\langle \delta_n | \psi_0 \rangle$ vanishes by Lemma 13.1.2. Collecting the non-vanishing terms we have

$$\langle \delta_n | (H - E_n)\delta_n \rangle + 2\langle \delta_n | (E_0 - H_0)\delta_n \rangle, \tag{13.15}$$

which gives the stated formula. □

Theorem 13.2.2. If the Brillouin–Wigner approximants E_n and ψ_n converge then their limits E and ψ satisfy $H\psi = E\psi$.

Proof. If we assume that E_n and ψ_n converge then, taking the limit of Theorem 13.2.1 as $n \to \infty$, their limits E and ψ will satisfy

$$(E_0 - H_0)(\psi - \psi) = (H - E)\psi - \langle \psi_0 | (H - E)\psi \rangle \psi_0, \tag{13.16}$$

so that

$$(H - E)\psi = \langle \psi_0 | (H - E)\psi \rangle \psi_0. \tag{13.17}$$

Taking the inner product with ψ, and recalling the normalization of Lemma 13.1.2, gives

$$\langle \psi | (H - E)\psi \rangle = \langle \psi_0 | (H - E)\psi \rangle. \tag{13.18}$$

Now the left-hand side is the limit of $\langle \psi_n | (H - E_{n+1})\psi_n \rangle$, and therefore vanishes, so that equation (13.17) reduces to $(H - E)\psi = 0$, as required. □

Theorem 13.2.3. Let E_n and ψ_n be the Brillouin–Wigner approximations, and assume that E and ψ are a true eigenvalue and eigenvector. Then

$$\begin{aligned} E - E_{n+1} &= \langle \psi - \psi_n | (E - H)(\psi - \psi_n) \rangle / \|\psi_n\|^2 \\ \psi - \psi_{n+1} &= (E_0 - H_0)^{-1} Q \left[(H' + E_0 - E_{n+1})(\psi - \psi_n) \right. \\ &\qquad \left. + (E_{n+1} - E)\,\psi \right]. \end{aligned}$$

Proof. We start by setting $\psi_n = \psi + \delta$ in the formula

$$(E_{n+1} - E)\|\psi_n\|^2 = \langle \psi_n | (H - E)\psi_n \rangle, \tag{13.19}$$

to get

$$\langle\delta|(H-E)\delta\rangle + \langle\psi|(H-E)\delta\rangle + \langle\delta|(H-E)\psi\rangle + \langle\psi|(H-E)\psi\rangle. \quad (13.20)$$

Since $H\psi = E\psi$ the last three terms vanish, leaving the stated formula.

For the wave function we may use Lemma 13.1.1 together with the recursive definition 13.1.1 to obtain

$$\begin{aligned}\psi - \psi_{n+1} &= (E_0 - H_0)^{-1}Q\,(H' + E_0 - E)\psi \\ &\quad - (E_0 - H_0)^{-1}Q\,(H' + E_0 - E_{n+1})\psi_n \\ &= (E_0 - H_0)^{-1}Q\,[(H' + E_0 - E_{n+1})\,(\psi - \psi_n) \\ &\quad + (E_{n+1} - E)\psi], \end{aligned} \quad (13.21)$$

as stated. □

The first inequality shows that if the wave function converges then the energy converges quadratically. This is what makes this scheme rather faster than some of the apparently simpler alternatives. In general, it can be more powerful than the Rayleigh–Schrödinger theory, as well as avoiding the assumption that E_u and ψ_u have power series expansions.

Theorem 13.2.4. Let ψ_n and E_n be the Brillouin–Wigner approximations to the wave function and energy for $H_u = H_0 + uH'$, and suppose that the true energy E and wave function ψ depend analytically on u. Then, under the assumptions of Theorem 13.2.3, ψ_n is accurate to order u^n, and E_n is accurate to order u^{2n-1}.

Proof. Since $\psi - \psi_0$ vanishes when $u = 0$ it is divisible by u. We can now prove by induction that $\psi - \psi_n$ is divisible by u^{n+1} and $E - E_n$ by u^{2n}, using

$$\psi - \psi_n = (E_0 - H_0)^{-1}Q\,[(uH' + E_0 - E_n)\,(\psi - \psi_{n-1}) + (E_n - E)\psi], \quad (13.22)$$

and also

$$E - E_n = \langle\psi - \psi_{n-1}|(E - H)(\psi - \psi_{n-1})\rangle / \|\psi_{n-1}\|^2 \quad (13.23)$$

from Theorem 13.2.3. This means that ψ_n is already correct to order u^n and E_n is correct to order u^{2n-1}. □

13.3. The Dalgarno–Lewis method

Impressive though the formula for ψ' in Theorem 12.4.3 looks, it is usually impractical to calculate an infinite series of inner products. Of course, one may be lucky and find that most of the terms vanish, so that the series collapses down to just a few terms, but an alternative approach to finding a first-order wave function, suggested by A. Dalgarno and J.T. Lewis, is often easier.

Since the vector ψ_n should be close to ψ_0 we try $\psi_n = (1+\Phi_n)\psi_0$ where Φ_n is an operator that remains to be determined. (It will be useful to make the obvious convention that Φ_0 vanishes.)

By Theorem 13.2.1 the approximate wave functions satisfy

$$\begin{aligned} Q\,(H' + E_0 - E_{n+1})\,(1+\Phi_n)\psi_0 &= (E_0 - H_0)(1+\Phi_{n+1})\psi_0 \\ &= (1+\Phi_{n+1})H_0\psi_0 - H_0(1+\Phi_{n+1})\psi_0 \\ &= [1+\Phi_{n+1}, H_0]\psi_0 \\ &= [\Phi_{n+1}, H_0]\psi_0. \end{aligned} \tag{13.24}$$

We have thus proved the following result:

Theorem 13.3.1. The sequence of vectors $\psi_n = (1+\Phi_n)\psi_0$ defined by $\Phi_0 = 0$ and

$$[\Phi_{n+1}, H_0]\psi_0 = Q\,(H' + E_0 - E_{n+1})\,(1+\Phi_n)\psi_0$$

satisfies the Brillouin–Wigner equations for the approximate eigenstate. In particular, the first-order approximation $\psi_1 = (1+\Phi_1)\psi_0$ can be obtained by solving the equation

$$(H' + E_0 - E_1)\psi_0 = [\Phi_1, H_0]\psi_0.$$

At this point it is useful to recall that ψ and ψ_0 are wave functions and that a plausible form for the operator Φ_n would be multiplication by a function ϕ_n, that is

$$(\Phi_n\psi_0)\,(\mathbf{x}) = \phi_n(\mathbf{x})\psi_0(\mathbf{x}). \tag{13.25}$$

Moreover, in practice we usually start with

$$H_0 = -\frac{\hbar^2}{2m}\nabla^2 + V. \tag{13.26}$$

Then multiplication by V commutes with multiplication by ϕ_n so that

$$\begin{aligned}[H_0, \Phi_n]\psi_0 &= -\frac{\hbar^2}{2m}[\nabla^2, \phi_n]\psi_0 \\ &= -\frac{\hbar^2}{2m}\left[\nabla^2(\phi_n\psi_0) - \phi_n\nabla^2\psi_0\right] \\ &= -\frac{\hbar^2}{2m}\left[(\nabla^2\phi_n)\psi_0 + 2\nabla\phi_n.\nabla\psi_0\right]. \end{aligned} \tag{13.27}$$

Thus we arrive at a differential equation for ϕ_{n+1}:

$$\frac{\hbar^2}{2m}\left[(\nabla^2\phi_{n+1})\psi_0 + 2\nabla\phi_{n+1}.\nabla\psi_0\right] = (H' + E_0 - E_{n+1})\,(1+\phi_n)\psi_0. \tag{13.28}$$

This can be further refined into the following theorem:

Theorem 13.3.2. When $H_0 = -\hbar^2\nabla^2/2m + V$ and H' is a multiplication operator, Φ_n can be expressed as multiplication by the solution ϕ_n of the differential equation

$$\frac{\hbar^2}{2m}\mathrm{div}(\psi_0^2\,\mathrm{grad}\,\phi_{n+1}) = (H' + E_0 - E_{n+1})\,(1+\phi_n)\psi_0^2.$$

Proof. We multiply equation (13.28) by ψ_0 and simplify to get the stated result. □

Corollary 13.3.3. In one dimension, when $H_0 = P^2/2m + V$ and H' is a multiplication operator then Φ_n can be expressed as multiplication by the function

$$\phi_{n+1}(x) = \frac{2m}{\hbar^2}\int \psi_0^{-2}\int (H' + E_0 - E_{n+1})\,(1+\phi_n)\psi_0^2.$$

Proof. The one-dimensional version of the theorem gives ϕ_{n+1} as the solution of the differential equation

$$\frac{\hbar^2}{2m}\left(\psi_0^2\,\phi'_{n+1}\right)' = (H' + E_0 - E_{n+1})\,(1+\phi_n)\psi_0^2, \tag{13.29}$$

which can be integrated to give the result. □

13.4. Example

As an example of these procedures we consider a one-dimensional oscillator in a uniform field F so that the potential is $V = \frac{1}{2}m\omega^2 x^2 + Fx$. This can be solved directly by noting that $V = \frac{1}{2}m\omega^2(x + F/m\omega^2)^2 - F^2/2m\omega^2$. By changing the origin we see that the energy levels are just $(n+\frac{1}{2})\hbar\omega - F^2/2m\omega^2$. For the purposes of illustration, however, we shall regard the term $H' = Fx$ as a perturbation of the usual oscillator Hamiltonian, H_0, and investigate what happens to the ground state. The first approximation to the energy is given by

$$\begin{aligned} E_1 &= \langle\psi_0|H\psi_0\rangle \\ &= \langle\psi_0|H_0\psi_0\rangle + F\langle\psi_0|X\psi_0\rangle \\ &= \tfrac{1}{2}\hbar\omega + F\int_{\mathbf{R}} x|\psi_0(x)|^2\,dx. \end{aligned} \tag{13.30}$$

Since the integrand in the last term is odd the integral vanishes, showing that $E_1 = E_0$. Applying the Dalgarno–Lewis method to find the first-order wave function leads us to the equation

$$\frac{\hbar^2}{2m}\left(\psi_0^2\phi_1'\right)' = (H' + E_0 - E_1)\,\psi_0^2 = Fx\psi_0^2. \tag{13.31}$$

Substituting $\psi_0(x) = N\exp(-m\omega x^2/2\hbar)$ and simplifying, we obtain

$$\phi_1'' - 2\frac{m\omega x}{\hbar}\phi_1' = \frac{2mF}{\hbar^2}x, \tag{13.32}$$

which has the obvious solution $\phi_1 = -(F/\hbar\omega)x$. The first-order wave function can therefore be written in terms of the ground and first excited state wave functions ψ_0 and φ_1 as

$$\begin{aligned} \psi_1 &= (1+\phi_1)\psi_0 \\ &= \psi_0 - \frac{F}{\hbar\omega}x\psi_0 \\ &= \psi_0 - \frac{F}{\hbar\omega}\sqrt{\frac{\hbar}{2m\omega}}\varphi_1. \end{aligned} \tag{13.33}$$

To evaluate the second Wigner–Brillouin approximation to the energy using Theorem 13.2.1 we need

$$\delta_1 = \psi_1 - \psi_0 = -\frac{F}{\hbar\omega}\sqrt{\frac{\hbar}{2m\omega}}\varphi_1. \tag{13.34}$$

Since $E_1 = E_0$ and $H - H_0 = FX$, this means that

$$\langle \delta_1 | \left[(H - E) + 2(E_0 - H_0) \right] \delta_1 \rangle = \frac{F^2}{2m\hbar\omega^3} \langle \varphi_1 | \left[FX + (E_0 - H_0) \right] \varphi_1 \rangle . \tag{13.35}$$

Now $\langle \varphi_1 | FX \varphi_1 \rangle$ vanishes, because the integrand is odd, and $(H_0 - E_0)\varphi_1 = \hbar\omega\varphi_1$, so this expression reduces to give

$$(E_2 - E_1) \, \| \psi_1 \|^2 = -\frac{F^2}{2m\hbar\omega^3} \hbar\omega . \tag{13.36}$$

Combining this with $\| \psi_1 \|^2 = 1 + F^2/m\hbar\omega^3$, we obtain

$$\begin{aligned} E_2 &= \tfrac{1}{2}\hbar\omega \left[1 - \left(1 + \frac{F^2}{m\hbar\omega^3} \right)^{-1} \frac{F^2}{m\hbar\omega^3} \right] \\ &= \tfrac{1}{2}\hbar\omega \left(1 + \frac{F^2}{m\hbar\omega^3} \right)^{-1} . \end{aligned} \tag{13.37}$$

This agrees with the exact solution up to terms in F^3.

13.5. The Born approximation

In Chapter 5 we investigated what happens when a particle moving in one dimension encounters a potential barrier. The freedom of real particles to move in three dimensions means that their behaviour can be much more complicated. Consider what happens in the presence of a 'short range force' for which the potential tends to zero at large distances. Ideally each particle starts so far from the region of interaction that the potential is virtually zero and the particle effectively free. It is fired with known energy towards the area where the potential is large and emerges from its encounter in a new direction, eventually arriving once more into the 'asymptotic region' where the potential is nearly zero and particles move almost freely.

This suggests that, as in perturbation theory, one must compare motion under two different Hamiltonians. In the idealized asymptotic region where the particle starts and finishes its journey, it moves freely with a Hamiltonian $H_0 = -\hbar^2\nabla^2/2m$. During the encounter the Hamiltonian is

$$H = -\frac{\hbar^2}{2m}\nabla^2 + V, \tag{13.38}$$

so that in this case the whole of the potential energy is regarded as a perturbation. This time we are not interested in calculating the energy E

of the particle, since that is under the control of the experimenter and can be assumed to be known. We do, however, want to know the form of the scattered wave function, and for this the Brillouin–Wigner formula,

$$\begin{aligned}\psi &= \psi_0 + (E - H_0)^{-1}Q(H - H_0)\psi \\ &= \psi_0 + (E - H_0)^{-1}QV\psi,\end{aligned} \tag{13.39}$$

is invaluable. Here we interpret ψ_0 as the wave function appropriate to the incoming particle as it starts on its journey.

In these scattering experiments the particle has positive energy, for it must be able to escape from the potential. We shall therefore write $E = \hbar^2k^2/2m$. We take $\psi_0 = A\exp(i\mathbf{k}.\mathbf{r})$ as the wave function appropriate to the incident particle. Recalling that $H_0 = -\hbar^2\nabla^2/2m$, we see that the main problem is to invert $(k^2 + \nabla^2)$. In effect we must solve

$$(\nabla^2 + k^2)f(\mathbf{r}) = \rho(\mathbf{r}). \tag{13.40}$$

If $k = 0$ then this is the Poisson equation, and there is a generalization of the Poisson integral formula that solves this problem for functions ρ that decay fast enough at ∞:

$$f(\mathbf{r}) = -\frac{1}{4\pi}\int_{\mathbf{R}^3} \frac{e^{ik|\mathbf{r}-\mathbf{r}'|}}{|\mathbf{r}-\mathbf{r}'|}\rho(\mathbf{r}')\,d^3\mathbf{r}'. \tag{13.41}$$

This can be proved by the same methods as the Poisson formula once one has noticed that $g(r) = \exp(ikr)/r$ is a rotationally invariant solution of the equation $(\nabla^2+k^2)g = 0$. (Since g is spherically symmetric the equation reduces to

$$\frac{\partial^2(rg)}{\partial r^2} + k^2(rg) = 0, \tag{13.42}$$

from which it is clear that $rg = \exp(ikr)$ is a solution.)

The projection Q is not needed in this case since it serves only to project onto the subspace where the integral formula makes sense. The Brillouin–Wigner formula therefore becomes the explicit identity

$$\psi(\mathbf{r}) = Ae^{i\mathbf{k}.\mathbf{r}} - \frac{2m}{4\pi\hbar^2}\int_{\mathbf{R}^3} \frac{e^{ik|\mathbf{r}-\mathbf{r}'|}}{|\mathbf{r}-\mathbf{r}'|}V(\mathbf{r}')\psi(\mathbf{r}')\,d^3\mathbf{r}'. \tag{13.43}$$

Definition 13.5.1. The *Born approximation* consists of taking the successive Brillouin–Wigner approximations to the solution of this problem:

$$\psi_n(\mathbf{r}) = Ae^{i\mathbf{k}.\mathbf{r}} - \frac{2m}{4\pi\hbar^2}\int_{\mathbf{R}^3} \frac{e^{ik|\mathbf{r}-\mathbf{r}'|}}{|\mathbf{r}-\mathbf{r}'|}V(\mathbf{r}')\psi_{n-1}(\mathbf{r}')\,d^3\mathbf{r}'.$$

Example 13.5.1. For example, the first-order Born approximation is

$$\begin{aligned}\psi_1(\mathbf{r}) &= Ae^{i\mathbf{k}.\mathbf{r}} - \frac{m}{2\pi\hbar^2}\int_{\mathbf{R}^3}\frac{e^{ik|\mathbf{r}-\mathbf{r}'|}}{|\mathbf{r}-\mathbf{r}'|}V(\mathbf{r}')\psi_0(\mathbf{r}')\,d^3\mathbf{r}' \\ &= \left(Ae^{i\mathbf{k}.\mathbf{r}} - \frac{mA}{2\pi\hbar^2}\int_{\mathbf{R}^3}\frac{e^{ik|\mathbf{r}-\mathbf{r}'|}}{|\mathbf{r}-\mathbf{r}'|}V(\mathbf{r}')e^{i\mathbf{k}.\mathbf{r}'}\,d^3\mathbf{r}'\right) \\ &= \left(1 - \frac{m}{2\pi\hbar^2}\int_{\mathbf{R}^3}\frac{e^{ik|\mathbf{r}-\mathbf{r}'|-i\mathbf{k}.(\mathbf{r}-\mathbf{r}')}}{|\mathbf{r}-\mathbf{r}'|}V(\mathbf{r}')\,d^3\mathbf{r}'\right)Ae^{i\mathbf{k}.\mathbf{r}}.\end{aligned} \tag{13.44}$$

We would expect this to be a reasonable approximation provided that

$$\frac{m}{2\pi\hbar^2}\left|\int_{\mathbf{R}^3}\frac{e^{ik|\mathbf{r}-\mathbf{r}'|-i\mathbf{k}.(\mathbf{r}-\mathbf{r}')}}{|\mathbf{r}-\mathbf{r}'|}V(\mathbf{r}')\,d^3\mathbf{r}'\right| \ll 1. \tag{13.45}$$

When the particle has again escaped to large distances r we have $|\mathbf{r}-\mathbf{r}'| \sim r - \mathbf{u}.\mathbf{r}'$, where $\mathbf{u} = \mathbf{r}/r$, and so, dropping terms of order r^{-2},

$$\begin{aligned}\psi_1(\mathbf{r}) &\sim \psi_0(\mathbf{r}) - \frac{mA}{2\pi\hbar^2}\int_{\mathbf{R}^3}\frac{e^{ikr}}{r}e^{-i(k\mathbf{u}-\mathbf{k}).\mathbf{r}'}V(\mathbf{r}')\,d^3\mathbf{r}' \\ &= \psi_0(\mathbf{r}) - \frac{mA}{2\pi\hbar^2}\frac{e^{ikr}}{r}\int_{\mathbf{R}^3}e^{i(\mathbf{k}-k\mathbf{u}).\mathbf{r}'}V(\mathbf{r}')\,d^3\mathbf{r}'.\end{aligned} \tag{13.46}$$

This is the sum of the incident wave ψ_0 and a scattered wave, which is a multiple of $\exp(ikr)/r$. For a given direction $\mathbf{u} = \mathbf{r}/r$ the scattering is governed by the coefficient

$$B(\mathbf{u}) = \frac{m}{2\pi\hbar^2}\int_{\mathbf{R}^3}e^{i(\mathbf{k}-k\mathbf{u}).\mathbf{r}'}V(\mathbf{r}')\,d^3\mathbf{r}'. \tag{13.47}$$

The current density of the scattered wave (Definition 2.5.1) is

$$\begin{aligned}\mathbf{j}_s &= \frac{\hbar|AB|^2}{2mi}\left[\frac{e^{-ikr}}{r}\left(\frac{ike^{ikr}}{r} - \frac{e^{ikr}}{r^2}\right) - \frac{e^{ikr}}{r}\left(\frac{-ike^{-ikr}}{r} - \frac{e^{-ikr}}{r^2}\right)\right]\mathbf{u} \\ &= \frac{\hbar k|A|^2|B(\mathbf{u})|^2}{mr^2}\mathbf{u},\end{aligned} \tag{13.48}$$

and its flux through an element $r^2 d\Omega$ of the surface of a sphere of radius r is therefore

$$\frac{\hbar k|A|^2}{m}|B(\mathbf{u})|^2 d\Omega. \tag{13.49}$$

This is just $|B|^2 d\Omega$ times the magnitude of the current density of the incoming wave, $\mathbf{j}_0 = (\hbar |A|^2/m)\mathbf{k}$, and leads us to make the following definition:

Definition 13.5.2. The differential

$$d\sigma = |B(\mathbf{u})|^2 d\Omega$$

is called the *differential scattering cross-section.*

Exercises

13.1 A harmonic oscillator with Hamiltonian

$$H_0 = \frac{P^2}{2m} + \frac{1}{2}m\omega^2 X^2$$

is perturbed by the addition of the term $H' = \lambda m X^2/2$. Show that the first-order correction to the ground state energy is $\lambda\hbar/4\omega$ and that the first-order correction to the wave function is a multiple of the second excited state wave function.

Calculate the second-order energy approximation in Rayleigh–Schrödinger and in Brillouin–Wigner perturbation theory, and compare your answers to the exact ground state energy.

13.2° A hydrogen atom in its ground state is acted on by a uniform weak electric field $\mathbf{F}$ of magnitude F. Assume that the Hamiltonian of the system has an eigenvector ψ that, referred to spherical polar coordinates (r, θ, ϕ) with Oz in the direction of $\mathbf{F}$, is given by

$$\psi(r, \theta, \phi) = \psi_0(r)[1 - F\cos\theta R(r)]$$

where ψ_0 is the ground state eigenvector of the hydrogen atom Hamiltonian. Prove that $R(r)$ satisfies the differential equation

$$\frac{d^2 r}{dr^2} + 2\left(\frac{1}{r} - \frac{1}{a}\right)\frac{dR}{dr} - \frac{2R}{r^2} = -\frac{2r}{ea}$$

where $a = \hbar^2/me^2$, m is the electron mass, and $-e$ is the electron charge. Verify that

$$R(r) = \frac{a^2}{e}\left[\frac{r}{a} + \frac{1}{2}\left(\frac{r}{a}\right)^2\right]$$

is a solution satisfying the necessary boundary conditions. Calculate the energy of the atom in the state described by ψ to second order in F, and deduce that the polarizability of the atom, for weak electric fields, is $9a^3/2$.

13.3° Show that the first Born approximation to the total cross-section for particles of momentum $\hbar k$ scattered by the Yukawa potential, $V = C\exp(-r/a)/r$, is

$$\frac{4m^2C^2a^4}{\hbar^4\left[1 + 4k^2a^2\sin^2(\theta/2)\right]^2}.$$

By taking the limit as $a \to \infty$, deduce Rutherford's formula for the differential scattering cross-section due to a Coulomb potential, C/r,

$$\frac{m^2C^2}{4\hbar^4k^4}\operatorname{cosec}^4(\theta/2).$$

13.4° Show that the first Born approximation to the differential cross-section for the scattering of particles of momentum $\hbar k$ by the spherically symmetric potential

$$V = Ce^{-r^2/a^2}$$

is proportional to

$$\exp\left[-2k^2a^2\sin^2(\theta/2)\right]$$

where θ is the scattering angle.

13.5° Derive the first Born approximation to the differential cross-section for the potential $V = Cr^2\exp(-r^2/a^2)$.

13.6° Show that the first Born approximation to the differential cross-section for the potential

$$V = \begin{cases} \hbar\lambda/2m & r < a \\ 0 & r > a \end{cases}$$

is

$$\frac{\lambda^2}{K^6}(\sin Ka - Ka\cos Ka)^2,$$

where $K = 2k\sin\frac{1}{2}\theta$ with θ the scattering angle.

13.7 Assume that ψ decays at least as fast as R^{-2} for large R. By applying Green's formula

$$\int_D (\psi\nabla^2\phi - \phi\nabla^2\psi)\,dV = \int_{\partial D}(\psi\nabla\phi - \phi\nabla\psi).d\mathbf{S}$$

to the volume D where $\epsilon < |\mathbf{r} - \mathbf{r}'| < R$, with

$$\phi = e^{ik|\mathbf{r}-\mathbf{r}'|}/|\mathbf{r} - \mathbf{r}'|,$$

and letting $\epsilon \to 0$ and $R \to \infty$, prove formula (13.41).

14 Variational methods

I am convinced that the spectrum of all chemical elements can be obtained ... from quantum theory in a unique manner without physics by bone-headed calculation.

WERNER HEISENBERG, letter to Pascual Jordan, 28 July 1926

14.1. Rayleigh quotients

Variational methods exploit an alternative characterization of eigenvectors and eigenvalues which lends itself better to approximation schemes. This is motivated by the geometrical fact that the points on a quadric, $\mathbf{x}.\mathbf{A}.\mathbf{x} = 1$, that are closest to or furthest from the centre lie on the principal axes, and these point along eigenvectors of the matrix, $\mathbf{A}$, defining the quadric. The inverse square of the distance $|\mathbf{x}|^{-2}$, which can be rewritten as $\mathbf{x}.\mathbf{A}.\mathbf{x}/|\mathbf{x}|^2$, is also stationary for such vectors. This suggests that we should consider how the expectation, $\mathsf{E}_\psi(H)$, of a self-adjoint operator H depends on the non-zero vector ψ. To emphasize the role of ψ let us introduce the function

$$f_H(\psi) = \mathsf{E}_\psi(H) = \frac{\langle\psi|H\psi\rangle}{\|\psi\|^2}. \tag{14.1}$$

Definition 14.1.1. The function $f_H(\psi)$ is called the *Rayleigh quotient.*

Theorem 14.1.1. The function $f_H(\psi)$ is stationary with respect to the addition of vectors in a subspace $\mathcal{K}$ if and only if $[H - f_H(\psi)]\psi$ is orthogonal to $\mathcal{K}$. The stationary values of f_H for all changes in ψ occur when ψ is an eigenvector of H, and they are the associated eigenvalues.

Proof. Let us suppose that ψ gives a stationary value of f_H. Choose a vector $\phi \in \mathcal{K}$ and consider the family $\{\psi_u = \psi + u\phi\}$ for $u \in [0, 1]$, for which

$$f_H(\psi + u\phi)\|\psi + u\phi\|^2 = \langle\psi + u\phi|H(\psi + u\phi)\rangle. \tag{14.2}$$

Expanding both sides we get

$$f_H(\psi + u\phi)\left[\|\psi\|^2 + u(\langle\phi|\psi\rangle + \langle\psi|\phi\rangle) + u^2\|\phi\|^2\right]$$
$$= \langle\psi|H\psi\rangle + u(\langle\phi|H\psi\rangle + \langle\psi|H\phi\rangle) + u^2\langle\phi|H\phi\rangle, \quad (14.3)$$

which may be differentiated at $u = 0$ to obtain

$$\frac{df_H(\psi)}{du}\|\psi\|^2 + f_H(\psi)\left(\langle\phi|\psi\rangle + \langle\psi|\phi\rangle\right) = \langle\phi|H\psi\rangle + \langle\psi|H\phi\rangle. \quad (14.4)$$

Rearranging terms, we have

$$\frac{df_H(\psi)}{du}\|\psi\|^2 = \left(\langle\phi|H\psi\rangle + \langle H\psi|\phi\rangle\right) - f_H(\psi)\left(\langle\phi|\psi\rangle + \langle\psi|\phi\rangle\right)$$
$$= \langle\phi|\left[H - f_H(\psi)\right]\psi\rangle + \langle\left[H - f_H(\psi)\right]\psi|\phi\rangle. \quad (14.5)$$

We choose ϕ to be the projection of $[H - f_H(\psi)]\,\psi$ into $\mathcal{K}$, so that we may write $[H - f_H(\psi)]\,\psi = \phi + \chi$ with $\chi \in \mathcal{K}^\perp$. Then

$$\frac{df_H(\psi)}{du}\|\psi\|^2 = \langle\phi|\phi + \chi\rangle + \langle\phi + \chi|\phi\rangle$$
$$= 2\|\phi\|^2, \quad (14.6)$$

from which we deduce that ϕ must vanish for a stationary value. This means that $[H - f_H(\psi)]\,\psi = \chi$ is orthogonal to $\mathcal{K}$, as asserted. When we allow arbitrary variations, $\mathcal{K}$ is the whole space and $\mathcal{K}^\perp = \{0\}$, so that

$$H\psi = f_H(\psi)\psi, \quad (14.7)$$

from which the second result follows immediately. □

Remark 14.1.1. For most interesting physical systems the function $f_H(\psi)$ is bounded below, corresponding to the physical fact that there is a ground state of the least possible energy.

Theorem 14.1.2. If $f_H(\psi)$ is bounded below and achieves its lower bound then $\min_\psi f_H(\psi)$ is the ground state energy of H, and any ψ for which it is attained is a ground state wave function. Conversely on a ground state wave function f_H attains its lower bound.

Proof. By the preceding result the minimum value of a function is automatically stationary, so the corresponding vector ψ must be an eigenvector of H, and $f_H(\psi)$ is an eigenvalue. Being the lower bound $f_H(\psi)$ must be the least eigenvalue, that is the ground state energy. The converse is immediate. □

Remark 14.1.2. Since $f_H(\psi/\|\psi\|) = f_H(\psi)$, it makes no difference whether we take a minimum over all vectors or just over normalized vectors. If ψ is not allowed to range over all vectors, but only over some preselected subset, then we shall probably not be able to attain the true minimum of $f_H(\psi)$, but we shall certainly get an upper bound on the eigenvalue.

The virial theorem 14.1.3. Let $H = T + V(\mathbf{r})$, where $T = -\hbar^2\nabla^2/2m$, and suppose that $H\psi = E\psi$. Then
(i) $2\mathsf{E}_\psi(T) = \mathsf{E}_\psi(\mathbf{X}.\nabla V)$.
(ii) If V is homogeneous of degree N then

$$\mathsf{E}_\psi(T) = \frac{N}{N+2}E \quad \text{and} \quad \mathsf{E}_\psi(V) = \frac{2}{N+2}E.$$

Proof. Suppose that ψ is a normalized eigenfunction of H. Consider the normalized trial wave functions $\psi_u(\mathbf{r}) = \exp(3u/2)\psi(\exp(u)\mathbf{r})$. Applying the chain rule

$$(\nabla\psi_u)(\mathbf{r}) = e^u e^{\frac{3}{2}u}(\nabla\psi)(e^u\mathbf{r}), \tag{14.8}$$

we see that $(P\psi_u)(\mathbf{r}) = \exp(u)(P\psi)(\exp(u)\mathbf{r})$, and making obvious changes of variable in the integrals, we have

$$\begin{aligned}\langle\psi_u|H\psi_u\rangle &= \frac{e^{2u}}{2m}\int \overline{\psi(e^u\mathbf{r})}(P^2\psi)(e^u\mathbf{r})e^{3u}d^3\mathbf{r} + \int V(\mathbf{r})|\psi(e^u\mathbf{r})|^2e^{3u}d^3\mathbf{r} \\ &= e^{2u}\langle\psi|T,\psi\rangle + \int V(e^{-u}\mathbf{r})|\psi(\mathbf{r})|^2e^{3u}d^3\mathbf{r} \\ &= e^{2u}\mathsf{E}_\psi(T) + \mathsf{E}_\psi\big(V\left(e^{-u}\mathbf{X}\right)\big).\end{aligned} \tag{14.9}$$

Since ψ_0 is a true eigenvector $f_H(\psi_u)$ must have a stationary value at $u = 0$, which gives the condition that

$$2e^{2u}\mathsf{E}_{\psi_u}(T) + d\mathsf{E}_\psi\big(V\left(e^{-u}\mathbf{X}\right)\big)/du \tag{14.10}$$

must vanish at $u = 0$. Applying the chain rule to calculate the derivative of V, this yields the virial theorem

$$2\mathsf{E}_\psi(T) = \mathsf{E}_\psi((\mathbf{X}.\nabla V)). \tag{14.11}$$

If V is homogeneous of degree N, then Euler's theorem gives $\mathbf{r}.\nabla V = NV$, so that we have

$$2\mathsf{E}_\psi(T) = N\mathsf{E}_\psi(V), \tag{14.12}$$

which combined with the obvious relation, $\mathsf{E}_\psi(T)+\mathsf{E}_\psi(V) = E$, gives (ii). □

Remark 14.1.3. Exercise 7.7 outlined a more elementary derivation of this result in one dimension, which can easily be extended to the general case, and also suggested useful applications of the result to calculate dispersions for the harmonic oscillator, and to show that the bound state energies of the hydrogen atom must be negative.

14.2. The ground state of helium

The Hamiltonian for the helium atom, which has already been discussed in Section 12.3, is

$$H = -\frac{\hbar^2}{2m}\left(\nabla_1^2 + \nabla_2^2\right) - \frac{2e^2}{4\pi\epsilon_0}\left(\frac{1}{r_1} + \frac{1}{r_2}\right) + \frac{e^2}{4\pi\epsilon_0|\mathbf{r}_1 - \mathbf{r}_2|}. \tag{14.13}$$

Proposition 14.2.1. The ground state energy of the helium atom is less than or equal to

$$-\left(\frac{27}{16}\right)^2 \frac{e^2}{4\pi\epsilon_0 a}.$$

Proof. Physical intuition might suggest that the presence of the second electron partly shields the positive charge on the nucleus. We therefore select the trial functions

$$\psi_Z(\mathbf{r}_1, \mathbf{r}_2) = \left(\frac{Z^3}{\pi a^3}\right) \exp\left(-\frac{Z}{a}(r_1 + r_2)\right), \tag{14.14}$$

which are the wave functions appropriate to a nuclear charge Z, and minimize $\langle\psi_Z|H\psi_Z\rangle/\|\psi_Z\|^2$ to find the best value of Z, and so to estimate the ground state energy.

Now, as in Section 12.3, we know that ψ_Z is the ground state of

$$H_Z = -\frac{\hbar^2}{2m}\left(\nabla_1^2 + \nabla_2^2\right) - \frac{Ze^2}{4\pi\epsilon_0}\left(\frac{1}{r_1} + \frac{1}{r_2}\right) = T + \frac{Z}{2}V, \tag{14.15}$$

with ground state energy, $E_Z = -Z^2e^2/4\pi\epsilon_0 a$. According to the remark following the virial theorem 14.1.3 we have

$$\begin{aligned} \mathsf{E}_{\psi_Z}(T) &= -E_Z = \frac{Z^2e^2}{4\pi\epsilon_0 a}, \\ \mathsf{E}_{\psi_Z}\left(\frac{ZV}{2}\right) &= 2E_Z = -\frac{2Z^2e^2}{4\pi\epsilon_0 a}, \end{aligned} \tag{14.16}$$

or

$$\mathsf{E}_{\psi_Z}(V) = -\frac{Ze^2}{\pi\epsilon_0 a}. \tag{14.17}$$

Finally the electronic repulsion $\langle\psi_Z|\left(e^2/4\pi\epsilon_0|\mathbf{r}_1 - \mathbf{r}_2|\right)\psi_Z\rangle$ was shown in Section 12.3 to be $5Ze^2/32\pi\epsilon_0 a$. Combining the various pieces we have

$$\begin{aligned} \langle\psi_Z|H\psi_Z\rangle &= \frac{Z^2e^2}{4\pi\epsilon_0 a} - \frac{Ze^2}{\pi\epsilon_0 a} + \frac{5Ze^2}{32\pi\epsilon_0 a} \\ &= \frac{Z^2e^2}{4\pi\epsilon_0 a} - \frac{27Ze^2}{32\pi\epsilon_0 a} \\ &= \frac{e^2}{4\pi\epsilon_0 a}\left[\left(Z - \frac{27}{16}\right)^2 - \left(\frac{27}{16}\right)^2\right]. \end{aligned} \tag{14.18}$$

When $Z = 27/16$ this achieves its minimum of

$$-\left(\frac{27}{16}\right)^2 \frac{e^2}{4\pi\epsilon_0 a} = -0.712\frac{e^2}{\pi\epsilon_0 a}, \tag{14.19}$$

which differs by less than 2% from the experimental value of $-0.73e^2/\pi\epsilon_0 a$, and is a considerable improvement on the $-0.6875e^2/\pi\epsilon_0 a$ of first-order perturbation theory. The wave function is that for a nuclear charge $27e/16$, suggesting that the actual charge of $2e$ has been reduced by $5e/16$ owing to the screening effects of the other electron. □

It is, of course, no accident that the variational method has produced a more accurate result than first-order perturbation theory. It came about because the family of trial functions included the true ground state wave function of the unperturbed Hamiltonian.

Proposition 14.2.2. If the family of trial vectors $\{\psi\}$ includes the ground state ψ_0 for the unperturbed Hamiltonian H_0 then

$$E_1 \geq \min_{\psi} \frac{\langle\psi|H\psi\rangle}{\|\psi\|^2} \geq E,$$

where E_1 denotes the first-order perturbation theoretic estimate of the ground state energy, and E denotes the true ground state energy.

Proof. We already know that the energy E gives a lower bound for the Rayleigh quotients. In perturbation theory we had for normalized ψ_0:

$$E_1 = \langle\psi_0|H\psi_0\rangle, \tag{14.20}$$

so clearly the minimum over ψ is less than or equal to E_1. □

14.3. Excited states

The excited states also satisfy extremal properties, which can be investigated inductively. Suppose that we have already found the k lowest lying energy levels for H, $k \geq 1$, and the eigenvectors span a subspace $\mathcal{H}_k$.

Theorem 14.3.1. If $\inf\left\{f_H(\psi) : \psi \in \mathcal{H}_k^\perp\right\}$ is attained for some ψ then this is an eigenvector corresponding to the $(k+1)$-th lowest energy level, and the infimum is the energy.

Proof. The subspace $\mathcal{H}_k$ is invariant under H and so therefore is $\mathcal{H}_k^\perp$. (For $\psi \in \mathcal{H}_k^\perp, \phi \in \mathcal{H}_k$, $\langle H\psi|\phi\rangle = \langle\psi|H\phi\rangle$ vanishes.) Applying Theorem 14.1.1 to $H|_{\mathcal{H}_k^\perp}$ we see that if $\inf\left\{f_H(\psi) : \psi \in \mathcal{H}_k^\perp\right\}$ is attained then it is the lowest energy level in $\mathcal{H}_k^\perp$. Since the k lowest energy levels were in $\mathcal{H}_k$ this is the $(k+1)$-th lowest energy overall. □

Hilbert's minimax theorem provides a more sophisticated formulation of the same result.

Theorem 14.3.2. If the infimum,

$$\inf\left\{\max\left\{f_H(\psi):\psi\in\mathcal{K}\right\}:\dim(\mathcal{K})=k\right\},$$

over all k-dimensional subspaces $\mathcal{K}$ of $\mathcal{H}$ is attained then it is the k-th lowest energy level of H and the vector ψ is the corresponding eigenvector.

Proof. We leave this reformulation to the reader. □

14.4. The Rayleigh–Ritz variational theory

Often it is convenient to take not just a set of trial functions but the space that they span. Let us therefore suppose that the trial functions lie in a finite-dimensional subspace $\mathcal{K}$. Rayleigh–Ritz theory uses the variational principle to reduce the general problem to the finite-dimensional case.

Theorem 14.4.1. Let $\{v_j\}$ denote a basis of the finite-dimensional subspace, $\mathcal{K}$. Then the k-th lowest root E of the equation

$$\det\left(\langle v_j|Hv_k\rangle - E\langle v_j|v_k\rangle\right)=0$$

provides an upper bound to the k-th lowest eigenvalue of H for $k\leq\dim\mathcal{K}$.

Proof. From the first part of Theorem 14.1.1 we know that $f_H(\psi)$ is stationary with respect to the addition of vectors in $\mathcal{K}$ when $[H-f_H(\psi)]\psi$ is orthogonal to $\mathcal{K}$. Writing $E=f_H(\psi)$, we see that

$$\langle v_j|(H-E)\psi\rangle=0, \tag{14.21}$$

for all j. If ψ is itself in $\mathcal{K}$ then it can be expanded as $\psi = \sum c_k v_k$, and then we have

$$0 = \sum_k \langle v_j|(H - E)v_k\rangle c_k = \sum_k (\langle v_j|Hv_k\rangle - E\langle v_j|v_k\rangle)c_k. \tag{14.22}$$

The condition for these linear equations to have a non-trivial solution for the coefficients c_k is

$$\det\left(\langle v_j|Hv_k\rangle - E\langle v_j|v_k\rangle\right) = 0, \tag{14.23}$$

and the assertion of the theorem follows. □

Definition 14.4.1. The matrix $(\langle v_j|v_k\rangle)$ is called the *Gramian* or *overlap matrix.* The equation

$$\det\left(\langle v_j|Hv_k\rangle - E\langle v_j|v_k\rangle\right) = 0$$

is called the *secular equation.*

For an orthonormal basis the overlap matrix is the identity, and the secular equation reduces to the characteristic equation. Many of the most accurate calculations of atomic ground state energies have used variational methods. As an illustration, in 1994, S.P. Goldman presented a simple calculation of the ground state energy of helium with a relative error of 3 parts in 10^8, by using a Rayleigh–Ritz method with 393 basis functions. These were written in terms of $r_M = \max(r_1, r_2)$ and $r_m = \min(r_1, r_2)$ as well as r_1 and r_2, themselves, and each was the sum of a term of the form

$$e^{-Z_1r_1 - Z_2r_2 - Z_Mr_M - Z_mr_m} r_1^{N_1} r_2^{N_2} r_M^{N_M} r_m^{N_m} \Lambda, \tag{14.24}$$

with the corresponding term with the particle positions interchanged, and each Λ a carefully chosen function of the angular terms alone.

14.5. Historical notes

Variational methods have been used for various problems since the early nineteenth century, and were used in particular by William Strutt, later Lord Rayleigh, in his investigations of wave motion. Following the pioneering work of Ivar Fredholm on integral equations in 1902, Hilbert started the rigorous mathematical study of variational procedures. One of his students, Walther Ritz, who worked on the problem shortly before his death from tuberculosis in 1909, independently rediscovered parts of Rayleigh's work.

One instructive use of variational methods in wave theory can be illustrated by considering the equation for stationary waves on a string, $y_{xx} = -\omega^2 c^2 y$, which can be written in quantum mechanical notation as $P^2 y = \hbar^2 \omega^2 c^2 y$. The fundamental frequency of this string is $\sqrt{E/\hbar c}$, where E is the lowest eigenvalue of P^2, or equivalently the minimum value of $\langle y|P^2 y\rangle$. If a finger stops the string, some previously admissible waves will be suppressed and this may raise the minimum, causing the frequency to rise. This provides a simple qualitative understanding of why open strings on a violin or guitar give lower notes than stopped strings.

Exercises

14.1° Estimate the ground state energy of the harmonic oscillator by taking trial functions of the form $\psi(x) = (a + bx)\exp(-\frac{1}{2}cx^2)$, where a, b, and c are real and c is positive.

14.2° A harmonic oscillator has Hamiltonian

$$H = \frac{P^2}{2m} + \frac{1}{2}m\omega^2 X^2.$$

Show that the expectation value of H in the state described by the wave function $x^n \exp(-\frac{1}{2}cx^2)$ (for n a non-negative integer and c a positive real number) is

$$(n + \tfrac{1}{2})\frac{m\omega^2}{2c} + \left(\frac{n - \frac{1}{4}}{n - \frac{1}{2}}\right)\frac{\hbar^2 c}{2m}.$$

By varying c obtain upper bounds on the ground state energy of the harmonic oscillator. What estimate of this energy does one obtain if one allows n to be non-integral and varies n as well as c?

14.3° Calculate a bound for the ground state energy of a hydrogen atom in a weak uniform magnetic field of strength F along the z-axis, using trial functions of the form $\psi(\mathbf{r}) = (1 + Az)\psi_0(\mathbf{r})$, with ψ_0 being the normalized ground state wave function of the hydrogen atom in the absence of the field.
[You may use the identities

$$\int_{\mathbf{R}^3} z^2 \psi_0^2 d^3\mathbf{r} = a^2, \qquad \int_{\mathbf{R}^3} \frac{z^2}{r}\psi_0^2 d^3\mathbf{r} = \frac{a}{2}.]$$

14.4° The total angular momentum operator in spherical polar coordinates has the form

$$L^2 = -\frac{\hbar^2}{\sin^2\theta}\left[\sin\theta\frac{\partial}{\partial\theta}\left(\sin\theta\frac{\partial}{\partial\theta}\right) + \frac{\partial^2}{\partial\phi^2}\right].$$

Using 1 and $\cos^2\theta$ as a basis for a space of trial wave functions on the unit sphere with inner product

$$\langle\psi_1|\psi_2\rangle = \int \overline{\psi_1}\psi_2 \sin\theta \, d\theta d\phi,$$

obtain Rayleigh–Ritz estimates for the lowest eigenvalues of L^2.

14.5° The Hamiltonian for a particle of mass M moving in one dimension is given by

$$H = \frac{P^2}{2M} - \frac{a\hbar^2}{2M}e^{-\kappa^2 X^2},$$

where a and κ are positive constants. For α positive, find the expectation value of H with respect to the state vector ψ_α defined by

$$\psi_\alpha(x) = e^{-\frac{1}{2}\alpha^2 x^2}.$$

Show that the best upper bound that can be put on the ground state energy using trial functions of this form occurs when $\alpha = \alpha_0$ is the positive root of

$$\alpha^2(\kappa^2 + \alpha^2)^3 = a^2\kappa^4.$$

14.6° A particle of mass m moves in the central potential $\hbar^2 U(r)/2m$, where

$$U(r) = \frac{b^2}{r^2} - \frac{a^2}{r},$$

and a and b are real constants. Find the expectation value of the Hamiltonian when the wave function of the particle is given by $\phi = \exp(-\kappa r)$ and show that the ground state energy is bounded above by

$$\frac{-\hbar^2}{2m}\frac{a^4}{4(1+2b^2)}.$$

A particle of mass m moves in the Coulomb potential

$$V(r) = -\frac{K}{r}, \quad K > 0.$$

Using the trial wave function $\psi_\beta = \exp(-\beta r)$ $(\beta > 0)$, obtain the best upper bound to the ground state energy.

14.7° A particle of mass m moves along the positive x-axis under the influence of a potential V that is infinite for negative x and $\mathcal{E}x$ for positive x. By calculating the expectation of H for trial functions of the form $\psi(x) = x^{\frac{1}{2}c}\exp(-\kappa x)$, as a function of κ and c, show that the ground state energy is less than

$$\left(\frac{27\mathcal{E}^2\hbar^2}{4m}\right)^{\frac{1}{3}}.$$

[You may assume that $\int_0^\infty x^n \exp(-\lambda x)\,dx = n!\lambda^{-(n+1)}$.]

14.8 A particle of mass m moves along the x-axis under the influence of a potential

$$V = \frac{1}{2}m\omega^2x^2 + \frac{\kappa(\kappa^2-1)}{6\hbar}m^2\omega^3x^4,$$

where $\kappa > 1$. Show that for trial wave functions of the form $\psi_\alpha(x) = N\exp(-\alpha m\omega x^2/2\hbar)$

$$F(\psi_\alpha) = \frac{\hbar\omega}{4}\left(\alpha + \alpha^{-1} + \frac{\kappa(\kappa^2-1)}{2\alpha^2}\right).$$

Show that this is minimum when $\alpha = \kappa$, and hence that the ground state energy is less than or equal to $(3\kappa^2+1)\hbar\omega/8\kappa$.

14.9° A particle of mass m moves in a potential

$$V(r) = \frac{-\hbar^2 C}{m}\frac{e^{-\kappa r}}{r}, \qquad r > 0,$$

where κ and C are positive constants. For $\alpha > 0$, find an expression for $F(\psi)$ where

$$\psi_\alpha(r) = e^{-\alpha r/2}, \qquad r > 0.$$

Show that stationary values of F occur when

$$(\alpha+\kappa)^3 = 2C\alpha(\alpha+3\kappa).$$

Show further that when $C = \frac{27}{20}\kappa$ the best estimate for the ground state energy occurs when $\alpha = 2\kappa$, and find it.

15 The semi-classical approximation

Just now I am teaching the foundations of poor deceased mechanics, which is so beautiful. What will her successor look like? With that question I torment myself incessantly.

ALBERT EINSTEIN, letter to Heinrich Zangger, 14 November 1911

15.1. The semi-classical approximation

There is another very useful sort of approximation which compares a quantum system not with a simpler quantum system, but with the corresponding classical system. We start by putting the wave function into polar form $\psi = a\exp(iS/\hbar)$.

Theorem 15.1.1. The wave function $\psi = a\exp(iS/\hbar)$ satisfies Schrödinger's equation

$$i\hbar\frac{\partial\psi}{\partial t} = -\frac{\hbar^2}{2m}\nabla^2\psi + V\psi$$

if and only if a and S satisfy the coupled equations

$$\frac{\partial S}{\partial t} + \frac{|\nabla S|^2}{2m} + V = \frac{\hbar^2}{2m}\frac{\nabla^2 a}{a},$$

$$\frac{\partial(a^2)}{\partial t} + \operatorname{div}\left(\frac{a^2}{m}\nabla S\right) = 0.$$

Proof. We immediately calculate that

$$\nabla\psi = \left(\frac{\nabla a}{a} + \frac{i}{\hbar}\nabla S\right)\psi, \qquad \frac{\partial\psi}{\partial t} = \left(\frac{1}{a}\frac{\partial a}{\partial t} + \frac{i}{\hbar}\frac{\partial S}{\partial t}\right)\psi, \tag{15.1}$$

and

$$\begin{aligned}\nabla^2\psi &= \left(\frac{\nabla^2 a}{a} - \frac{|\nabla a|^2}{a^2} + \frac{i}{\hbar}\nabla^2 S\right)\psi + \left(\frac{\nabla a}{a} + \frac{i}{\hbar}\nabla S\right).\left(\frac{\nabla a}{a} + \frac{i}{\hbar}\nabla S\right)\psi \\ &= \left(\frac{\nabla^2 a}{a} + \frac{i}{\hbar}\nabla^2 S + 2\frac{i}{\hbar}\frac{\nabla a}{a}.\nabla S - \frac{1}{\hbar^2}|\nabla S|^2\right)\psi.\end{aligned} \tag{15.2}$$

Schrödinger's equation now reads

$$\left(\frac{i\hbar}{a}\frac{\partial a}{\partial t} - \frac{\partial S}{\partial t}\right)\psi$$
$$= \left[-\frac{1}{2m}\left(\hbar^2\frac{\nabla^2 a}{a} + i\hbar\nabla^2 S + 2i\hbar\frac{\nabla a}{a}.\nabla S - |\nabla S|^2\right) + V\right]\psi, \tag{15.3}$$

and since we are supposing that both a and S are real we may separate the real and imaginary multiples of ψ to obtain the two equations

$$\frac{\partial S}{\partial t} + \frac{|\nabla S|^2}{2m} + V = \frac{\hbar^2}{2m}\frac{\nabla^2 a}{a}, \tag{15.4}$$

and

$$\frac{1}{a}\frac{\partial a}{\partial t} + \frac{1}{2m}\nabla^2 S + \frac{1}{m}\frac{\nabla a}{a}.\nabla S = 0. \tag{15.5}$$

On multiplying through by $2a^2$ the second of these can be rewritten as

$$\frac{\partial(a^2)}{\partial t} + \text{div}\left(\frac{a^2}{m}\nabla S\right) = 0. \qquad \square$$

Remark 15.1.1. One may readily check that when $\psi = a\exp(iS/\hbar)$, the probability density is $\rho = a^2$ and the probability current is $\mathbf{j} = a^2\nabla S/m$, so that the second condition is just the continuity equation of Proposition 2.5.1.

So far everything has been exact, but if we look at the first condition we see that the right-hand side is of order $\hbar^2$, whilst the left-hand side is independent of $\hbar$. This suggests the following approximation:

Definition 15.1.1. The *semi-classical approximation* uses the wave function $\psi = a\exp(iS/\hbar)$ where a and S satisfy the following coupled equations:

$$\frac{\partial S}{\partial t} + \frac{|\nabla S|^2}{2m} + V = 0,$$
$$\frac{\partial(a^2)}{\partial t} + \text{div}\left(\frac{a^2}{m}\nabla S\right) = 0.$$

The semi-classical approximation is also known as the WKB approximation after Wentzel, Kramers, and Brillouin who introduced it into quantum theory. Sometimes the initial J is added, for Jeffreys, to commemorate an earlier pioneer of the method. In fact, a closely related technique for solving differential equations was introduced as early as 1837 by Liouville and independently by Green. It is really part of an asymptotic approximation to the wave function for very small values of $\hbar$. One could go further and find a series,

$$\psi = e^{iS/\hbar}\left(a + \sum_{n=1}^{\infty} a_n \hbar^n\right), \tag{15.6}$$

which is asymptotic in the sense that

$$\hbar^{-N}\left[e^{-iS/\hbar}\psi - \left(a + \sum_{n=1}^{N} a_n \hbar^n\right)\right] \tag{15.7}$$

tends to 0 as $\hbar \to 0$. (Even though the series $\sum a_n \hbar^n$ may be divergent for non-zero $\hbar$, the asymptotic approximation, which at any stage uses only a finite number of terms, can nonetheless be very accurate.)

The particular advantage of the semi-classical approximation is that the first equation involves only S, and in fact it is the Hamilton–Jacobi equation of classical mechanics. It is the equation satisfied by the classical action

$$S = \int L\, dt \tag{15.8}$$

where L is the Lagrangian for the classical system, and the integral is taken from some fixed starting point, $\mathbf{y}$, to $\mathbf{x}$, and calculated for a journey taking time t. The continuity equation is also classical in form (being the equation for a classical fluid), so that we have approximated quantum theory by a classical theory.

The connection between Schrödinger's equation and Hamilton–Jacobi theory has inspired some attempts to reinterpret quantum theory using classical hidden variables. David Bohm and Imre Fényes independently introduced two such examples in 1952. Both relied on the observation that the true equation for S can be written in the form

$$\frac{\partial S}{\partial t} + \frac{|\nabla S|^2}{2m} + V - \frac{\hbar^2}{2m}\frac{\nabla^2 a}{a} = 0, \tag{15.9}$$

which is the Hamilton–Jacobi equation for a potential $V - \hbar^2\nabla^2 a/2ma$. This suggests that one is dealing with classical mechanics with an additional potential, $-\hbar^2\nabla^2 a/2ma$, which depends on the probability density, $a = \sqrt{\rho}$. Bohm's example postulated this, together with some rules designed to remove other counterintuitive features of the new potential. Fényes' theory is based on probabilistic ideas. In a later refinement, due to Edward Nelson, the new potential appears to be the exact analogue of a term which appears when one considers the classical Brownian motion of a pollen grain, buffeted by surrounding molecules.

However, despite their obvious appeal, such ideas do not really return us to a familiar kind of classical theory. As we have seen in Section 10.4, the new theory cannot be local, that is there must be long range interdependence. The equations we have given are non-relativistic equations, and it is more difficult to find convincing relativistic analogues. Moreover, there are other, more subtle, peculiarities, which mean that one has simply swapped one set of problems for another, and it is a matter of personal judgement as to which difficulties one prefers. At a purely practical level, it is often easier to solve the linear Schrödinger equation than the coupled non-linear equations for S and a.

15.2. Semi-classical examples

We noted above that the argument, S, of the semi-classical wave function can be calculated directly from solutions of the classical equations of motion, and a from the continuity equation. As a matter of fact the situation is even better, since solutions for a can be calculated directly from S, as the following result shows.

Theorem 15.2.1. (Van Vleck) Let $S(\mathbf{x}, \mathbf{y})$ be a solution of the Hamilton–Jacobi equation where $\mathbf{y}$ denotes the position of the starting point for the action (or constants of integration, if one simply solves the equation). Then

$$a^2 = \det\left(\frac{\partial^2 S}{\partial x_j \partial y_k}\right)$$

gives a solution of the continuity equation.

Proof. We shall only deal with the case of one spatial dimension leaving the general case to the reader. Then

$$a^2 = \frac{\partial^2 S}{\partial x \partial y}, \tag{15.10}$$

so that

$$\begin{aligned}\frac{\partial(a^2)}{\partial t} &= \frac{\partial^3 S}{\partial x \partial y \partial t}\\ &= -\frac{\partial^2}{\partial x \partial y}\left[\frac{1}{2m}\left(\frac{\partial S}{\partial x}\right)^2 + V\right]\\ &= -\frac{\partial}{\partial x}\left(\frac{1}{m}\frac{\partial S}{\partial x}\frac{\partial^2 S}{\partial x \partial y}\right)\\ &= -\frac{1}{m}\frac{\partial}{\partial x}\left(\frac{\partial S}{\partial x}a^2\right),\end{aligned} \tag{15.11}$$

which is the continuity condition. (In fact this result still holds for any parameters y_1, y_2, y_3, and not just the starting coordinates.) □

For a free particle there is a solution of the Hamilton–Jacobi equation of the form $S = m|\mathbf{x} - \mathbf{y}|^2/2t$. (To see this we note that

$$\nabla S = m(\mathbf{x} - \mathbf{y})/t \qquad \text{and} \qquad \partial S/\partial t = -m|\mathbf{x} - \mathbf{y}|^2/2t^2, \tag{15.12}$$

from which it easily follows that

$$\frac{\partial S}{\partial t} + \frac{|\nabla S|^2}{2m} = 0, \tag{15.13}$$

so that S does satisfy the equation.) The corresponding probability density is

$$a^2 = \det\left(\frac{\partial^2 S}{\partial x_j \partial y_k}\right) = \det\left(\frac{-m}{t}1\right) = -\left(\frac{m}{t}\right)^3. \tag{15.14}$$

The semi-classical (unnormalized) wave function can therefore be written as

$$\psi = \left(\frac{m}{t}\right)^{\frac{3}{2}} \exp\left(\frac{im}{2\hbar t}|\mathbf{x} - \mathbf{y}|^2\right). \tag{15.15}$$

We can now adapt some of the other results of classical Hamiltonian mechanics to our present needs.

Proposition 15.2.2. The Hamilton–Jacobi equation has solutions of the form

$$S(\mathbf{x},t) = W(\mathbf{x}) - Et$$

provided that

$$\frac{|\nabla W|^2}{2m} + V = E.$$

The corresponding semi-classical wave function

$$\psi = a(\mathbf{x}) \exp\left(\frac{iW(\mathbf{x})}{\hbar}\right) \exp\left(-i\frac{Et}{\hbar}\right)$$

is a solution of Schrödinger's time-independent equation with energy E.

Proof. This follows immediately on substituting $S = W - Et$ into the previous equations, and then calculating $i\hbar\partial\psi/\partial t$. □

Corollary 15.2.3. In one dimension the Hamilton–Jacobi equation has solutions of the form

$$S(x,t) = \pm \int \sqrt{2m[E - V(x)]}dx - Et,$$

and the continuity equation has solutions of the form

$$a(x) = A\,[E - V(x)]^{-\frac{1}{4}},$$

where A is a constant.

Proof. In one dimension the equation for W becomes

$$\frac{1}{2m}\left(\frac{dW}{dx}\right)^2 + V(x) = E, \tag{15.16}$$

which can be rearranged and integrated to give the stated formula for S. The amplitude can be found directly from the continuity equation, which in one dimension becomes

$$\frac{d}{dx}\left(\frac{a^2}{m}\frac{dW}{dx}\right) = 0. \tag{15.17}$$

Clearly the solution of this gives

$$a^2 = B\left(\frac{dW}{dx}\right)^{-1} = \pm\frac{B}{\sqrt{2m[E-V(x)]}}, \tag{15.18}$$

for some constant B. The result now follows on taking square roots and changing constants. (One can also deduce the formula for a from van Vleck's solution, as in the following example.) □

Example 15.2.1. (The free particle in one dimension) For the free particle $V = 0$ so one has simply

$$S = \pm\sqrt{2mE}x - Et. \tag{15.19}$$

This is very different in form from the solution $m(x-y)^2/2t$ we would expect from the three-dimensional case. It is easy to check that both are solutions of the Hamilton–Jacobi equation, whose solutions can take many different forms. This time E plays the role of the constant of integration, so that

$$a^2 = \frac{\partial^2 S}{\partial x \partial E} = \pm\sqrt{m/2E}, \tag{15.20}$$

and ψ is just a constant multiple of the plane wave

$$\exp\left[i(\pm\sqrt{2mE}x - Et)/\hbar\right]. \tag{15.21}$$

Remark 15.2.1. We expect the semi-classical approximation to be useful where the term that we have dropped, $\hbar^2\nabla^2 a/2ma$, is small with respect to the terms that we retained. The form of a in one dimension suggests that this will be reasonable provided that $E - V$ is not too small, that is we stay away from the classical turning points.

15.3. The Bohr–Sommerfeld condition

Owing to the appearance of the square root in the formula for S in Corollary 15.2.3 there are two independent semi-classical solutions, which we write as

$$A_\pm[E-V(x)]^{-\frac{1}{4}}\exp\left(\pm\frac{i}{\hbar}\int_b^x \sqrt{2m[E-V(x)]}\,dx\right), \tag{15.22}$$

where c is some arbitrary basepoint. We shall now show that there can be other constraints that force us to use a particular linear combination of these two approximate solutions.

The square root makes good sense as long as one is dealing with the *classical* region where $E > V$. In the non-classical region, where $E < V(x)$, it is natural to make the complex substitution $E - V(x) = |E - V(x)| \exp(\pm i\pi)$, so that our two semi-classical solutions can then be written as

$$A_{\pm} e^{\mp i\frac{\pi}{4}} |E - V(x)|^{-\frac{1}{4}} \exp\left(\pm i e^{\pm i\frac{\pi}{2}} \int_b^x \sqrt{2m|E - V(x)|}\, dx/\hbar \right). \quad (15.23)$$

It is easy to check directly that these are valid approximate solutions. To pass from the classical to the non-classical regions one must traverse a turning point where $V = E$, and we may as well choose this as the reference point, c. For definiteness let us suppose that it is at the extreme right-hand edge of the permitted classical region.

To the right of the classical region the solution should be given by a decaying exponential, so the appropriate form of solution is

$$C|E - V(x)|^{-\frac{1}{4}} \exp\left(- \int_c^x \sqrt{2m|E - V(x)|}\, dx/\hbar \right), \quad (15.24)$$

which we can obtain by consistently choosing either the upper or the lower signs in our non-classical solutions and taking

$$A_{\pm} = C e^{\pm i\frac{\pi}{4}}. \quad (15.25)$$

This means that the appropriate combination to use in the classical region is

$$2C[E - V(x)]^{-\frac{1}{4}} \cos\left(\frac{1}{\hbar} \int_c^x \sqrt{2m[E - V(x)]}\, dx + \frac{\pi}{4} \right)$$
$$= 2C[E - V(x)]^{-\frac{1}{4}} \cos\left(\frac{[W(x) - W(c)]}{\hbar} + \frac{\pi}{4} \right). \quad (15.26)$$

For a turning point, b, on the left-hand edge of the classical region one obtains by a similar argument the semi-classical wave function of the form

$$2B[E - V(x)]^{-\frac{1}{4}} \cos\left(\frac{[W(x) - W(b)]}{\hbar} - \frac{\pi}{4} \right), \quad (15.27)$$

for some constant, B. However, when there is a single classical region bounded by the two points b and c, these two expressions must coincide, and, in particular, we must have

$$\frac{W(x) - W(b)}{\hbar} - \frac{\pi}{4} = \frac{W(x) - W(c)}{\hbar} + \frac{\pi}{4} + n\pi, \quad (15.28)$$

for some integer n, or, on simplifying,

$$W(c) - W(b) = \left(n + \frac{1}{2}\right)\hbar\pi. \tag{15.29}$$

In fact, since the left-hand side is positive that integer must be non-negative, and we have the following result:

The Bohr–Sommerfeld condition 15.3.1. Suppose that the potential V is less than E just on the interval (b, c), with equality $V = E$ at the endpoints Then for a consistent semi-classical solution we require that

$$\int_b^c \sqrt{2m[E - V(s)]}ds = (n + \tfrac{1}{2})\pi\hbar$$

for some non-negative integer n.

The Bohr–Sommerfeld rule is more usually expressed in terms of an integral round a closed curve in phase space, that is from b to c using the positive square root and back again using the negative root, which doubles the integral and gives

$$\oint \sqrt{2m[E - V(s)]}ds = (n + \tfrac{1}{2})2\pi\hbar. \tag{15.30}$$

This rule was used as an *ad hoc* way to do quantum calculations before the development of a consistent quantum theory by Heisenberg and Schrödinger. Not surprisingly, since it is a patchwork of approximations, it often leads to incorrect conclusions.

The above argument hinged on the behaviour of the semi-classical solutions near a turning point, but we have already noted that the whole basis for the approximation breaks down, because the term that we dropped to get the Hamilton–Jacobi equation can no longer be safely ignored. Fortunately, another approximation, which we shall now describe, comes to the rescue near the turning points, and can be used to justify our formulae. Suppose that c is a turning point. Then, to first order,

$$V(x) \sim V(c) + (x - c)V'(c) = E + (x - c)V'(c). \tag{15.31}$$

This gives as an approximation to Schrödinger's time-independent equation:

$$-\frac{\hbar^2}{2m}\frac{d^2\psi}{dx^2} + (x - c)V'(c)\psi = 0. \tag{15.32}$$

This *Airy equation* can be solved by Fourier transforming as in Section 3.5, and we shall assume, for definiteness, that $V'(c)$ is positive. Using the solution given by equation (3.44) with $eF = V'(c)$ and $E = eFc = cV'(c)$, we have

$$\widehat{\psi}(p) = N \exp\left(i\frac{p^3}{6m\hbar V'(c)} - i\frac{cp}{\hbar}\right). \tag{15.33}$$

Inverting the transform and introducing a new constant B we obtain

$$\psi = B \int_{\mathbf{R}} \exp\left[\frac{i}{\hbar}\left(p(x-c) + \frac{p^3}{6mV'(c)}\right)\right] dp. \tag{15.34}$$

We are interested in the behaviour of this integral as one moves away from the turning point and attempts to weld it to the semi-classical solution. For large x the integrand oscillates very rapidly and, long before the advent of quantum mechanics, Lord Kelvin had realized that interference effects would mean that most of the contributions would cancel. Only near the extrema of the exponential would the oscillations be slower, so that the dominant contribution to the integral comes from there. (This technique for finding asymptotic forms of integrals is known as the *method of stationary phase.*)

Now the phase $\Phi(x,p) = p(x-c) + p^3/6mV'(c)$ has its extrema when

$$\frac{p^2}{2m} + V'(c)(x-c) = 0, \tag{15.35}$$

which is the linear approximation to

$$\frac{p^2}{2m} + V(x) = E. \tag{15.36}$$

Let us first look at the classical region where $x < c$, and this equation has real solutions for p. At these points, $p_\pm$, the phase takes the value

$$\Phi(x, p_\pm) = \pm\tfrac{2}{3}\sqrt{2mV'(c)}(x-c)^{\frac{3}{2}}. \tag{15.37}$$

By integrating the expression for $E - V$ we see that near the turning point this is approximately

$$\pm W(x) = \pm \int_c^x \sqrt{2m(E-V)}\, dx, \tag{15.38}$$

so that we have recovered the semi-classical expression for the phase.

The amplitude can also be recovered by changing variables near the stationary values $p_\pm$ to $p = p_\pm + u$. Dropping higher order terms from the Taylor series, this gives

$$\Phi(x,p) = \Phi(x,p_\pm) + \tfrac{1}{2}\Phi''(x,p_\pm)u^2, \tag{15.39}$$

where the prime denotes a derivative with respect to p. We therefore have contributions to the integral of

$$B \int_{\mathbf{R}} \exp\left[\frac{i}{\hbar}\left(\Phi(x,p_\pm) + \tfrac{1}{2}\Phi''(x,p_\pm)u^2\right)\right] du. \tag{15.40}$$

The integrand resembles a normal distribution function whose variance is $-i\hbar/\Phi''$. We would expect this to integrate to give us

$$\sqrt{2\pi\hbar\Phi''(x,p_\pm)}e^{\pm i\frac{\pi}{4}}, \tag{15.41}$$

and this may be justified with appropriate contour integrals. Introducing a new constant C, we are left with

$$Ce^{\pm i\frac{\pi}{4}}\left(\Phi''\right)^{-\frac{1}{2}} e^{i\Phi(x,p_\pm)}. \tag{15.42}$$

We can directly calculate that

$$\Phi''(x,p_\pm)^{-\frac{1}{2}} = \sqrt{V'(c)/p_\pm} = \sqrt{\frac{mV'(c)}{2(c-x)}}, \tag{15.43}$$

which is approximately a multiple of the semi-classical expression for the amplitude, a. We have already noted that $\Phi(x,p_\pm)$ is approximately given by the same integral as W, so this confirms that we can obtain the same result by a method that is valid near the turning points.

For the non-classical region, $x > c$, we allow p to become complex in the integral in equation (15.34). The integrand is holomorphic, and by standard complex variable techniques the integral can be shown to be identical to that along the hyperbolic contour on which

$$p = -i\sqrt{2mV'(c)(x-c)^3}\left(\omega e^t + \omega^2 e^{-t}\right), \tag{15.44}$$

where $\omega = \exp(2\pi i/3)$. A straightforward substitution shows that on this contour the integral reduces to a multiple of

$$e^{-\frac{2}{3}\sqrt{2mV'(c)(x-c)^3}} \int e^{-\frac{2}{3}\sqrt{2mV'(c)(x-c)^3}[\cosh(3t)-1]}\left(\omega e^t - \omega^2 e^{-t}\right) dt. \tag{15.45}$$

This shows immediately that one has a decaying exponential of exactly the form derived earlier. More careful analysis of the integral gives the other factors of the previous calculation.

In recent years the semi-classical approximation has been given a firm foundation in differential geometry. The behaviour near the classical turning points is of particular interest, since it is linked with the mathematical study of singularities.

Exercises

15.1° Use the WKB approximation to obtain estimates of the energy levels of the harmonic oscillator with potential $\frac{1}{2}m\omega^2x^2$.

15.2 Calculate the additional potential, $-\hbar^2 a''/2ma$, in the case of an harmonic oscillator eigenstate. Show that this tends to a constant if the energy is large.

15.3° Derive the equations for the WKB approximation for a stationary state of a one-dimensional system with energy E. Solve the equations and deduce that the WKB approximation gives an exact solution in this case if and only if the potential $V(x)$ is given by

$$V(x) = E - (\alpha x + \beta)^{-4},$$

for some constants α and β. For which potentials does the WKB approximation give an exact solution of Schrödinger's equation for all energies E?

16 Systems of several particles

One now ceases to understand the quantum theory at all.

WERNER HEISENBERG, letter to Wolfgang Pauli, 9 October 1923

16.1. Identical particles

One often wants to study systems, such as the helium atom of Sections 12.3 and 14.2, that contain more than one particle. We describe two-particle systems by a wave function, $\psi(\mathbf{r}_1, \mathbf{r}_2)$, where $\mathbf{r}_1$ is the position of the first particle, and $\mathbf{r}_2$ that of the second. If we interchange the two particles, then the wave function becomes $\psi(\mathbf{r}_2, \mathbf{r}_1)$. Suppose, however, that the two particles are absolutely identical, and that there is no way of distinguishing whether the first is at $\mathbf{r}_1$ and the second at $\mathbf{r}_2$ or vice versa. (So far as is known, this is the case for electrons and many other subatomic particles.) The two wave functions $\psi(\mathbf{r}_1, \mathbf{r}_2)$ and $\psi(\mathbf{r}_2, \mathbf{r}_1)$ must then represent the same physical state, and so must be multiples of one another. In other words, we can write

$$\psi(\mathbf{r}_1, \mathbf{r}_2) = \lambda\psi(\mathbf{r}_2, \mathbf{r}_1), \tag{16.1}$$

for some complex number λ. Since we have no way of knowing which particle is where, we must also have $\psi(\mathbf{r}_2, \mathbf{r}_1) = \lambda\psi(\mathbf{r}_1, \mathbf{r}_2)$, which means that

$$\psi(\mathbf{r}_1, \mathbf{r}_2) = \lambda^2\psi(\mathbf{r}_1, \mathbf{r}_2), \tag{16.2}$$

and tells us that $\lambda = \pm 1$. Moreover, as we shall now show, all wave functions must give the same choice of sign for λ. For, if ψ_- changes sign when its arguments are interchanged and ψ_+ does not, then $\psi_+ + \psi_-$ changes to $\psi_+ - \psi_-$ when its arguments are transposed. These can only be multiples of each other if one of ψ_+ and ψ_- vanishes, which means that one or other sign cannot occur.

It is now a simple matter to consider systems with an arbitrary number, n, of indistinguishable particles. We cannot tell which of the particles is at which point, so for any permutation $\pi \in S_n$, any physically acceptable wave function must satisfy

$$\psi(\mathbf{r}_1, \mathbf{r}_2, \ldots, \mathbf{r}_n) = \lambda(\pi)\psi(\mathbf{r}_{\pi(1)}, \mathbf{r}_{\pi(2)}, \ldots, \mathbf{r}_{\pi(n)}) \tag{16.3}$$

for some $\lambda(\pi) \in \mathbf{C}$. If we follow one permutation, π, by another, σ, then we see that

$$\lambda(\sigma\pi) = \lambda(\sigma)\lambda(\pi). \tag{16.4}$$

(In the terminology of Definition 9.1.1, λ defines a one-dimensional representation of the symmetric group S_n.) From this we deduce that

$$\lambda(\sigma\pi\sigma^{-1}) = \lambda(\sigma)\lambda(\pi)\lambda(\sigma)^{-1} = \lambda(\pi), \tag{16.5}$$

and so we see that conjugate elements give the same scalar $\lambda(\pi)$. Now, any transposition is conjugate to the interchange of 1 and 2, since, for example,

$$(r\,s) = (1\,r)(2\,s)(1\,2)(2\,s)^{-1}(1\,r)^{-1}, \tag{16.6}$$

and we therefore deduce that either every transposition, τ, gives $\lambda(\tau) = 1$, or they all give $\lambda(\tau) = -1$. As librarians know only too well, the symmetric group is generated by transpositions, so in the first case every permutation, π, has $\lambda(\pi) = 1$, whilst in the second $\lambda(\pi) = 1$ for even permutations (those which are products of even numbers of transpositions), but $\lambda(\pi) = -1$ for odd permutations. This proves the following result:

Proposition 16.1.1. If the wave function ψ is sent to a multiple of itself by every permutation π in S_n, then either

$$\psi(\mathbf{r}_1, \mathbf{r}_2, \ldots, \mathbf{r}_n) = \psi(\mathbf{r}_{\pi(1)}, \mathbf{r}_{\pi(2)}, \ldots, \mathbf{r}_{\pi(n)})$$

or

$$\psi(\mathbf{r}_1, \mathbf{r}_2, \ldots, \mathbf{r}_n) = \epsilon(\pi)\psi(\mathbf{r}_{\pi(1)}, \mathbf{r}_{\pi(2)}, \ldots, \mathbf{r}_{\pi(n)})$$

where

$$\epsilon(\pi) = \begin{cases} 1 & \text{for even permutations } \pi \\ -1 & \text{for odd permutations } \pi. \end{cases}$$

Both these possibilities occur in nature and they divide the known particles into two classes.

Definition 16.1.1. If $\psi(\mathbf{r}_1, \mathbf{r}_2, \ldots, \mathbf{r}_n) = \psi(\mathbf{r}_{\pi(1)}, \mathbf{r}_{\pi(2)}, \ldots, \mathbf{r}_{\pi(n)})$ for all permutations π we call the individual particles *bosons*, or say that they satisfy *Bose–Einstein statistics.* If for odd permutations, π, the wave function satisfies $\psi(\mathbf{r}_1, \mathbf{r}_2, \ldots, \mathbf{r}_n) = -\psi(\mathbf{r}_{\pi(1)}, \mathbf{r}_{\pi(2)}, \ldots, \mathbf{r}_{\pi(n)})$, then we call the individual particles *fermions*, or say that they satisfy *Fermi–Dirac statistics.*

Example 16.1.1. Of the known particles, electrons, protons, neutrons, and quarks are fermions, whilst photons and mesons are bosons. Whether a particle is a fermion or a boson is linked to its intrinsic spin. (Just as the electron is described by two-component wave functions corresponding to the value $l = \frac{1}{2}$ of the angular momentum, so in general one requires $(2l+1)$-component wave functions and talks of the particle having spin l.) It can be shown that particles with integral spin should be bosons, and those with non-integer spin should be fermions, a result known as the *spin-statistics theorem.* This was proved by Jacobus de Wet in 1940 and the first published proof given by Pauli ten years later.

16.2. Bosons and fermions

The behaviour under permutations places strong restrictions on the sort of wave functions that can represent boson or fermion states, but it is nonetheless quite easy to construct wave functions of the appropriate kind, by averaging.

Proposition 16.2.1. For any wave function, ψ, on $\mathbf{R}^{3n}$, the wave function

$$(Q_\lambda\psi)(\mathbf{r}_1, \mathbf{r}_2, \ldots, \mathbf{r}_n) = \frac{1}{n!}\sum_{\pi\in S_n} \lambda(\pi)\psi(\mathbf{r}_{\pi(1)}, \mathbf{r}_{\pi(2)}, \ldots, \mathbf{r}_{\pi(n)})$$

satisfies $(Q_\lambda\psi)(\mathbf{r}_1, \mathbf{r}_2, \ldots, \mathbf{r}_n) = \lambda(\pi)(Q_\lambda\psi)(\mathbf{r}_{\pi(1)}, \mathbf{r}_{\pi(2)}, \ldots, \mathbf{r}_{\pi(n)})$ for all $\pi \in S_n$. Q_λ is a projection.

Proof. This is proved by direct calculation. For any permutation $\sigma \in S_n$ we have

$$(Q_\lambda\psi)(\mathbf{r}_{\sigma(1)}, \mathbf{r}_{\sigma(2)}, \ldots, \mathbf{r}_{\sigma(n)}) = \frac{1}{n!}\sum_{\pi\in S_n} \lambda(\pi)\psi(\mathbf{r}_{\pi\sigma(1)}, \mathbf{r}_{\pi\sigma(2)}, \ldots, \mathbf{r}_{\pi\sigma(n)}), \tag{16.7}$$

and as π runs over all permutations $\pi\sigma$ does too (albeit in a different order), so we may introduce $\rho = \pi\sigma$ and rewrite the sum on the right as

$$\frac{1}{n!}\sum_{\rho\in S_n} \lambda(\rho\sigma^{-1})\psi(\mathbf{r}_{\rho(1)}, \mathbf{r}_{\rho(2)}, \ldots, \mathbf{r}_{\rho(n)}) = \lambda(\sigma)^{-1}(Q_\lambda\psi)(\mathbf{r}_1, \mathbf{r}_2, \ldots, \mathbf{r}_n). \tag{16.8}$$

We also note that if each permutation, π, of the arguments of a wave function ϕ merely introduces a factor $\lambda(\pi)$, then, since $\lambda(\pi)^2 = 1$, each factor in the sum defining $Q_\lambda\phi$ is identical, and since there are $n!$ terms we have $Q_\lambda\phi = \phi$. Applying this to $\phi = Q_\lambda\psi$, we see that $Q_\lambda^2\psi = Q_\lambda\psi$, showing that Q_λ is a projection. □

By taking $\lambda \equiv 1$, we see that if

$$(Q_1\psi)(\mathbf{r}_1, \mathbf{r}_2, \ldots, \mathbf{r}_n) = \frac{1}{n!}\sum_{\pi\in S_n} \psi(\mathbf{r}_{\pi(1)}, \mathbf{r}_{\pi(2)}, \ldots, \mathbf{r}_{\pi(n)}) \tag{16.9}$$

is not identically zero then it is a bosonic wave function, and, similarly, taking $\lambda = \epsilon$

$$(Q_\epsilon\psi)(\mathbf{r}_1, \mathbf{r}_2, \ldots, \mathbf{r}_n) = \frac{1}{n!}\sum_{\pi\in S_n} \epsilon(\pi)\psi(\mathbf{r}_{\pi(1)}, \mathbf{r}_{\pi(2)}, \ldots, \mathbf{r}_{\pi(n)}) \tag{16.10}$$

is a fermionic wave function, provided it is not the zero function. If ψ is already bosonic, or fermionic, then $Q_1\psi = \psi$ or $Q_\epsilon\psi = \psi$, respectively.

When $n = 2$ the formulae simplify considerably to give

$$\begin{aligned}(Q_1\psi)(\mathbf{r}_1, \mathbf{r}_2) &= \tfrac{1}{2}\left[\psi(\mathbf{r}_1, \mathbf{r}_2) + \psi(\mathbf{r}_2, \mathbf{r}_1)\right] \\ (Q_\epsilon\psi)(\mathbf{r}_1, \mathbf{r}_2) &= \tfrac{1}{2}\left[\psi(\mathbf{r}_1, \mathbf{r}_2) - \psi(\mathbf{r}_2, \mathbf{r}_1)\right].\end{aligned} \tag{16.11}$$

In this case we can express the original wave function as the sum of $Q_1\psi$ and $Q_\epsilon\psi$, but this fails for larger n. There have been various attempts to build theories that include other kinds of particle statistics, but, although these have other uses, they do not seem to be needed for the basic particles observed in nature. (In some parts of solid state physics, where electrons are effectively confined to a surface, it is useful to allow the possibility that transposition of the particle positions changes the wave function by a factor, λ, which is neither 1 nor -1. This gives rise to so-called *anyon* statistics.)

One interesting case of the above expressions for bosonic and fermionic wave functions occurs for a separable function $\psi_1(\mathbf{r}_1)\psi_2(\mathbf{r}_2)$. Its fermionic projection is

$$Q_\epsilon\psi(\mathbf{r}_1, \mathbf{r}_2) = \tfrac{1}{2}\left[\psi_1(\mathbf{r}_1)\psi_2(\mathbf{r}_2) - \psi_1(\mathbf{r}_2)\psi_2(\mathbf{r}_1)\right] = \tfrac{1}{2}\begin{vmatrix}\psi_1(\mathbf{r}_1) & \psi_2(\mathbf{r}_1)\\ \psi_1(\mathbf{r}_2) & \psi_2(\mathbf{r}_2)\end{vmatrix}. \tag{16.12}$$

It is easy to generalize this to n particles, to see that if $\psi(\mathbf{r}_1, \mathbf{r}_2, \ldots, \mathbf{r}_n) = \prod_{j=1}^n \psi_j(\mathbf{r}_j)$. Then

$$Q_\epsilon\psi(\mathbf{r}_1, \mathbf{r}_2, \ldots, \mathbf{r}_n) = \frac{1}{n!}\begin{vmatrix}\psi_1(\mathbf{r}_1) & \psi_2(\mathbf{r}_1) & \ldots & \psi_n(\mathbf{r}_1)\\ \psi_1(\mathbf{r}_2) & \psi_2(\mathbf{r}_2) & \ldots & \psi_n(\mathbf{r}_2)\\ \vdots & \vdots & & \vdots \\ \psi_1(\mathbf{r}_n) & \psi_2(\mathbf{r}_n) & \ldots & \psi_n(\mathbf{r}_n)\end{vmatrix}. \tag{16.13}$$

This form is sometimes called a Slater determinant. It immediately leads us to Pauli's exclusion principle:

Theorem 16.2.2. If any two wave functions ψ_j are the same then $Q_\epsilon(\psi_1\psi_2\ldots\psi_n)$ vanishes.

Proof. A determinant vanishes if two of its columns are the same. □

Pauli's exclusion principle may be paraphrased to say that two fermions cannot simultaneously occupy the same state. (The reader may be wondering why we did not mention this when we discussed the helium atom, and be uneasy that we used a zeroth-order wave function in which both electrons were in the same ground state. This is permissible because we neglected the electron spin. Had we included it we should have had one electron in one spin state and the other in an orthogonal state, thereby avoiding a clash with the Pauli principle.) In fact it is possible to extract a more powerful result by the same procedure. Suppose that the individual wave functions, ψ_j, have to be selected from an N-dimensional space, such as the space of wave functions corresponding to a particular, degenerate, energy level. The properties of a determinant mean that it clearly vanishes if any of its columns are linear combinations of the others.

Theorem 16.2.3. Suppose that n fermions can each be described by wave functions in an N-dimensional space. Then the number of independent fermion states of the n particles is $\binom{N}{n}$.

Proof. Let $\{\psi_1, \psi_2, \ldots, \psi_N\}$ be an orthonormal basis for the space of single fermions. Any state of n fermions can be constructed as a combination of Slater determinants of size n. The order of the columns is unimportant since it can only affect the sign, and there are $\binom{N}{n}$ ways of choosing a set of n columns from the N basis vectors. □

We may similarly count states of n bosons.

Theorem 16.2.4. Suppose that n bosons can each be described by wave functions in an N-dimensional space. Then the number of independent boson states of the n particles is $\binom{N+n-1}{n}$.

Proof. In this case the bosonic part of the product $\psi_1 \dots \psi_n$ is

$$\sum_{\pi} \psi_1(\mathbf{r}_{\pi(1)})\psi_2(\mathbf{r}_{\pi(1)}) \dots \psi_n(\mathbf{r}_{\pi(n)}). \tag{16.14}$$

There are again N independent choices for each ψ_j, but this time repetitions are permitted, so, typically, ψ_j might occur k_j times. The number of independent choices is therefore the same as the number of ways of taking products of monomials in N variables, $x_1, \dots, x_N$, that is expressions of the form

$$x_1^{k_1} \dots x_N^{k_N} \tag{16.15}$$

whose total degree $k_1 + k_2 + \dots + k_N$ is n. It is easy to see that these terms arise as the coefficient of the s^n term in

$$\sum_{k_1,\dots,k_N} (sx_1)^{k_1}(sx_2)^{k_2} \dots (sx_N)^{k_n} = \prod_j \sum (sx_j)^{k_j} = \prod_j (1 - sx_j)^{-1}. \tag{16.16}$$

On setting $x_j = 1$ for all j, the right-hand side becomes just $(1 - s)^{-N}$, whilst the coefficient of s^n on the left-hand side counts the number of possible terms of total degree n. The binomial theorem tells us that

$$(1 - s)^{-N} = \sum \binom{N + n - 1}{n} s^n, \tag{16.17}$$

from which the result follows. □

16.3. The periodic table

Despite its simplicity Pauli's exclusion principle is of crucial importance in the physical world, for it prevents the fermions of which most ordinary matter is composed from accumulating together in the lowest energy state. Without it solid matter could not exist in stable forms. Moreover, it determines the nature of the elements of which ordinary matter is composed.

Let us first consider a hydrogen-like atom, but taking into account electron spin and the exclusion principle, both of which were neglected previously. As we observed in Section 8.8 the electron has spin $\frac{1}{2}$, and its wave function should really be $\mathbf{C}^2$ valued. (We may think of the two basis vectors as representing states with spin up and spin down.) There is now a two-dimensional space of possible ground states for the electron, spanned, in the notation of Proposition 4.3.1, by

$$\psi_{100}(\mathbf{r})\begin{pmatrix}1\\0\end{pmatrix} \qquad \text{and} \qquad \psi_{100}(\mathbf{r})\begin{pmatrix}0\\1\end{pmatrix}. \tag{16.18}$$

Similarly there are now $2n^2$ states with energy $Ze^2/8\pi\epsilon_0 n^2 a$, since each of the n^2 independent wave functions can be combined with two possible spin states. As we add more electrons they tend naturally to fill the lowest energy states first. However, Pauli's exclusion principle stops them from all occupying the ground state, as there is room for only two. After the first two electrons, the first excited state will be filled, where there is room for $2 \times 2^2 = 8$ electrons, and so on. We thus see the periodic table building up, with hydrogen and helium in the first row, followed by the eight elements lithium, beryllium, boron, carbon, nitrogen, oxygen, fluorine, and neon in the second, and so on. Of course, we have oversimplified by ignoring the interactions between electrons and by assuming that the energies of the states are independent of the electron spin, which is not quite true. With such refinements it is possible to understand the whole of the periodic table in terms of the above principles.

Moreover, we can also understand the notion of valence. For example, oxygen has room for two of its eight electrons in the ground state whilst the remainder can occupy six of the eight available places amongst the first excited states. This leaves two more places amongst the first excited states for electrons poached from other atoms, and is responsible for oxygen's valency of 2. Of course, the above discussion is rather vague and does not, for example, immediately explain why two oxygen atoms can bind together to form an oxygen molecule. Nonetheless, it can be refined with a more precise mathematical analysis based upon detailed Schrödinger equations coupled with the requirement that the n-electron state vector must be antisymmetric. Although in practice the resulting equations can only be solved numerically, the principles involved seem to be correct, so that one might almost say that chemistry is an exploration of the consequences of Pauli's exclusion principle.

The periodic table is not the only important manifestation of Pauli's exclusion principle. One much more extreme example is provided by neutron stars. In younger and middle-aged stars like our sun, fusion processes convert hydrogen nuclei into those of heavier elements, and the radiation pressure from these thermonuclear reactions is sufficient to stop the star from collapsing under its own gravitational attraction. Eventually, however, when all the possible thermonuclear fuel is exhausted and after a complicated series of transformations, the gravitational forces start to win. The star's atoms, ionized aeons earlier in its thermonuclear furnace, are now subjected to conditions so extreme that their nuclei and electrons interact to form neutrons (and other particles that are radiated away). These are compressed by the gravitational forces into a neutron star. However, unless the mass is so great or confined within so small a radius as to form a black hole, there is then a respite, because the neutrons, being fermions, cannot all occupy the state of least gravitational energy, to which they would otherwise be inexorably pulled.

16.4. Bose–Einstein condensation

By contrast with fermions, bosons are gregarious and occupy the same state more often than one might have expected. For example, photons are bosons, and lasers create large numbers of them in the same state. At sufficiently low temperatures large numbers of bosons will all fill the lowest energy state, in a process known as Bose–Einstein condensation. In 1995 a team of physicists in Boulder, Colorado, succeeded in cooling around 2000 rubidium atoms to a temperature of 2×10^{-7} degrees above absolute zero, so that they all condensed into a single clump of around 10^{-4} metres diameter. Even earlier, however, the effects of Bose–Einstein condensation had made themselves apparent in an indirect way in the anomalous behaviour of liquid helium. Helium is found in two isotopes, ^{3}He and ^{4}He. The commoner isotope ^{4}He, of atomic weight 4, has a nucleus made of two protons and two neutrons. Since all these constituents have spin $\frac{1}{2}$, the nucleus has integral spin and is a boson. At temperatures below 2 K (well below its boiling point of 4 K) it becomes a *superfluid*, able to climb up the walls of any container, and having virtually no viscosity. The nucleus of the rarer ^{3}He contains only one neutron, and so it is a fermion. It exhibits anomalous behaviour only below about 0.002 K, when the nuclei pair up to form bosons.

In 1911 Kamerlingh Onnes discovered that some materials appear to lose all resistance to the flow of electric currents at low enough temperatures, a phenomenon known as *superconductivity*. Leon Cooper suggested in 1956 that this was the result of a mechanism, similar to that in ^{3}He, whereby the electrons carrying the currents pair up. These *Cooper pairs*, being bosons, can show Bose–Einstein condensation, which enables them to carry the current freely. It was long believed that the electron pairing was caused by interactions with the crystal lattice of the superconductor. This was the BCS theory, elaborated by Bardeen, Cooper, and Schrieffer, and for which they received the 1972 Nobel Prize in Physics. However, this theory suggested that no material could be superconducting above about 25 K. The discovery of high temperature superconductors in 1985, whose structure was far more complicated than the previously known materials, showed that this could not be the whole story. It is believed, however, that the idea of some kind of pairing of charge carriers (not necessarily the electrons) is still valid.

In 1962 Brian Josephson, then a student at Cambridge, realized that Cooper pairs would be able to tunnel through a sufficiently thin insulator placed in a superconducting circuit. These 'Josephson junctions' form the basis of superconducting quantum interference device or SQUID, an extremely sensitive tool for measuring magnetic fields. SQUIDs exhibit quantum effects on an everyday scale, and have been proposed for some extremely sensitive tests of quantum mechanical effects.

16.5*. Tensor products

So far our description of many-particle systems has been limited to wave functions, but often a more algebraic approach is needed. In the general case where the two quantum states are described by vectors in spaces $\mathcal{H}_1$ and $\mathcal{H}_2$ we want to construct an inner product space $\mathcal{H}$, whose vectors represent states of the combined system. The most obvious approach is to choose orthonormal bases $\{\xi_j\}_{j\in J}$ and $\{\eta_k\}_{k\in K}$ for $\mathcal{H}_1$ and $\mathcal{H}_2$, and then pick any inner product space $\mathcal{H}$ with an orthonormal basis of the form $\{\zeta_{jk} : (j,k) \in J \times K\}$. This is consistent with the approach for wave functions, since in that case we could simply take for $\zeta_{jk}(\mathbf{r}_1, \mathbf{r}_2)$ the separable wave function, $\xi_j(\mathbf{r}_1)\eta_k(\mathbf{r}_2)$. Although not every wave function of two variables is separable, we expect that they can all be expanded as an (infinite) sum of separable solutions, so we would expect $\mathcal{H}$ to include all two-particle wave functions. In order to distinguish the kind of product appearing in separable solutions, $\zeta_{jk}(\mathbf{r}_1, \mathbf{r}_2) = \xi_j(\mathbf{r}_1)\eta_k(\mathbf{r}_2)$, from the ordinary pointwise product where we multiply the values of ξ_j and η_k at the same point, we shall henceforth write it as $\zeta_{jk} = \xi_j \otimes \eta_k$. (The symbol $\otimes$ is read as *tensor.*)

Unfortunately our construction of $\mathcal{H}$ is quite arbitrary. Had we chosen different bases for $\mathcal{H}_1$ and $\mathcal{H}_2$ it is far from obvious that we should have ended up with the same space $\mathcal{H}$ for the combined system. It is therefore more useful to try to characterize the space intrinsically without using bases. Intuitively we expect $\mathcal{H}$ to incorporate both the single-particle spaces, and to be the smallest space with this property (that is, any other space that contains both the single-particle spaces should also include $\mathcal{H}$). These two requirements correspond to the two conditions of the following definition. (We recall that a bilinear map from $\mathcal{H}_1 \times \mathcal{H}_2$ to $\mathcal{K}$ is a map that is linear in each of its two arguments.)

Definition 16.5.1. Let $\mathcal{H}_1$ and $\mathcal{H}_2$ be inner product spaces. Suppose that there exists a space $\mathcal{H}_1 \otimes \mathcal{H}_2$ with the following properties:
(i) there is a bilinear map from $\mathcal{H}_1 \times \mathcal{H}_2$ to $\mathcal{H}_1 \otimes \mathcal{H}_2$ written as $(\psi_1, \psi_2) \mapsto \psi_1 \otimes \psi_2$;
(ii) for any bilinear map, β, from $\mathcal{H}_1 \times \mathcal{H}_2$ to an inner product space $\mathcal{K}$, there is also a unique linear map $\hat{\beta}$ from $\mathcal{H}_1 \otimes \mathcal{H}_2$ to $\mathcal{K}$ satisfying

$$\hat{\beta}(\psi_1 \otimes \psi_2) = \beta(\psi_1, \psi_2).$$

Then $\mathcal{H}_1 \otimes \mathcal{H}_2$ is called the *tensor product* of $\mathcal{H}_1$ and $\mathcal{H}_2$, and its elements are called tensors. The inner product on $\mathcal{H}_1 \otimes \mathcal{H}_2$ is defined by

$$\langle \phi_1 \otimes \phi_2 | \psi_1 \otimes \psi_2 \rangle = \langle \phi_1 | \psi_1 \rangle \langle \phi_2 | \psi_2 \rangle.$$

It is fairly clear that the space $\mathcal{H}$ constructed in terms of bases has these properties. Any vectors in $\mathcal{H}_1$ and $\mathcal{H}_2$ can be expanded as $\psi_1 = \sum x_j \xi_j$ and $\psi_2 = \sum y_k \eta_k$ and we have the bilinear map

$$(\psi_1, \psi_2) \mapsto \sum x_j y_k \zeta_{jk}. \tag{16.19}$$

This means that $\zeta_{jk} = \xi_j \otimes \eta_k$. Given $\beta : \mathcal{H}_1 \times \mathcal{H}_2 \to \mathcal{K}$ we need a linear transformation $\hat{\beta} : \mathcal{H} \to \mathcal{K}$ which must in particular satisfy $\hat{\beta}(\xi_j \otimes \eta_k) = \beta(\xi_j, \eta_k)$. This forces us to take

$$\hat{\beta}\left(\sum z_{jk} \zeta_{jk}\right) = \sum z_{jk} \beta(\xi_j, \eta_k), \tag{16.20}$$

so that $\hat{\beta}$ is uniquely defined on a general element of $\mathcal{H}$. Moreover, it is easy to check that it satisfies all the requirements of the definition. This shows that our definition is not vacuous, but we must now check that there is only one possible space $\mathcal{H}_1 \otimes \mathcal{H}_2$.

Proposition 16.5.1. Any two tensor product spaces satisfying the requirements of the definition are isomorphic.

Proof. Suppose that besides $\mathcal{H}_1 \otimes \mathcal{H}_2$ the space $\mathcal{H}_1 \odot \mathcal{H}_2$ also satisfies the provisions of the definition. By the first of the two conditions there is a bilinear map from $\mathcal{H}_1 \times \mathcal{H}_2$ to $\mathcal{H}_1 \odot \mathcal{H}_2$, which we denote by $\odot : (\psi_1, \psi_2) \mapsto \psi_1 \odot \psi_2$. But then, by the second part of the definition of $\mathcal{H}_1 \otimes \mathcal{H}_2$, there exists a linear map $\hat{\odot}$ from $\mathcal{H}_1 \otimes \mathcal{H}_2$ to $\mathcal{H}_1 \odot \mathcal{H}_2$. Reversing the roles of the two contenders for the tensor product we also have a linear map $\hat{\otimes}$ from $\mathcal{H}_1 \odot \mathcal{H}_2$ to $\mathcal{H}_1 \otimes \mathcal{H}_2$. Since each is unique, it is easy to see that $\hat{\odot}$ and $\hat{\otimes}$ must be mutually inverse maps, and so set up the required isomorphism between the two spaces. □

The construction of $\mathcal{H} = \mathcal{H}_1 \otimes \mathcal{H}_2$ in terms of bases immediately gives us its dimension.

Proposition 16.5.2. If $\mathcal{H}_1$ and $\mathcal{H}_2$ are finite dimensional then the dimension of the tensor product is

$$\dim(\mathcal{H}_1 \otimes \mathcal{H}_2) = \dim \mathcal{H}_1 \dim \mathcal{H}_2.$$

We can also easily prove some other useful properties.

Proposition 16.5.3. For any spaces $\mathcal{H}_1$, $\mathcal{H}_2$ and $\mathcal{H}$, $\mathcal{H}^*$, the dual of $\mathcal{H}$, and $\mathcal{L}(\mathcal{H}_1, \mathcal{H}_2)$, the linear transformations from $\mathcal{H}_1$ to $\mathcal{H}_2$, we have
(i) $\mathbf{C} \otimes \mathcal{H} \cong \mathcal{H} \cong \mathcal{H} \otimes \mathbf{C}$;
(ii) $\mathcal{H}_1 \otimes (\mathcal{H}_2 \otimes \mathcal{H}) \cong (\mathcal{H}_1 \otimes \mathcal{H}_2) \otimes \mathcal{H}$;
(iii) $\mathcal{H} \otimes (\mathcal{H}_1 \oplus \mathcal{H}_2) \cong (\mathcal{H} \otimes \mathcal{H}_1) \oplus (\mathcal{H} \otimes \mathcal{H}_2)$;
(iv) $\mathcal{H}_2 \otimes \mathcal{H}_1^* \cong \mathcal{L}(\mathcal{H}_1, \mathcal{H}_2)$.

Proof. (i) Now that we know that the tensor product space is essentially unique we need only check that $\mathcal{H}$ satisfies the properties required of $\mathbf{C} \otimes \mathcal{H}$ to know that they are isomorphic. Now the scalar product that takes (λ, ψ) to $\lambda\psi$ is a bilinear map from $\mathbf{C} \times \mathcal{H}$ to $\mathcal{H}$. Given any bilinear map β from $\mathbf{C} \times \mathcal{H}$ to $\mathcal{K}$ we may define a linear map $\hat{\beta}$ from $\mathcal{H}$ to $\mathcal{K}$ by

$$\hat{\beta}(\psi) = \beta(1, \psi). \tag{16.21}$$

Since

$$\hat{\beta}(\lambda\psi) = \beta(1, \lambda\psi) = \lambda\beta(1, \psi) = \beta(\lambda, \psi), \tag{16.22}$$

this map satisfies the compatibility condition, and it is easily seen to be unique, so the result follows. The proofs of the other parts follow the same lines as these. The isomorphism in the last part sends $\xi \otimes \eta \in \mathcal{H}_2 \otimes \mathcal{H}_1^*$ to the linear transformation, $\psi \in \mathcal{H}_1 \mapsto \eta(\psi)\xi$. □

The above proofs show that the real purpose of the second condition in the definition is really to enable us to extend to the whole of the tensor product space maps which are initially defined only on elements of the form $\psi_1 \otimes \psi_2$, and this is why we only needed to define the inner product for tensors of the form $\psi_1 \otimes \psi_2$.

Definition 16.5.2. Vectors that can be written in the form $\psi_1 \otimes \psi_2$ are called *decomposable tensors.*

As already mentioned decomposable tensors are the algebraic analogue of separable solutions of a differential equation. Just as most solutions are not separable, most tensors are not themselves decomposable, but are linear combinations of decomposable tensors and this fact leads to the phenomenon known as *entanglement*, whereby the properties of one particle depend to some extent on those of another. We saw an example of this in the case of the two Einstein–Podolsky–Rosen photons in Section 10.4. Entanglement posed one of the major obstacles to the calculation of the energy levels of helium and more complex atoms before the development of quantum theory. Although perturbation and variational methods start with separable wave functions, it is easy to check that the true wave functions do not decompose in that way.

We can, of course extend the notion of tensor product to more than just two particles.

Definition 16.5.3. Let $\mathcal{H}_1, \mathcal{H}_2, \ldots, \mathcal{H}_n$ and $\mathcal{K}$ be inner product spaces.

A map from $\mathcal{H}_1 \times \mathcal{H}_2 \times \ldots \times \mathcal{H}_n$ to $\mathcal{K}$ which is linear in each argument is said to be *n-linear.*

Definition 16.5.4. In the same notation as the previous definition suppose that there is a space, $\mathcal{H}_1 \otimes \mathcal{H}_2 \otimes \ldots \otimes \mathcal{H}_n$, that satisfies the two conditions that
(i) there is an n-linear map from

$$\mathcal{H}_1 \times \mathcal{H}_2 \times \ldots \times \mathcal{H}_n \to \mathcal{H}_1 \otimes \mathcal{H}_2 \otimes \ldots \otimes \mathcal{H}_n$$

which takes $(\psi_1, \psi_2, \ldots, \psi_n)$ to the vector $\psi_1 \otimes \psi_2 \otimes \ldots \otimes \psi_n$;
(ii) for any n-linear map β from $\mathcal{H}_1 \times \mathcal{H}_2 \times \ldots \times \mathcal{H}_n$ to $\mathcal{K}$ there is a linear map $\hat{\beta}$ from $\mathcal{H}_1 \otimes \mathcal{H}_2 \otimes \ldots \otimes \mathcal{H}_n$ to $\mathcal{K}$ such that

$$\hat{\beta}(\psi_1 \otimes \ldots \otimes \psi_n) = \beta(\psi_1, \psi_2, \ldots, \psi_n);$$

then $\mathcal{H}_1 \otimes \mathcal{H}_2 \otimes \ldots \otimes \mathcal{H}_n$ is said to be the *tensor product* of the spaces $\mathcal{H}_1, \mathcal{H}_2, \ldots, \mathcal{H}_n$.

The inner products are given by the obvious product

$$\langle \phi_1 \otimes \ldots \otimes \phi_n | \psi_1 \otimes \ldots \otimes \psi_n \rangle = \prod \langle \phi_j | \psi_j \rangle.$$

We also need to consider observables in quantum mechanics, which means looking at operators on tensor products. If A_j is a linear operator on $\mathcal{H}_j$, for $j = 1, \ldots, n$, then there is an obvious operator $A_1 \otimes A_2 \otimes \ldots \otimes A_n$ on $\mathcal{H}_1 \otimes \mathcal{H}_2 \otimes \ldots \otimes \mathcal{H}_n$, defined by

$$(A_1 \otimes A_2 \otimes \ldots \otimes A_n)(\psi_1 \otimes \psi_2 \otimes \ldots \otimes \psi_n) = A_1\psi_1 \otimes A_2\psi_2 \otimes \ldots \otimes A_n\psi_n. \quad (16.23)$$

It is common to write $A(k)$ for this product when $A_k = A$ and $A_j = 1$ for $j \neq k$. (Thus, for example, one writes $A(1) = A \otimes 1$ and $A(2) = 1 \otimes A$.)

16.6*. Symmetric and antisymmetric tensors

Tensor products provide an algebraic description of systems of several particles, whether they may be distinguished or not. We shall now consider how to describe fermions and bosons algebraically. There are two equivalent, but slightly different, ways of doing this. The first is to build the permutation symmetries into the definition by using only those multilinear forms that have the correct behaviour under permutations. Since they are indistinguishable each individual particle is described by vectors in the same inner product space $\mathcal{H}_0$.

Definition 16.6.1. Let $\mathcal{H}_0$ and $\mathcal{K}$ be inner product spaces. An n-linear map

$$\beta : \mathcal{H}_0 \times \mathcal{H}_0 \times \ldots \times \mathcal{H}_0 \to \mathcal{K}$$

is said to be *symmetric* if $\beta(\xi_1, \xi_2, \ldots, \xi_n) = \beta(\xi_{\pi(1)}, \xi_{\pi(2)}, \ldots, \xi_{\pi(n)})$, for all $\pi \in S_n$, and *antisymmetric* if

$$\beta(\xi_1, \xi_2, \ldots, \xi_n) = \epsilon(\pi)\beta(\xi_{\pi(1)}, \xi_{\pi(2)}, \ldots, \xi_{\pi(n)}),$$

where ϵ is defined as in Proposition 16.1.1.

Definition 16.6.2. Let $\mathcal{H}_0$ be an inner product space. Suppose that there exists a space $\bigotimes_S^n \mathcal{H}_0 = \mathcal{H}_0 \otimes_S \ldots \otimes_S \mathcal{H}_0$ with the following properties:
(i) there is a symmetric n-linear map from $\mathcal{H}_0 \times \ldots \times \mathcal{H}_0$ to $\bigotimes_S^n \mathcal{H}_0$ written as $(\psi_1, \ldots, \psi_n) \mapsto \psi_1 \otimes_S \ldots \otimes_S \psi_n$;
(ii) for any symmetric n-linear map β from $\mathcal{H}_0 \times \ldots \times \mathcal{H}_0$ to a space $\mathcal{K}$, there is a unique linear map $\hat{\beta}$ from $\bigotimes_S^n \mathcal{H}_0$ to $\mathcal{K}$ satisfying $\hat{\beta}(\psi_1 \otimes_S \ldots \otimes_S \psi_2) = \beta(\psi_1, \ldots, \psi_n)$; then $\bigotimes_S^n \mathcal{H}_0$ is called the n-fold *symmetric tensor product* of $\mathcal{H}_0$.

Definition 16.6.3. Let $\mathcal{H}_0$ be an inner product space. Suppose that there exists a space $\bigwedge^n \mathcal{H}_0 = \mathcal{H}_0 \wedge \ldots \wedge \mathcal{H}_0$ with the following properties:
(i) there is an antisymmetric n-linear map from $\mathcal{H}_0 \times \ldots \times \mathcal{H}_0$ to $\bigwedge^n \mathcal{H}_0$ written as $(\psi_1, \ldots, \psi_n) \mapsto \psi_1 \wedge \ldots \wedge \psi_n$;
(ii) for any antisymmetric n-linear map β from $\mathcal{H}_0 \times \ldots \times \mathcal{H}_0$ to a space $\mathcal{K}$ there is a unique linear map $\hat{\beta}$ from $\bigwedge^n \mathcal{H}_0$ to $\mathcal{K}$ satisfying $\hat{\beta}(\psi_1 \wedge \ldots \wedge \psi_2) = \beta(\psi_1, \ldots, \psi_n)$; then $\bigwedge^n \mathcal{H}_0$ is called the n-fold *exterior product* of $\mathcal{H}_0$.

We have set things up so that bosons and fermions can be described by elements of the symmetric tensor product and exterior product, respectively. They can also be described by elements of the ordinary tensor product that have a particular symmetry. To see this we note that there is an obvious action of the permutation $\pi \in S_n$ on decomposable tensors, given by

$$U(\pi)\,(\xi_1 \otimes \xi_2 \otimes \ldots \otimes \xi_n) = \xi_{\pi(1)} \otimes \xi_{\pi(2)} \otimes \ldots \otimes \xi_{\pi(n)}. \tag{16.24}$$

Since the tensor product space is spanned by decomposable tensors, we can extend $U(\pi)$ linearly to the whole space. We leave it to the reader to check the following result, which follows from the fact that, like the ordinary tensor product, the symmetric and exterior product are unique up to isomorphism.

Theorem 16.6.1. The symmetric tensor product can be identified with the subspace of $\psi \in \mathcal{H}_1 \otimes \ldots \otimes \mathcal{H}_n$ that satisfies $U(\pi)\psi = \psi$, for all permutations π, and the exterior product with the subspace of ψ that satisfies $U(\pi)\psi = -\psi$ for odd permutations.

We can now deduce that the other properties of tensor products given in Proposition 16.5.3, such as associativity and distributivity over direct sums, carry over to the exterior and symmetric tensor product.

16.7*. Tensor products of group representations

One often has a symmetry group G acting on the spaces of the individual particles and one wishes to know how this symmetry is reflected in the combined system.

Definition 16.7.1. If U_1 and U_2 are unitary representations of G on $\mathcal{H}_1$ and $\mathcal{H}_2$ respectively, then there is a *tensor product* representation $U = U_1 \otimes U_2$ on $\mathcal{H} = \mathcal{H}_1 \otimes \mathcal{H}_2$ defined by

$$U(x)\,(\psi_1 \otimes \psi_2) = U_1(x)\psi_1 \otimes U_2(x)\psi_2$$

for x in G, ψ_1 in $\mathcal{H}_1$, and ψ_2 in $\mathcal{H}_2$.

This definition can similarly be extended to products of more than two spaces. There is a very simple formula for the character of a tensor product representation.

Proposition 16.7.1. Let χ_1 and χ_2 be the characters of the representations U_1 and U_2; then the character of $U_1 \otimes U_2$ is given by $\chi(g) = \chi_1(g)\chi_2(g)$, for all $g \in G$

Proof. From orthonormal bases ξ_j and η_k for $\mathcal{H}_1$ and $\mathcal{H}_2$ we may construct the orthonormal basis $\xi_j \otimes \eta_k$ for $\mathcal{H}_1 \otimes \mathcal{H}_2$. Then

$$\begin{aligned} \operatorname{tr}(U_1(g) \otimes U_2(g)) &= \sum_{jk} \langle \xi_j \otimes \eta_k | U_1(g)\xi_j \otimes U_2(g)\eta_k \rangle \\ &= \sum_{jk} \langle \xi_j | U_1(g)\xi_j \rangle \langle \eta_k | U_2(g)\eta_k \rangle \\ &= \sum_j \langle \xi_j | U_1(g)\xi_j \rangle \sum_k \langle \eta_k | U_2(g)\eta_k \rangle \\ &= \chi_1(g)\chi_2(g) \end{aligned} \tag{16.25}$$

as asserted. □

Remark 16.7.1. By Corollary 9.8.3 we already know that any finite-dimensional representation U of the rotation group can be decomposed as a direct sum of irreducible subrepresentations. Its character is the sum of irreducible characters and so the coefficients n_l will just give the multiplicity or number of times that D^l occurs in the decomposition of U. This must apply in particular when U is the tensor product of representations, for example when U is the tensor product of two irreducible representations $D^k \otimes D^l$.

Theorem 16.7.2. (The Clebsch–Gordan series) The tensor product of irreducible representations of the rotation group decomposes as

$$D^k \otimes D^l \cong D^{|k-l|} \oplus D^{|k-l|+1} \oplus \ldots \oplus D^{k+l}.$$

Proof. Recalling the expression in Theorem 9.8.1 for the irreducible characters of the rotation group, the character of the right-hand side is

$$\sum_{j=|k-l|}^{k+l} \Delta_j(\theta) = \operatorname{cosec}^2\left(\tfrac{1}{2}\theta\right) \sum_{j=|k-l|}^{k+l} \sin\left(\tfrac{1}{2}\theta\right) \sin\left[\left(j+\tfrac{1}{2}\right)\theta\right]$$

$$
\begin{aligned}
&= \tfrac{1}{2}\operatorname{cosec}^2\left(\tfrac{1}{2}\theta\right) \sum_{j=|k-l|}^{k+l} \{\cos(j\theta) - \cos[(j+1)\theta]\} \\
&= \tfrac{1}{2}\operatorname{cosec}^2\left(\tfrac{1}{2}\theta\right) \{\cos\left(|k-l|\theta\right) - \cos\left[(k+l+1)\,\theta\right]\} \\
&= \operatorname{cosec}^2\left(\tfrac{1}{2}\theta\right) \sin\left[\left(k+\tfrac{1}{2}\right)\theta\right] \sin\left[\left(l+\tfrac{1}{2}\right)\theta\right], \qquad (16.26)
\end{aligned}
$$

which is just $\Delta_k(\theta)\Delta_l(\theta)$, the character of $D^k \otimes D^l$, as required. □

Example 16.7.1. $D^1 \otimes D^l \cong D^{l+1} \oplus D^l \oplus D^{l-1}$.

Example 16.7.2. $D^{\frac{1}{2}} \otimes D^{\frac{1}{2}} \cong D^1 \oplus D^0$. This shows that two spin $\frac{1}{2}$ particles have combined angular momentum 0 or 1. This is in line with intuition, which suggests that their spins either line up to give $\frac{1}{2}+\frac{1}{2}=1$ or are in opposite directions giving $\frac{1}{2}-\frac{1}{2}=0$.

Theorem 16.7.3. If U_1 and U_2 are representations of $\mathbf{R}$ with infinitesimal generators H_1 and H_2 respectively then the infinitesimal generator of $U_1 \otimes U_2$ is $H_1 \otimes 1 + 1 \otimes H_2$.

Proof. The generator is

$$
\begin{aligned}
i\hbar\frac{d}{dt}\left(U_1 \otimes U_2\right)\Big|_{t=0} &= i\hbar\frac{d}{dt}U_1 \otimes U_2\Big|_{t=0} + i\hbar U_1 \otimes \frac{d}{dt}U_2\Big|_{t=0} \\
&= H_1 \otimes 1 + 1 \otimes H_2, \qquad (16.27)
\end{aligned}
$$

as stated. □

16.8*. Tensor operators

Sometimes it is possible to extract information from group representations even when the quantum system is not symmetric. For example, although, as we have seen, a uniform electric field breaks the rotational symmetry of an atomic Hamiltonian, it does so in a rather controlled way.

Definition 16.8.1. Let U, W, and D be unitary representations of G on the spaces $\mathcal{H}$, $\mathcal{K}$, and $\mathcal{L}$, respectively. A *tensor operator* for (U, W, D) is a linear map T from $\mathcal{L}$ to the space of linear transformations $\mathcal{L}(\mathcal{K}, \mathcal{H})$, such that for all v in $\mathcal{L}$ and x in G

$$U(x)T(v)W(x)^{-1} = T(D(x)v).$$

Example 16.8.1. If D is the trivial representation on the space $\mathcal{L} = \mathbf{C}$, then $T(v) = vT(1)$ is determined by the single operator $T(1)$. The defining relation becomes

$$U(x)T(1) = T(1)W(x), \tag{16.28}$$

so that $T(1)$ is just an intertwining operator for U and W.

Example 16.8.2. Let $U = W$ be the representation of $SO(3)$ on wave functions defined in Example 9.1.2, and $D = D^1$ the three-dimensional irreducible representation. Then for $\mathbf{v}$ in $\mathbf{C}^3$ we may define $X(\mathbf{v}) = \mathbf{X}.\mathbf{v}$, $P(\mathbf{v}) = \mathbf{P}.\mathbf{v}$, $L(\mathbf{v}) = \mathbf{L}.\mathbf{v}$, where $\mathbf{X}$, $\mathbf{P}$, and $\mathbf{L}$ are the usual position, momentum, and angular momentum operators. Each of these is a tensor operator for (U, U, D^1). For example,

$$\begin{aligned}(U(A)X(\mathbf{v})U(A)^{-1}\psi)(\mathbf{r}) &= \left(X(\mathbf{v})U(A)^{-1}\psi\right)(A^{-1}\mathbf{r})\\ &= \left(\mathbf{v}.(A^{-1}\mathbf{r})\right)\left(U(A)^{-1}\psi\right)(A^{-1}\mathbf{r})\\ &= (A\mathbf{v}.\mathbf{r})\psi(\mathbf{r})\\ &= (X(A\mathbf{v})\psi)(\mathbf{r}),\end{aligned} \tag{16.29}$$

so that

$$U(A)X(\mathbf{v})U(A)^{-1} = X(A\mathbf{v}). \tag{16.30}$$

If we set $A = R_{\mathbf{n}}(t)$ and differentiate this relationship we get the commutation relations

$$[L_j, X_k] = i\hbar\epsilon_{jkl}X_l \tag{16.31}$$

proved in Proposition 8.1.1.

Example 16.8.3. Let $U = W$ be the representation of the translations of $\mathbf{R}^3$ on wave functions described in Example 9.1.4. For a continuous function f let $M(f) = f(\mathbf{X})$, that is

$$(M(f)\psi)(\mathbf{x}) = f(\mathbf{x})\psi(\mathbf{x}), \tag{16.32}$$

and let D be defined on functions f by exactly the same formula as U is defined on wave functions. Then

$$\begin{aligned}(U(\mathbf{a})M(f)\psi)(\mathbf{x}) &= (M(f)\psi)(\mathbf{x}-\mathbf{a}) \\ &= f(\mathbf{x}-\mathbf{a})\psi(\mathbf{x}-\mathbf{a}) \\ &= (D(\mathbf{a})f)(\mathbf{x})(U(\mathbf{a})\psi)(\mathbf{x}),\end{aligned} \tag{16.33}$$

so that M defines a tensor operator. If we take $t\mathbf{a}$ in place of $\mathbf{a}$ and differentiate with respect to t then we obtain the relation

$$[P(\mathbf{a}), f(\mathbf{X})] = \frac{\hbar}{i}(\mathbf{a}.\nabla f)(\mathbf{X}). \tag{16.34}$$

On taking $f(\mathbf{x}) = x_j$ we easily recover the commutation relations between momentum and position.

Arthur Wightman and George Mackey independently suggested that this is the real origin of the commutation relations. If quantum theory is to be able to describe local interactions at a point then a position operator $\mathbf{X}$ must exist. If it is to be independent of the choice of origin then $M(f) = f(\mathbf{X})$ must be a tensor operator for the translations, and then the commutation relations follow.

Lemma 16.8.1. Let T be a tensor operator for (U, W, D). Then there exists an operator $\hat{T}$ from $\mathcal{L} \otimes \mathcal{K}$ to $\mathcal{H}$ defined for v in $\mathcal{L}$ and ψ in $\mathcal{K}$ by

$$\hat{T}(v \otimes \psi) = T(v)\psi.$$

Moreover, $\hat{T}$ intertwines $D \otimes W$ and U.

Proof. There is a bilinear map from $\mathcal{L} \times \mathcal{K}$ to $\mathcal{H}$ defined by

$$(v, \psi) \mapsto T(v)\psi, \tag{16.35}$$

so by Definition 16.5.1(ii) there must exist a linear map $\hat{T}$ from $\mathcal{L} \otimes \mathcal{K}$ to $\mathcal{H}$ satisfying

$$\hat{T}(v \otimes \psi) = T(v)\psi. \tag{16.36}$$

Moreover, for any x in G,

$$\begin{aligned}U(x)\hat{T}(v \otimes \psi) &= U(x)T(v)\psi \\ &= T(D(x)v)W(x)\psi \\ &= \hat{T}(D(x)v \otimes W(x)\psi) \\ &= \hat{T}(D(x) \otimes W(x))(v \otimes \psi),\end{aligned} \tag{16.37}$$

from which the result now follows. □

Theorem 16.8.2. (Wigner–Eckart theorem) Let U be an irreducible representation of G. Suppose that $D \otimes W$ can be decomposed into irreducibles, and that U occurs at most once in the decomposition. Let T be a tensor operator for (U, W, D); then for any ξ in $\mathcal{H}$, η in $\mathcal{K}$, and v in $\mathcal{L}$

$$\langle \xi | T(v)\eta \rangle = \lambda_T \, c(\xi, v, \eta),$$

where λ_T is independent of the vectors ξ, η, and v, and c is independent of T.

Proof. Suppose that U occurs exactly once in the decomposition of $D \otimes W$, so that we may write

$$D \otimes W = U_0 \oplus U^{\perp}, \tag{16.38}$$

where U_0 is equivalent to U and $U^{\perp}$ does not contain any irreducible equivalent to U. Accordingly the linear operator $\hat{T}$ must split into $T_U \oplus T_{\perp}$, where T_U denotes the restriction to the space on which U_0 operates and $T_{\perp}$ the restriction to the space of $U^{\perp}$. Since U is irreducible Schur's lemma tells us that T_U is either 0 or an isomorphism. Moreover, if S denotes any non-zero intertwining operator then $S^{-1}T_U$ intertwines U with itself and so for some scalar λ_T

$$S^{-1}T_U = \lambda_T. \tag{16.39}$$

On the other hand, since $U^{\perp}$ contains no irreducible equivalent to U, $T_{\perp} = 0$, so we have

$$\hat{T} = \lambda_T S \oplus 0 = \lambda_T S. \tag{16.40}$$

If $D \otimes W$ contains no representation equivalent to U then, by the same argument, $\hat{T} = 0$. From this relation we see that

$$\begin{aligned} \langle \xi | T(v)\eta \rangle &= \langle \xi | \hat{T}(v \otimes \eta) \rangle \\ &= \langle \xi | \lambda_T S(v \otimes \eta) \rangle \\ &= \lambda_T \langle \xi | S(v \otimes \eta) \rangle. \end{aligned} \tag{16.41}$$

The answer now follows on writing $c(\xi, v, \eta) = \langle \xi | S(v \otimes \eta) \rangle$. □

Definition 16.8.2. The numbers $c(\xi, v, \eta)$ are called *Clebsch–Gordan coefficients.*

Remark 16.8.1. The point of this result is that, since the Clebsch–Gordan coefficients are independent of T they can be tabulated for interesting representations. Then one needs only to find the single scalar λ_T in order to know all the matrix elements $\langle \xi | T(v)\eta \rangle$. This can be done by a single direct calculation for some convenient triple (ξ, v, η).

By Proposition 16.7.2 we have

$$D^k \otimes D^l = D^{|k-l|} \oplus D^{|k-l|+1} \oplus \ldots \oplus D^{k+l}, \tag{16.42}$$

so that the hypotheses of the theorem hold for the representations $U = D^j$, $W = D^l$, and $D = D^k$ of the rotation group. The perturbing term for the Stark effect, $\mathbf{F}.\mathbf{X} = X(\mathbf{F})$, is a tensor operator. So, if we take ξ and η to be wave functions of angular momentum j and l, respectively, then the theorem tells us that $\langle \xi | \mathbf{F}.\mathbf{X}\eta \rangle$ vanishes unless $|j - l| < 1$, and can provide information about the non-vanishing values too. A useful tool for this end is the following:

Corollary 16.8.3. Under the hypotheses of the theorem let S be a non-zero tensor operator for (U, W, D). Then there exists a complex number μ_T such that for any ξ in $\mathcal{H}$, η in $\mathcal{K}$, and v in $\mathcal{L}$

$$\langle \xi | T(v)\eta \rangle = \mu_T \langle \xi | S(v)\eta \rangle.$$

Proof. Since S is non-zero the scalar λ_S does not vanish, and therefore

$$\langle \xi | T(v)\eta \rangle = \lambda_T \, c(\xi, v, \eta) = \frac{\lambda_T}{\lambda_S} \langle \xi | S(v)\eta \rangle, \tag{16.43}$$

so that we may take $\mu_T = \lambda_T / \lambda_S$. □

This allows us to calculate all the inner products involving a given sort of tensor operators, from those for a single tensor operator, S, such as $\mathbf{L}$ or $\mathbf{X}$. This circumvents direct calculation and deals with a whole class of such problems at the same time.

Exercises

16.1° Explain the behaviour of the wave function for two identical particles moving in one dimension when the particles are interchanged if they are bosons. What happens if the particles are fermions?

Suppose that particles, moving in one dimension, of 'charge' e_1 and e_2 interact according to the potential

$$-\tfrac{1}{2}e_1e_2(X_1 - X_2)^2.$$

Write down the Hamiltonian for two spinless particles, each of mass m and 'charge' -1, moving in the field of an infinitely massive 'nucleus' of 'charge' $K > 0$ placed at the origin. By rotating coordinates, or otherwise, show that the Hamiltonian only has a ground state if $K > 2$. If the particles are bosons find the energies and multiplicities of the lowest two energy levels when $K = 8/3$.

16.2° Show that the states of n distinguishable spin l particles can be described quantum mechanically on a space of dimension $(2l + 1)n$. How does the description change if the particles are indistinguishable bosons?

Show that if the components of $\mathbf{L}^{(k)}$ give the angular momentum of the k-th particle for $k = 1, \ldots, n$, then the components of $\mathbf{L}^{(1)} + \mathbf{L}^{(2)}$ also satisfy the commutation relations for angular momentum.

Two spin l particles are placed in a uniform magnetic field of magnitude B and the interaction energy is represented by

$$\tfrac{1}{2}\gamma B\mathbf{L}^{(1)}.\mathbf{L}^{(2)},$$

where the components of $\mathbf{L}^{(k)}$ give the angular momentum of the k-th particle for $k = 1, 2$. Find the energy levels of the system and their degeneracies when the particles are (i) distinguishable, (ii) bosons.

16.3° Two independent spin $\frac{1}{2}$ systems have spin operators $\mathbf{s}(1) = \frac{1}{2}\hbar\boldsymbol{\sigma}(1)$ and $\mathbf{s}(2) = \frac{1}{2}\hbar\boldsymbol{\sigma}(2)$. Show that the eigenvalues of $\boldsymbol{\sigma}(1).\boldsymbol{\sigma}(2)$ are 1 and -3 and find the corresponding eigenvectors.

State the property of the state vectors representing the physical states of identical particles under identical particle interchange.

The Hamiltonian for a system of two spin $\frac{1}{2}$ particles, labelled 1 and 2, is given by

$$H = h(1) + h(2) - \kappa\boldsymbol{\sigma}(1).\boldsymbol{\sigma}(2).$$

The self-adjoint operator h has non-degenerate eigenvalues $E_0 < E_1 < E_2 < \ldots$ with eigenfunctions $\phi_0(\mathbf{x}), \phi_1(\mathbf{x}), \phi_2(\mathbf{x}), \ldots$, respectively, and $h(j)$ acts on the space of particle j. What are the possible energy eigenvalues for the two-particle system (i) when the particles are not identical and (ii) when the particles are identical? Show that the ground states in the two cases have the same degeneracy if $4\kappa > (E_1 - E_0)$.

17 Relativistic wave equations

Henceforth space by itself, and time by itself, are doomed to fade away into mere shadows, and only a kind of union of the two will preserve an independent reality.

HERMANN MINKOWSKI, lecture at Cologne, 21 September 1908

17.1. Minkowski space

One serious objection to the whole of quantum theory as we have developed it so far, is that it is inconsistent with the theory of relativity. In order to see how we may overcome this defect, at least as far as the special theory of relativity is concerned, we shall first recall some of its basic features. Shortly after Einstein's original 1905 papers on the subject, Minkowski realized that the physical ideas could also be understood geometrically, by combining space and time into a single four-dimensional real vector space, M, whose elements we shall call four-vectors. Each observer can assign coordinates (x^0, x^1, x^2, x^3) to a point x in M, and interpret $\mathbf{x} = (x^1, x^2, x^3)$ as normal spatial coordinates and x^0 as ct, where t is time and c is the speed of light. (By convention, these indices are written as superscripts rather than subscripts.) The constancy of the speed of light gives the bilinear form

$$g(x, y) = x^0y^0 - x^1y^1 - x^2y^2 - x^3y^3 = x^0y^0 - \mathbf{x}.\mathbf{y} \tag{17.1}$$

a special significance, since the associated quadratic form $g(x, x) = c^2t^2 - |\mathbf{x}|^2$ vanishes for points on a light ray through the origin.

Definition 17.1.1. The four-dimensional real vector space, M, with the bilinear form, g, is known as *Minkowski space*.

The group of linear transformations, Λ, of M such that

$$g(\Lambda x, \Lambda x) = g(x, x),$$

is called the *Lorentz group*, and its elements are known as *Lorentz transformations*. The *proper orthochronous Lorentz group* is the subgroup of Lorentz transformations, Λ, with positive determinant such that $g(x, \Lambda x)$ is positive whenever $g(x, x)$ is positive.

The last constraint ensures that proper orthochronous Lorentz transformations do not reverse the direction of time, and the former then rules out spatial reflections, just as the corresponding determinant condition picks out the rotations from the orthogonal group. In fact, any rotation R can be identified with a Lorentz transformation that affects only the spatial coordinates of an observer, sending $(x^0, \mathbf{x})$ to $(x^0, R\mathbf{x})$. The coordinate systems of different observers are connected by Lorentz transformations. We shall refer to the elements of M as four-vectors. Each observer and object traces out a curve, its worldline, in Minkowski space as time passes. (For example, the path through M of an observer stationary at the origin of a reference frame is the curve $(ct, \mathbf{0})$.) At each point the curve has a tangent, which, parametrizing the curve by τ, is in the direction $U = dx/d\tau$. Since observers move more slowly than light $g(U, U)$ is positive, and the parameter τ can be chosen so that $g(U, U) = c^2$.

Definition 17.1.2. The parameter τ that ensures that $U = dx/d\tau$ satisfies $g(U, U) = c^2$ is known as the *proper time* along the curve and U is called thc *four-velocity* of the observer.

In the case of the static observer $x = (ct, \mathbf{0})$ can be differentiated with respect to t, to give the tangent $V = (c, \mathbf{0})$. This is already appropriately normalized so that, in this case, the proper time is t.

Schrödinger's equation was motivated by Planck's law for energy and the de Broglie relation for momentum, so we need to investigate the corresponding relativistic concepts.

Definition 17.1.3. The *rest mass* of a body is its mass as measured in a frame in which it is at rest. The *four-momentum* of a body whose rest mass is m and four-velocity U is $p = mU$.

By definition, we have

$$g(p, p) = g(mU, mU) = m^2 g(U, U) = m^2 c^2. \tag{17.2}$$

As we shall now see the four-momentum combines both the momentum and the energy of the body and this is the relativistic formula linking them. With respect to a coordinate system it says that ${p^0}^2 - |\mathbf{p}|^2 = m^2c^2$, which can be rewritten as

$$(p^0 - mc) = |\mathbf{p}|^2/(p^0 + mc). \tag{17.3}$$

For a static body on the curve $(ct, \mathbf{0})$ the four-momentum is $(mc, \mathbf{0})$, so we expect that for slow moving bodies $\mathbf{p}$ should be small and p^0 close to mc. Our previous relation then gives

$$(p^0 - mc) \sim |\mathbf{p}|^2/2mc, \tag{17.4}$$

suggesting that we should interpret $c(p^0 - mc)$ as the kinetic energy of the body and $\mathbf{p}$ as its spatial momentum.

Definition 17.1.4. The *energy* of a body with four-momentum p as seen by an observer with four-velocity W is $g(W, p)$.

For the 'static' observer with four-velocity $V = (c, \mathbf{0})$ this relativistic energy is $g(V, p) = cp^0$. It is made up of the famous constant term, mc^2, and additional kinetic terms, which, for slowly moving bodies, are approximately $|\mathbf{p}|^2/2m$.

It is easy to express a plane wave $\psi(t, \mathbf{x}) = \exp[-i(\omega t - \mathbf{k}.\mathbf{x})]$ in relativistic form by introducing the *frequency four-vector* $\kappa = (\omega, c\mathbf{k})$ since then $\psi(x) = \exp[g(\kappa, x)/c]$. We can now recognize that the Planck and de Broglie relations are just the temporal and spatial components of a linear relationship between the four-momentum and four-frequency: $cp = \hbar\kappa$. In fact, it is sufficient to know the Planck relationship, which can be written as $g(V, p) = \hbar g(V, \kappa)/c$, since if this is true for one observer, then relativistic invariance means that the corresponding identity must be true for all uniformly moving observers. That is, $g(W, p) = \hbar g(W, \kappa)/c$ or $g(W, cp - \hbar\kappa) = 0$, for any four-velocity W. It is easy to see that this implies that $cp - \hbar\kappa = 0$, giving the de Broglie law as well.

Einstein was originally led to develop the special theory of relativity in order to harmonize Maxwell's electromagnetic equations and the laws of motion. The laws of electromagnetism can be expressed in terms of an electrostatic potential ϕ and a magnetic vector potential $\mathbf{A}$, from which the electric and magnetic fields can be recovered as

$$\mathbf{E} = -\frac{\partial \mathbf{A}}{\partial t} - \operatorname{grad} \phi, \qquad \mathbf{B} = \operatorname{curl} \mathbf{A}. \tag{17.5}$$

The potentials are not unique since it is possible to add a gradient, $\operatorname{grad} \chi$, to $\mathbf{A}$ and to subtract $\partial\chi/\partial t$ from ϕ without affecting the fields. Such changes are called *gauge transformations*, and play a crucial role in the modern understanding of physics. Not surprisingly, these concepts have an elegant reformulation in Minkowski space.

Definition 17.1.5. Let ϕ be the electrostatic potential and $\mathbf{A}$ the magnetic vector potential. The *electromagnetic four-potential* is $\Phi = (\phi, c\mathbf{A})$.

The significance of this is that any observer with four-velocity U will see an electrostatic potential $g(U, \Phi)$ and the component of Φ perpendicular to U will give the magnetic vector potential.

17.2. The Klein–Gordon equation

The Planck–de Broglie relations mean that the plane wave can also be expressed as $\psi(x) = \exp[-ig(p, x)/\hbar]$, and the constraint $g(p,p) = m^2c^2$ on the four-momentum leads to a wave equation much as in the non-relativistic case in Chapter 2. First we introduce a relativistic Fourier transform that decomposes any given wave function into plane waves

$$\psi(x) = (2\pi\hbar)^{-2} \int_M e^{ig(p,x)/\hbar} (\mathcal{F}\psi)(p)dp. \tag{17.6}$$

Exploiting the Einstein summation convention that repeated Greek indices are summed over the values 0, 1, 2, and 3, we can differentiate to get

$$i\hbar W^\mu \frac{\partial\psi}{\partial x^\mu} = (2\pi\hbar)^{-2} \int_M g(p, W) e^{-ig(p,x)/\hbar} (\mathcal{F}\psi)(p)dp, \tag{17.7}$$

giving

$$i\hbar\left(\mathcal{F}W^\mu \frac{\partial\psi}{\partial x^\mu}\right) = g(p, W)(\mathcal{F}\psi)(p)dp. \tag{17.8}$$

In particular $ic\hbar\partial/\partial x^0 = i\hbar\partial/\partial t$ transforms into the relativistic energy cp^0, and $i\hbar\nabla$ transforms into $-\mathbf{p}$. This agrees with the conventions used earlier for non-relativistic quantum theory. The constraint $g(p,p) = m^2c^2$ therefore gives the equation

$$-\frac{\hbar^2}{c^2}\frac{\partial^2\psi}{\partial t^2} + \hbar^2\nabla^2\psi = m^2c^2\psi. \tag{17.9}$$

Definition 17.2.1. The equation

$$\hbar^2\left(\frac{1}{c^2}\frac{\partial^2\psi}{\partial t^2} - \nabla^2\psi\right) + m^2c^2\psi = 0$$

is known as the free *Klein–Gordon equation*.

Remark 17.2.1. If the mass m vanishes then the Klein–Gordon equation reduces to a standard wave equation

$$\frac{1}{c^2}\frac{\partial^2\psi}{\partial t^2} - \nabla^2\psi = 0. \tag{17.10}$$

In this way one can regard the ordinary wave equation as a degenerate quantum equation although Planck's constant no longer appears in it.

17.3. The Yukawa potential

For wave functions that do not depend on time, the Klein–Gordon equation reduces to Yukawa's equation, $\nabla^2\psi = (mc/\hbar)^2\psi$, and when $\psi = \psi(r)$ is spherically symmetric this becomes

$$\frac{1}{r}\frac{\partial^2(r\psi)}{\partial r^2} = (mc/\hbar)^2\psi, \tag{17.11}$$

whose solutions are of the form

$$r\psi = Ae^{mcr/\hbar} + Be^{-mcr/\hbar} \tag{17.12}$$

(compare Section 13.5). As usual the exponentially increasing solution is physically unacceptable, leaving

$$\psi(r) = \frac{Be^{-mcr/\hbar}}{r}. \tag{17.13}$$

The importance of this wave function was first pointed out by Yukawa Hideki, who assigned it an important role in explaining the stability of atoms. It is known as the Yukawa potential.

Experiments between the wars showed fairly clearly that the nucleus of an atom consists of positively charged protons and uncharged neutrons. To explain why electrostatic repulsion between the protons did not cause the nucleus to explode, Yukawa proposed that there must be another force, now called the *strong force*, described by the Yukawa potential. If $B > Ze^2/4\pi\epsilon_0$, then this exceeds the electrostatic potential $-Ze^2/4\pi\epsilon_0 r$ at short distances, but owing to the exponential term it is smaller at large distances. Such an attractive force could overcome electrostatic potential at short distances (less than about $\hbar/mc$) whilst being small enough outside the atom to explain why it had hitherto eluded detection. Just as the photon is associated to the electromagnetic field, particles called *mesons* are associated to Yukawa's field, and the parameter m in Yukawa's equation can be interpreted as their mass. It can be estimated from the size of the nucleus. (In the case of the electromagnetic field the photon is massless, since $m = 0$.) When mesons were later detected experimentally it was found that there were many different sorts having different masses. The lightest is around 264 times more massive than an electron.

17.4. The Dirac equation

Unfortunately, despite its elegance, there are serious reasons for believing that the Klein–Gordon equation is not basic, but must be a consequence of other more fundamental equations. Schrödinger's non-relativistic equation is only first order in time, and so can be solved once the starting value of the wave function is known. Being second order, the Klein–Gordon equation requires initial values of both ψ and $\partial\psi/\partial t$ for its solution. To avoid this striking difference one would need a first-order relativistic equation as well. (Maxwell's equations for electromagnetism provide a good model for this. In empty space the fields satisfy second-order wave equations, but these are a consequence of the first-order dynamical equations, $\partial\mathbf{E}/\partial t = c^2 \operatorname{curl}\mathbf{B}$ and $\partial\mathbf{B}/\partial t = -\operatorname{curl}\mathbf{B}$.) Now, relativistic invariance means that space and time are on the same footing, so that the equation must have the same order in the time and space derivatives, and, by the above argument, that order should be 1.

Fourier transforming, this means that we need a linear constraint on the momentum of the form $\gamma(p) = \mu$. On the other hand this must be consistent with the known quadratic constraint $g(p,p) = m^2c^2$, or else we should be able to eliminate p^0 between the equations and get unphysical constraints involving only the spatial components of p. By rescaling γ we may as well assume that $\mu = mc$ and then the equations

$$\gamma(p)^2 = m^2c^2 = g(p,p) \tag{17.14}$$

tell us that for consistency we need $\gamma(p)^2 = g(p,p)$ for all $p \in M$.

Proposition 17.4.1. The following conditions are equivalent:
(a) $\gamma(p)^2 = g(p,p)$, for all $p \in M$;
(b) $\gamma(p)\gamma(q) + \gamma(q)\gamma(p) = 2g(p,q)$, for all $p, q \in M$.

Proof. The first condition clearly follows from the second by putting $p = q$. Conversely, by the linearity of γ we have $\gamma(p+q) = \gamma(p) + \gamma(q)$, so that

$$\begin{aligned}\gamma(p)\gamma(q) + \gamma(q)\gamma(p) &= \gamma(p+q)^2 - \gamma(p)^2 - \gamma(q)^2 \\ &= g(p+q,p+q) - g(p,p) - g(q,q) = 2g(p,q).\end{aligned} \tag{17.15}$$

□

Corollary 17.4.2. There are no solutions of these conditions for which $\gamma(p)$ and $\gamma(q)$ commute for all p and q.

Proof. If there were commuting solutions to these equations, then we should have $g(p,q) = \gamma(p)\gamma(q)$, and

$$g(p,q)^2 = \gamma(p)^2\gamma(q)^2 = g(p,p)g(q,q), \tag{17.16}$$

for all p and q in M. This equation is, however, violated whenever p and q are orthogonal unit spatial vectors such as $(0,\mathbf{i})$ and $(0,\mathbf{j})$. □

Fortunately, there are non-commuting matrices $\gamma(p)$ that satisfy this condition and, moreover, they are essentially unique.

Theorem 17.4.3. In 2×2 block form, the matrices

$$\gamma(p) = \begin{pmatrix} p_0 & -\boldsymbol{\sigma}.\mathbf{p} \\ \boldsymbol{\sigma}.\mathbf{p} & -p_0 \end{pmatrix}$$

satisfy the equation $\gamma(p)^2 = g(p,p)$ for all $p \in M$, where the components σ_1, σ_2, σ_3 of $\boldsymbol{\sigma}$ are the Pauli spin matrices.

Proof. This follows by direct calculation, using the relation $(\boldsymbol{\sigma}.\mathbf{p})^2 = |\mathbf{p}|^2$, since

$$\begin{aligned}\gamma(p)^2 &= \begin{pmatrix} p_0^2 - \boldsymbol{\sigma}.\mathbf{p}^2 & 0 \\ 0 & p_0^2 - \boldsymbol{\sigma}.\mathbf{p}^2 \end{pmatrix} \\ &= \begin{pmatrix} p_0^2 - |\mathbf{p}|^2 & 0 \\ 0 & p_0^2 - |\mathbf{p}|^2 \end{pmatrix} \\ &= g(p,p)\begin{pmatrix} 1 & 0 \\ 0 & 1 \end{pmatrix},\end{aligned} \tag{17.17}$$

as required. □

These matrices are known as the *Dirac matrices.* In fact we shall show later, in Section 17.8, that any operators $\gamma(p)$ depending linearly on $p \in M$ and satisfying this constraint are direct sums of operators equivalent to these. This is therefore the smallest set of matrices that can be used and all other 4×4 matrices satisfying the constraint are equivalent to these.

It is often useful to work in coordinates. We may choose an orthogonal basis e_μ, $\mu = 0, 1, 2, 3$, for M such that $g(e_0, e_0) = 1$ and $g(e_j, e_j) = -1$ for $j = 1, 2, 3$ and write $\gamma_\mu = \gamma(e_\mu)$, so that $\gamma(p) = \gamma_\mu p^\mu$. With the above choice of matrices we have

$$\gamma_0 = \begin{pmatrix} 1 & 0 \\ 0 & -1 \end{pmatrix} \qquad \gamma_j = \begin{pmatrix} 0 & -\sigma_j \\ +\sigma_j & 0 \end{pmatrix}. \tag{17.18}$$

Proposition 17.4.4. The operators γ_μ satisfy $\gamma_0^2 = 1$, $\gamma_j^2 = -1$, for $j = 1, 2, 3$, and

$$\gamma_\mu \gamma_\nu + \gamma_\nu \gamma_\mu = 0,$$

for $\mu \neq \nu$.

Proof. By definition of γ_μ we have

$$\gamma_\mu \gamma_\nu + \gamma_\nu \gamma_\mu = g(e_\mu, e_\nu) \tag{17.19}$$

and the result follows on giving the values of $g(e_\mu, e_\nu)$. □

We shall refer to the identities in this proposition and its earlier coordinate-free versions as the anticommutation relations for the γ matrices.

Definition 17.4.1. The four-dimensional space on which the γ matrices act is called the space of *Dirac spinors.*

This suggests that the appropriate relativistic equation for which we have been looking is, in momentum space form,

$$\gamma(p)(\mathcal{F}\psi)(p) = mc(\mathcal{F}\psi)(p), \tag{17.20}$$

where ψ must now be a spinor-valued wave function, with four components. Recalling the formulae following equation (17.2) for the transforms of the momenta we now make the following definition:

Definition 17.4.2. In terms of the four-momentum operators

$$P^0 = i\hbar\frac{\partial}{\partial x^0}, \quad P^j = -i\hbar\frac{\partial}{\partial x^j}$$

the *free Dirac equation* is given by

$$\gamma(P)\psi = mc\psi,$$

where the differential operator $D = \gamma(P) = \gamma_\mu P^\mu$ is called the *Dirac operator*.

In practice the different signs attached to the space and time derivatives (which derived ultimately from the different signs in g) are often a nuisance, so it is useful to remove them by a redefinition of the γ matrices. We simply set $\gamma^0 = \gamma_0$, but $\gamma^j = -\gamma_j$. (This is consistent with the conventions used in general relativity.) The Dirac operator may then be rewritten as

$$D = i\hbar\gamma^\mu\frac{\partial}{\partial x^\mu} = i\hbar\gamma^\mu\partial_\mu, \tag{17.21}$$

and the free Dirac equation becomes

$$i\hbar\gamma^\mu\partial_\mu\psi = mc\psi. \tag{17.22}$$

It is believed to be the equation appropriate to the description of an electron and also other light particles, such as the muon and tau particle, which seem to be distinguished from the electron only by their masses.

The Dirac equation can easily be converted to a Schrödinger equation. In fact, using the convention that repeated Roman indices are summed over 1, 2, and 3, we have

$$i\hbar\gamma^0\frac{\partial\psi}{\partial t} = c\left(mc - i\hbar\gamma^j\frac{\partial}{\partial x^j}\right)\psi, \tag{17.23}$$

or, on using the fact that ${\gamma^0}^2 = 1$,

$$i\hbar\frac{\partial\psi}{\partial t} = c\gamma^0\left(mc - i\hbar\gamma^j\frac{\partial}{\partial x^j}\right)\psi, \tag{17.24}$$

which inspires the following definition:

Definition 17.4.3. The operator

$$H_D = mc^2\gamma_0 - c\gamma_0\gamma_j P^j = c\gamma_0\left(mc - i\hbar\gamma^j \frac{\partial}{\partial x^j}\right)$$

is called the *free Dirac Hamiltonian.*

We recall that $\gamma(p)$ was defined to ensure that

$$g(p,p)\widehat{\psi}(p) = \gamma(p)^2\widehat{\psi}(p) = m^2c^2\widehat{\psi}(p), \tag{17.25}$$

where we have written $\mathcal{F}\psi = \widehat{\psi}$ for the Fourier transform. This means that $\widehat{\psi}(p)$ must vanish except when p is on the hyperboloid $g(p,p) = m^2c^2$. The inner product on Dirac spinors is given by

$$mc\int \widehat{\psi}(p)^\dagger\widehat{\psi}(p)\frac{dp^1dp^2dp^3}{|p^0|}, \tag{17.26}$$

where $\dagger$ denotes the conjugate transpose of the Dirac spinor. If the wave function is concentrated in the region where the spatial momentum $|\mathbf{p}|$ is small then $|p^0| \sim mc$ and this closely approximates the non-relativistic formula for momentum space inner products.

17.5. Antiparticles

Given the block form of the γ matrices, it is sensible to split each Dirac spinor into a pair of two-component vectors

$$\psi(x) = \begin{pmatrix}\psi_+\\ \psi_-\end{pmatrix}, \tag{17.27}$$

and similarly for its Fourier transform. Then the Dirac equation takes the explicit form

$$\begin{pmatrix} p_0 & -\boldsymbol{\sigma}.\mathbf{p}\\ \boldsymbol{\sigma}.\mathbf{p} & -p_0\end{pmatrix}\begin{pmatrix}\widehat{\psi}_+\\ \widehat{\psi}_-\end{pmatrix} = mc\begin{pmatrix}\widehat{\psi}_+\\ \widehat{\psi}_-\end{pmatrix}, \tag{17.28}$$

or, equivalently, as a pair of coupled two-component equations

$$\begin{aligned}(p^0 - mc)\widehat{\psi}_+ &= \boldsymbol{\sigma}.\mathbf{p}\widehat{\psi}_-\\ (p^0 + mc)\widehat{\psi}_- &= \boldsymbol{\sigma}.\mathbf{p}\widehat{\psi}_+.\end{aligned} \tag{17.29}$$

We expect p^0 to be positive, so that $p^0 \neq -mc$. Then the second equation defines $\widehat{\psi}_-$ as

$$\widehat{\psi}_- = \frac{\boldsymbol{\sigma}.\mathbf{p}}{p^0 + mc}\widehat{\psi}_+, \tag{17.30}$$

and the consistency condition $g(p,p) = m^2c^2$ ensures that this also satisfies the first equation. Thus the problem is completely solved once one knows the two-component wave function $\widehat{\psi}_+$. Formally, $\widehat{\psi}(p)$ appears in the inverse Fourier transform for $\psi(x)$ in the combination $\psi(x) = \exp(-ig(x,p))\,\widehat{\psi}(p)$, which has the form of a plane wave, so these are often called plane wave solutions.

Normally we should expect the spatial momenta to be small and $p^0 \sim mc$, so that the ψ_- components would be small with respect to those of ψ_-, but there is no mathematical reason why this should not be reversed. The problem is that the consistency condition $g(p,p) = m^2c^2$ is satisfied whenever

$$p^0 = \pm(m^2c^2 + |\mathbf{p}|^2)^{\frac{1}{2}}, \tag{17.31}$$

so that the energy could be positive or negative. In principle this is inherent in the relativistic relationship between energy and momentum, but without quantum theory one could argue that energy can only be lost continuously, making it impossible to jump the gap from the lowest positive energy, mc, to the highest negative energy, $-mc$. In quantum theory energy can be lost in discrete packets, and, since particles tend to lose energy, one would have expected all the particles to have settled into negative energy states long ago. Dirac realized, however, that since the equation describes spin $\frac{1}{2}$ particles, which are therefore fermions, this difficulty could be circumvented. The Pauli exclusion principle forbids multiple occupation of a state by two fermions, so that if all the negative energy states are already occupied then the remaining particles must have positive energies. Since one measures only energy differences this vast sea of negative energy particles could go unnoticed.

In practice one would expect that from time to time a particle of energy $E < -mc^2$ would absorb enough energy to give it positive energy and leave a vacancy or 'hole' in the sea of negative energy particles. The zero energy of such a hole would still exceed by $-E > mc^2$ the negative energy of the particle that was there before, and the hole could therefore be interpreted as a particle of rest mass m and positive energy $-E$. This is known as the antiparticle associated with the original particle. By boosting the previously unnoticed negative energy particle to a positive energy and leaving behind the hole, interpreted as an antiparticle, the original energy would therefore seem to have simultaneously created a particle and an antiparticle. Conversely the positive energy particle might drop back into the hole releasing the excess energy as it does so, and this would look like the particle and antiparticle annihilating each other in a burst of energy. We shall see, when we consider the Dirac equation in an electromagnetic field in Section 18.3, that these antiparticles must have the opposite charge to the original particles. A couple of years after Dirac published his equation Anderson discovered a positively charged particle otherwise identical to the electron. This particle, now called the *positron*, is the antiparticle to the electron. It is now thought that to each kind of particle (whether or not it is described by the Dirac equation), there corresponds an antiparticle, though in some cases such as the photon the particle may be its own antiparticle. Towards the end of 1995 a team at CERN in Geneva succeeded in constructing anti-atoms from antiprotons and positrons. So far these have been very short lived, but the hope is that new techniques will give them a long enough lifetime to start checking whether, for example, gravity affects antimatter in the same way as ordinary matter.

17.6. The Weyl equation

When the mass m in the Dirac equation vanishes, the equations take a simpler form

$$\begin{aligned} p^0\widehat{\psi}_+ &= \boldsymbol{\sigma}.\mathbf{p}\widehat{\psi}_- \\ p^0\widehat{\psi}_- &= \boldsymbol{\sigma}.\mathbf{p}\widehat{\psi}_+. \end{aligned} \tag{17.32}$$

On setting $\widehat{\psi}_L = \widehat{\psi}_+ - \widehat{\psi}_-$ and $\widehat{\psi}_R = \widehat{\psi}_+ + \widehat{\psi}_-$, the equations decouple to give two two-component equations

$$\begin{aligned} (p^0 + \boldsymbol{\sigma}.\mathbf{p})\,\widehat{\psi}_L &= 0 \\ (p^0 - \boldsymbol{\sigma}.\mathbf{p})\,\widehat{\psi}_R &= 0. \end{aligned} \tag{17.33}$$

It is, therefore, open to us to choose solutions in which one of the component two-vectors vanishes.

Definition 17.6.1. The equation $(p^0 - \boldsymbol{\sigma}.\mathbf{p})\,\widehat{\psi}_R = 0$ is known as the *Weyl equation.*

This equation has been widely used to describe neutrinos, massless spin $\frac{1}{2}$ particles, predicted on theoretical grounds by Pauli and Fermi long before the first was detected experimentally in 1956 by Fred Reines (who was awarded a share of the 1995 Nobel Prize for his discovery) and Cowan. (It is now known that there are distinct neutrinos associated with the electron, the muon, and the tau particle.) This equation was initially scorned by physicists, since it is not invariant under the parity operator. (The parity operator reverses all spatial directions and so interchanges solutions of $(p^0 - \boldsymbol{\sigma}.\mathbf{p})\,\widehat{\psi}_R = 0$ with those of $(p^0 + \boldsymbol{\sigma}.\mathbf{p})\,\widehat{\psi}_L = 0$. Having an equation with vanishing ψ_R, but no corresponding equation with vanishing ψ_L, implies that nature is not invariant under reflection. The suffix L on ψ refers to the fact that it is by convention a left-handed particle, and its right-handed equivalent does not appear in nature.) 'God could not be only weakly left-handed' was Pauli's reaction, referring to the role of the neutrino in the so-called weak interactions. The neutrino itself was later discovered and when, in 1957, experiments of Wu confirmed the suggestion of Lee and Yang that parity violation could occur in weak interactions, Weyl's equation was rehabilitated. Doubts of a different sort have arisen more recently, with suggestions that neutrinos might after all have a small mass, and that the three experimentally observed neutrinos might each be superpositions of neutrinos with different masses. There is, as yet, no firm experimental evidence for this, though it might help to reconcile astrophysical models of the interior of the sun with the paucity of solar neutrinos detected on the earth. In that case Weyl's equation would just be an approximation to the correct equation.

17.7. The angular momentum

The free Dirac Hamiltonian can be used exactly as in non-relativistic quantum mechanics to calculate energy levels and find constants of the motion. It is natural to start by considering angular momentum.

Theorem 17.7.1. The Dirac Hamiltonian, H_D, commutes with $\mathbf{L} + \frac{1}{2}\hbar\tilde{\boldsymbol{\sigma}}$, where

$$\tilde{\boldsymbol{\sigma}} = \begin{pmatrix} \boldsymbol{\sigma} & 0 \\ 0 & \boldsymbol{\sigma} \end{pmatrix},$$

but not with the orbital angular momentum $\mathbf{L}$.

Proof. We shall start by calculating the commutator of L_k with H_D. For this we notice that all the γ matrices have constant entries and so commute with L_k. We know the commutators between L_k and $P^j = -i\hbar\partial_j$ from Proposition 8.1.1, so that we have

$$[L_k, H_D] = -c\gamma^0\gamma_j[L_k, P^j] = -ic\hbar\gamma^0\gamma_j\epsilon_{kjl}P^l, \tag{17.34}$$

which shows that H_D does not commute with the orbital angular momentum operators.

On the other hand when we calculate the commutators of the Pauli spin matrices with H_D it is the γ matrices that give problems. We first note that from the explicit formulae

$$\gamma_0\gamma_l = \begin{pmatrix} 1 & 0 \\ 0 & -1 \end{pmatrix}\begin{pmatrix} 0 & -\sigma_l \\ \sigma_l & 0 \end{pmatrix} = \begin{pmatrix} 0 & \sigma_l \\ \sigma_l & 0 \end{pmatrix}. \tag{17.35}$$

This means that

$$[\tilde{\sigma}_k, H_D] = [\tilde{\sigma}_k, mc^2\gamma_0 - c\gamma_0\gamma_l P^l] = c\begin{pmatrix} 0 & [\sigma_k, \sigma_l] \\ [\sigma_k, \sigma_l] & 0 \end{pmatrix} P^l. \tag{17.36}$$

The commutation relations for the Pauli spin matrices enable us to reduce this to

$$2ic\epsilon_{klj}\begin{pmatrix} 0 & \sigma_j \\ \sigma_j & 0 \end{pmatrix} P^l = -2ic\epsilon_{klj}\gamma_0\gamma_j P^l. \tag{17.37}$$

Multiplying this by $\frac{1}{2}\hbar$, adding it to the previous commutator, and using the antisymmetry of ϵ_{kjl}, gives the result. It was not actually necessary to use the explicit matrix form of the γ matrices: it is perfectly possible to deduce the result directly from the anticommutation relations in Proposition 17.4.4. In that case the conserved quantity is

$$L_k + \frac{i\hbar}{4}\epsilon_{kjl}\gamma_j\gamma_l. \qquad \square$$

This result shows that one must add a spin term, two copies of $\frac{1}{2}\hbar\sigma$, to the orbital angular momentum before one obtains an expression that commutes with H_D and so is a constant of the motion. If we wish to retain the law of conservation of angular momentum then we are more or less forced to assign to the electron an intrinsic angular momentum or spin of $\frac{1}{2}\hbar\sigma$. This confirms the idea already suggested by the presence of the Pauli spin matrices that the Dirac equation describes a particle with spin $\frac{1}{2}$.

We conclude this section with another elementary property of the free Dirac equation. Knowing the Hamiltonian, it is easy to calculate the velocity of a Dirac particle.

Proposition 17.7.2. In the Heisenberg picture the velocity of a particle described by the Dirac equation is

$$\frac{dX^k}{dt} = c\gamma_0\gamma^k.$$

Proof. Since the position operators X^k commute with the γ matrices we have

$$i\hbar\frac{dX^k}{dt} = [X^k, H_D] = -c\gamma_0\gamma_j[X^k, P^j] = -ic\hbar\gamma_0\gamma_j\delta_{jk} = -ic\hbar\gamma_0\gamma_k,$$

which is equivalent to the stated result. □

The first surprising thing about this result is that the components of the velocity do not commute with each other and so cannot be simultaneously measured. Furthermore, we have

$$\left(\gamma_0\gamma^j\right)^2 = -{\gamma_0}^2{\gamma^j}^2 = 1, \tag{17.38}$$

so that the only eigenvalues of a velocity component are $\pm c$. Physically this can be understood as a result of the fact that velocity is calculated from successive precise position measurements, which cause a large uncertainty in the momentum. Relativistically a large spatial momentum implies a velocity close to that of light. In practice one could think of the particle continually changing direction to give a much lower net velocity. One can also calculate the motion of the particle (Exercise 17.7). This picture of the motion is sometimes called *Zitterbewegung* or *trembling motion.*

17.8. Uniqueness of the gamma matrices

We shall now prove our earlier assertion that the γ matrices in the chiral representation are unique up to equivalence.

Theorem 17.8.1. Let $\gamma(p)$ be operators on a finite-dimensional complex vector space V, which depend linearly on p and satisfy $\gamma(p)^2 = g(p,p)$ for all $p \in M$. Then V is a direct sum of four-dimensional subspaces in which, for a suitable choice of basis, $\gamma(p)$ takes the matrix form

$$\gamma(p) = \begin{pmatrix} p_0 & -\boldsymbol{\sigma}.\mathbf{p} \\ \boldsymbol{\sigma}.\mathbf{p} & -p_0 \end{pmatrix}.$$

Proof. We shall work with the operators $\gamma_\mu = \gamma(e_\mu)$. Using the anti-commutation relations of Proposition 17.4.4, it is easy to check that γ_0 and

$\gamma_1\gamma_2$ commute with each other. We may therefore find common eigenvectors for both. Let us write Ω_1 for one of these, so that for suitable complex numbers, α and β, we have

$$\gamma_0\Omega_1 = \alpha\Omega_1 \quad \text{and} \quad \gamma_1\gamma_2\Omega_1 = \beta\Omega_1. \tag{17.39}$$

Since $\gamma_0^2 = 1$ we see that $\alpha = \pm 1$, and similarly the identity $(\gamma_1\gamma_2)^2 = -\gamma_1^2\gamma_2^2 = -1$ means that $\beta = \pm i$. We readily check that

$$\gamma_0\gamma_3\Omega_1 = -\gamma_3\gamma_0\Omega_1 = -\alpha\gamma_3\Omega_1, \tag{17.40}$$

and that

$$\gamma_1\gamma_2\gamma_3\Omega_1 = \gamma_3\gamma_1\gamma_2\Omega_1 = \beta\gamma_3\Omega_1, \tag{17.41}$$

so that $\Omega_3 = -\gamma_3\Omega_1$ is also a common eigenvector with eigenvalues $-\alpha$ and β. Similarly, $\Omega_2 = \gamma_3\gamma_1\Omega_1$ and $\Omega_4 = \gamma_1\Omega_1$ are common eigenvectors with eigenvalues α and $-\beta$, and $-\alpha$ and $-\beta$, respectively. Since all four possible signs occur, we may permute the four vectors to ensure that $\alpha = 1$ and $\beta = -i$. On the subspace with basis Ω_1, Ω_2, Ω_3, and Ω_4 the operators γ_0 and $\gamma_1\gamma_2$ have the diagonal block matrix form

$$\gamma_0 = \begin{pmatrix} 1 & 0 \\ 0 & -1 \end{pmatrix} \qquad \gamma_1\gamma_2 = \begin{pmatrix} -i\sigma_3 & 0 \\ 0 & -i\sigma_3 \end{pmatrix}. \tag{17.42}$$

Moreover, the vectors Ω_j have been defined in such a way that

$$\gamma_3 = \begin{pmatrix} 0 & -\sigma_3 \\ \sigma_3 & 0 \end{pmatrix} \qquad \gamma_1 = \begin{pmatrix} 0 & -\sigma_1 \\ \sigma_1 & 0 \end{pmatrix}. \tag{17.43}$$

From these we may check that $\gamma_2 = -\gamma_1(\gamma_1\gamma_2)$ also has the desired form. The vectors Ω_j for $j = 1$ to 4 span an invariant subspace. If this is the whole space then we are finished. Otherwise we pick another linearly independent eigenvector Ω_1' and repeat the process. □

17.9. Lorentz covariance

The γ matrices that appear in the Dirac equation are unique up to equivalence, so that all observers should agree on the form of the Dirac equation. Indeed, if the coordinate systems of two observers are related by a Lorentz transformation Λ then, since that preserves g, we have

$$\gamma(\Lambda p)^2 = g(\Lambda p, \Lambda p) = g(p, p) \tag{17.44}$$

so that $p \to \gamma(\Lambda p)$ provides another possible choice of matrices, and by uniqueness this must be equivalent to the previous choice so that there exist operators, $\Gamma(\Lambda)$, such that

$$\gamma(\Lambda p) = \Gamma(\Lambda)\gamma(p)\Gamma(\Lambda)^{-1}. \tag{17.45}$$

(We could equally well use $\pm\Gamma(\Lambda)$, so there are actually two such operators for each Λ. In what follows we shall assume that one has been chosen arbitrarily.) We can now see how Dirac spinors transform under Lorentz transformations.

Theorem 17.9.1. Let

$$V(\Lambda)\widehat{\psi}(p) = \Gamma(\Lambda)\widehat{\psi}(\Lambda^{-1}p),$$

and let $\widehat{D}$ be the Fourier-transformed Dirac operator

$$(\widehat{D}\widehat{\psi})(p) = \gamma(p)\widehat{\psi}(p).$$

Then

$$V(\Lambda)\widehat{D}V(\Lambda)^{-1} = \widehat{D},$$

and if $\widehat{\psi}$ is a solution of the free Dirac equation, then so is $V(\Lambda)\widehat{\psi}$.

Proof. First the definitions give us

$$\begin{aligned}(V(\Lambda)\widehat{D}V(\Lambda)^{-1}\widehat{\psi})(p) &= \Gamma(\Lambda)(\widehat{D}V(\Lambda)^{-1}\widehat{\psi})(\Lambda^{-1}p)\\ &= \Gamma(\Lambda)\gamma(\Lambda^{-1}p)(V(\Lambda)^{-1}\widehat{\psi})(\Lambda^{-1}p)\\ &= \Gamma(\Lambda)\gamma(\Lambda^{-1}p)\Gamma(\Lambda)^{-1}(\widehat{\psi})(p)\\ &= \gamma(p)\widehat{\psi}(p) = (\widehat{D}\widehat{\psi})(p).\end{aligned} \tag{17.46}$$

Since this identity can be rewritten in the form $V(\Lambda)\widehat{D} = \widehat{D}V(\Lambda)$, when $\widehat{\psi}$ satisfies the free Dirac equation we have

$$\widehat{D}V(\Lambda)\widehat{\psi} = V(\Lambda)\widehat{D}\widehat{\psi} = V(\Lambda)mc\widehat{\psi} = mcV(\Lambda)\widehat{\psi}, \tag{17.47}$$

showing that $V(\Lambda)\widehat{\psi}$ also satisfies the equation. □

17.10. Explicit Lorentz transforms for spinors

In this section we shall obtain explicit formulae for the operator $\Gamma(\Lambda)$ which will also enable us to show why the Dirac equation describes a spin $\frac{1}{2}$ particle.

Lemma 17.10.1. For any vectors $p, q \in M$ such that $g(q,q) \neq 0$, we have

$$\gamma(q)\gamma(p)\gamma(q)^{-1} = \gamma(R_q p),$$

where

$$R_q p = 2\frac{g(q,p)}{g(q,q)}q - p.$$

The transformation R_q is in the Lorentz group, but not proper.

Proof. Since $\gamma(q)^2 = g(q,q)$ we have $\gamma(q)^{-1} = \gamma(q)/g(q,q)$. On multiplying the anticommutation relation

$$\gamma(q)\gamma(p) + \gamma(p)\gamma(q) = 2g(q,p) \tag{17.48}$$

on the right by $\gamma(q)^{-1}$ we obtain

$$\gamma(q)\gamma(p)\gamma(q)^{-1} + \gamma(p) = 2g(q,p)\gamma(q)/g(q,q) \tag{17.49}$$

from which the result follows by rearrangement, and the linearity of γ. One readily checks that $g(q, R_q p) = g(q,p)$ so that the component in the q direction is unchanged, and for similar reasons, the orthogonal components are reversed. A straightforward calculation shows that R_q is in the Lorentz group, though it is never proper (since the determinant is $(-1)^3$), and it need not be orthochronous. □

The transformation R_q leaves the component of p in the direction q unchanged, and reverses the components of p orthogonal to q. To illustrate this, we take the unit vector $q = e_0$ to obtain

$$R_q p = 2g(e_0, p)e_0 - p = p_0 e_0 - p_j e_j. \tag{17.50}$$

If we take $q = (0, \mathbf{q})$ with $\mathbf{q}$ a unit vector, then $g(q,q) = -|\mathbf{q}|^2 = -1$, and

$$R_q p = 2(\mathbf{q}.\mathbf{p})q - p = (p_0, 2(\mathbf{p}.\mathbf{q})\mathbf{q} - \mathbf{p}). \tag{17.51}$$

The spatial component along $\mathbf{q}$ is unchanged, but orthogonal components are reversed, which is precisely the effect of a rotation through π about the axis $\mathbf{q}$.

Lemma 17.10.2. Let Λ be a proper Lorentz transformation that fixes two linearly independent vectors v_1 and v_2, and take any non-null vector u orthogonal to v_1 and v_2. If $\Lambda u = -u$ take any w orthogonal to all three vectors u, v_1, and v_2, and otherwise set $w = (1 + \Lambda)u$. Then $\Lambda = R_w R_u$.

Proof. Since Λ fixes v_j and preserves g we have

$$g(\Lambda u, v_j) = g(\Lambda u, \Lambda v_j) = g(u, v_j) = 0, \tag{17.52}$$

showing that Λu, and also $w = u + \Lambda u$, are in the plane orthogonal to v_1 and v_2. Let us first deal with the case when $\Lambda u \neq -u$, so that $w = u + \Lambda u$, and

$$\begin{aligned} g(w, w) &= g(u, u) + 2g(u, \Lambda u) + g(\Lambda u, \Lambda u) \\ &= g(u, u) + 2g(u, \Lambda u) + g(u, u) \\ &= 2\left[g(u, u) + g(u, \Lambda u)\right] = 2g(u, w). \end{aligned} \tag{17.53}$$

This means that

$$R_w u = \frac{2g(u, w)}{g(w, w)} w - u = w - u = \Lambda u. \tag{17.54}$$

Since it is also obvious that R_u fixes u, we deduce that $\Lambda u = R_w R_u u$. Each of the two reflections R_w and R_u reverses the direction of orthogonal vectors, so their product must fix both v_1 and v_2. It therefore has the same action as Λ. We have already observed that the reflections R_w and R_u each have determinant -1, so their product is, like Λ, proper. It is easy to see that any proper Lorentz transformation that fixes three independent vectors must be the identity, and applying this to $\Lambda^{-1} R_w R_u$, we see that $R_w R_u = \Lambda$. The special case when $\Lambda u = -u$ is also easily checked. □

Theorem 17.10.3. Every proper orthochronous Lorentz transformation is the product of an even number of elements R_q and the product of the $g(q, q)$ over all q is 1.

Proof. It is shown in most books on special relativity that every proper orthochronous Lorentz transformation can be expressed as the product of two rotations and a standard Lorentz transformation (one that fixes two spatial axes). Any rotation fixes the time axis and a spatial axis, and so falls into the class of proper Lorentz transformations considered above, and so, by definition, do standard Lorentz transformations. □

Corollary 17.10.4. Suppose that the proper Lorentz transformation $\Lambda = R_{u_1} R_{u_2} \ldots R_{u_k}$. Then $\Gamma(\Lambda) = \gamma(u_1)\gamma(u_2)\ldots\gamma(u_k)$ satisfies

$$\gamma(\Lambda p) = \Gamma(\Lambda)\gamma(p)\Gamma(\Lambda)^{-1}$$

for all $p \in M$.

Proof. We know that we may write $\Lambda = R_{q_1} R_{q_2} \ldots R_{q_k}$, so that

$$\begin{aligned}\gamma(q_1)\gamma(q_2)\ldots\gamma(q_k)\gamma(p)\gamma(q_k)^{-1}\ldots\gamma(q_1)^{-1} &= \gamma(R_{q_1} R_{q_2} \ldots R_{q_k} p) \\ &= \gamma(\Lambda p), \qquad (17.55)\end{aligned}$$

as we asserted. □

One of the advantages of this method is that it not only confirms that $\Gamma(\Lambda)$ of this form exists but tells us how to find it. Consider, for example, a rotation R through θ about an axis $\mathbf{n}$. Choosing a vector $\mathbf{u}$ perpendicular to e_0 and $\mathbf{n}$, and noting that

$$R\mathbf{u} = \cos\theta\mathbf{u} + \sin\theta\mathbf{n}\times\mathbf{u}, \qquad (17.56)$$

we see that

$$\begin{aligned}(1+R)\mathbf{u} &= (1+\cos\theta)\mathbf{u} + \sin\theta\mathbf{n}\times\mathbf{u} \\ &= 2\cos(\tfrac{1}{2}\theta)\left[\cos(\tfrac{1}{2}\theta)\,\mathbf{u} + \sin(\tfrac{1}{2}\theta)\,\mathbf{n}\times\mathbf{u}\right]. \qquad (17.57)\end{aligned}$$

The reflection R_w is unaffected by the normalization of w, and it is more convenient to define $\mathbf{w}$ as the unit vector $\cos(\frac{1}{2}\theta)\mathbf{u} + \sin(\frac{1}{2}\theta)\mathbf{n}\times\mathbf{u}$. We now note that

$$\begin{aligned}\gamma(\mathbf{w})\gamma(\mathbf{u}) &= \begin{pmatrix} 0 & -\boldsymbol{\sigma}.\mathbf{w} \\ \boldsymbol{\sigma}.\mathbf{w} & 0 \end{pmatrix}\begin{pmatrix} 0 & -\boldsymbol{\sigma}.\mathbf{u} \\ \boldsymbol{\sigma}.\mathbf{u} & 0 \end{pmatrix} \\ &= \begin{pmatrix} -(\boldsymbol{\sigma}.\mathbf{w})(\boldsymbol{\sigma}.\mathbf{u}) & 0 \\ 0 & -(\boldsymbol{\sigma}.\mathbf{w})(\boldsymbol{\sigma}.\mathbf{u}) \end{pmatrix}. \qquad (17.58)\end{aligned}$$

Using Theorem 8.8.1 we have

$$(\boldsymbol{\sigma}.\mathbf{u})(\boldsymbol{\sigma}.\mathbf{w}) = \mathbf{u}.\mathbf{w} + i\boldsymbol{\sigma}.\mathbf{u} \times \mathbf{w} = \cos\tfrac{1}{2}\theta + i\sin\tfrac{1}{2}\theta\boldsymbol{\sigma}.\mathbf{n} = \exp\left(+\tfrac{1}{2}i\theta\boldsymbol{\sigma}.\mathbf{n}\right), \tag{17.59}$$

so that

$$\Gamma(R) = \gamma(\mathbf{u})\gamma(\mathbf{w}) = -\begin{pmatrix} \exp(\frac{1}{2}i\theta\boldsymbol{\sigma}.\mathbf{n}) & 0 \\ 0 & \exp(\frac{1}{2}i\theta\boldsymbol{\sigma}.\mathbf{n}) \end{pmatrix}. \tag{17.60}$$

This shows immediately that the two component spinors ψ_1 and ψ_2 transform as spinors under the rotation group.

Corollary 17.10.5. The infinitesimal generator of rotations about the axis $\mathbf{n}$ is given by

$$\mathbf{n}.\mathbf{L} + \tfrac{1}{2}\hbar\begin{pmatrix} \mathbf{n}.\boldsymbol{\sigma} & 0 \\ 0 & \mathbf{n}.\boldsymbol{\sigma} \end{pmatrix},$$

where $\mathbf{L}$ is the standard orbital angular momentum operator.

Proof. The infinitesimal generator is obtained by differentiating the action of rotations about $\mathbf{n}$ through an angle t,

$$i\hbar\frac{d}{dt}V(\Lambda)\widehat{\psi}(p) = i\hbar\frac{d}{dt}\Gamma(\Lambda)\widehat{\psi}(\Lambda^{-1}p). \tag{17.61}$$

For rotations we have

$$\Gamma(\Lambda) = \begin{pmatrix} \exp(-\frac{1}{2}it\boldsymbol{\sigma}.\mathbf{n}) & 0 \\ 0 & \exp(-\frac{1}{2}it\boldsymbol{\sigma}.\mathbf{n}) \end{pmatrix} \tag{17.62}$$

and

$$i\hbar\left.\frac{d\Gamma(\Lambda)}{dt}\right|_{t=0} = \begin{pmatrix} \frac{1}{2}\hbar(\boldsymbol{\sigma}.\mathbf{n}) & 0 \\ 0 & \frac{1}{2}\hbar(\boldsymbol{\sigma}.\mathbf{n}) \end{pmatrix}. \tag{17.63}$$

The derivative of $\widehat{\psi}$ just gives the momentum space version of the calculation that yielded the orbital angular momentum, $\mathbf{n}.\mathbf{L}$, so putting the pieces together we obtain the desired result. □

Expressing the free Dirac Hamiltonian as

$$H_D = c\big[P^0 + \gamma_0(mc - D)\big] \tag{17.64}$$

and using the fact that rotations leave P^0 and γ_0 unchanged and commute with D we obtain a second proof of the earlier result on the conservation of $L_k + \frac{1}{2}\hbar\sigma_k$.

17.11. Historical remarks

When Dirac first realized that his equation also predicted a positively charged version of the electron, it was natural to assume that it must be the proton, which was the only known positively charged particle at that time. Weyl gave strong grounds for insisting that the new particle must have the same mass as the electron, and so could not be the much heavier proton. In fact the proton is now regarded as a composite particle built out of quarks, for which the Dirac equation is not appropriate.

The equations that we have discussed are by no means the only wave equations for relativistic particles. There are others covering higher spin particles. There is also a totally different way to approach relativistic quantum theory: if one requires a Lorentz-invariant theory then one could simply investigate the irreducible representations of the Lorentz group. In fact, one could take the larger group, called the *Poincaré group*, which includes the translations in M as well (that is, the maps that take $m \in M$ to $m+a$ for $a \in M$). The irreducible representations of the Poincaré group were classified by Wigner in 1939. One family of these is characterized by a positive real number m and a half integer s, interpreted as mass and spin, respectively. Another family has zero mass and a half-integer spin-like parameter called *helicity*. When the helicity is 1, the particle can be interpreted as a photon and the two helicity states as its polarization. Other representations could describe particles travelling faster than light, or zero-mass particles with a continuous spin parameter, but there is no experimental evidence for any of these.

Exercises

17.1° Show that the Klein–Gordon equation

$$\frac{1}{c^2}\frac{\partial^2\psi}{\partial t^2} - \nabla^2\psi + \frac{m^2c^2}{\hbar^2}\psi = 0$$

has normalizable separable solutions of the form $\psi(t,\mathbf{r}) = T(t)R(r)$ provided that $T(t)$ has the form $\exp(-iEt/\hbar)$ with the energy $E < mc^2$.

17.2 The Dirac equation for a particle of rest mass m is written in the form

$$i\hbar\gamma^\mu\partial_\mu\psi = mc\psi,$$

where

$$\gamma^0 = \begin{pmatrix} 1 & 0 \\ 0 & -1 \end{pmatrix}, \qquad \gamma^j = \begin{pmatrix} 0 & \sigma^j \\ -\sigma^j & 0 \end{pmatrix}, \qquad j = 1,2,3,$$

and σ_1, σ_2, σ_3 are the usual Pauli spin matrices and $\partial_\mu = \partial/\partial x^\mu$, $\mu = 0, 1, 2, 3$. Show that the Dirac equation for a particle of rest mass m has solutions of the form

$$\psi(ct, \mathbf{r}) = e^{-iEt/\hbar}\begin{pmatrix} f(r)v \\ g(r)\boldsymbol{\sigma}.\mathbf{r}v \end{pmatrix},$$

where f and g are real-valued functions of r and $v \in \mathbf{C}^2$ is a fixed non-zero column vector, provided that

$$rg' + 3g = \frac{i}{c\hbar}(E - mc^2)f$$
$$f' = \frac{i}{c\hbar}(E + mc^2)g.$$

Deduce that for solutions to exist f must satisfy the differential equation

$$(rf)'' = \left(\frac{m^2c^4 - E^2}{c^2\hbar^2}\right)(rf).$$

Hence or otherwise show that there are no normalizable solutions of this form unless $|E| < mc^2$, and find the solutions in this case.

17.3 A particle of rest mass m satisfies the free particle Dirac equation

$$(\gamma.P - mc)\psi = 0,$$

where

$$\gamma^0 = \begin{pmatrix} 1 & 0 \\ 0 & -1 \end{pmatrix}, \quad \gamma^j = \begin{pmatrix} 0 & \sigma^j \\ -\sigma^j & 0 \end{pmatrix},$$

and σ_1, σ_2, σ_3 are the usual Pauli spin matrices. Show that

$$\mathbf{J} = \mathbf{X} \times \mathbf{P} + \frac{i\hbar}{4}\boldsymbol{\gamma} \times \boldsymbol{\gamma}$$

commutes with the relativistic Hamiltonian, where $\boldsymbol{\gamma} = (\gamma^1, \gamma^2, \gamma^3)$. Give a physical interpretation of $\mathbf{J}$ and the fact that it commutes with the Hamiltonian.

Show further that $\mathbf{J}.\mathbf{P}$ commutes with the Hamiltonian, and find the eigenvalues of $\mathbf{J}.\mathbf{P}$ when the particle is in a mutual eigenstate of energy with eigenvalue E and three-momentum with eigenvalue $(0, 0, p)$.

[You may assume that $\gamma^\mu\gamma^\nu + \gamma^\nu\gamma^\mu = 2g_{\mu\nu}$ and that $\boldsymbol{\sigma} \times \boldsymbol{\sigma} = 2i\boldsymbol{\sigma}$ where $\boldsymbol{\sigma} = (\sigma_1, \sigma_2, \sigma_3)$.]

17.4° The Dirac Hamiltonian for a free electron is given by

$$H = c(\boldsymbol{\alpha}.\mathbf{p}) + \beta mc^2$$

where

$$\boldsymbol{\alpha} = \begin{pmatrix} 0 & \boldsymbol{\sigma} \\ \boldsymbol{\sigma} & 0 \end{pmatrix}, \qquad \beta = \begin{pmatrix} 1 & 0 \\ 0 & -1 \end{pmatrix}$$

and $\boldsymbol{\sigma} = (\sigma_1, \sigma_2, \sigma_3)$ with components the Pauli spin matrices, and 1 the unit 2×2 matrix.
(i) Show that $\mathbf{J} = \mathbf{L} + \frac{1}{2}\hbar\tilde{\boldsymbol{\sigma}}$, where $\mathbf{L} = \mathbf{X} \times \mathbf{P}$ and

$$\tilde{\boldsymbol{\sigma}} = \begin{pmatrix} \boldsymbol{\sigma} & 0 \\ 0 & \boldsymbol{\sigma} \end{pmatrix},$$

commutes with the Hamiltonian, and comment briefly on this result.
(ii) If $\hbar K = (\tilde{\boldsymbol{\sigma}}.\mathbf{L} + \hbar)$, prove that K^2 is a constant of the motion.
(iii) Show that the Hamiltonian may be written

$$H = c\alpha_r p_r + i\hbar c|\mathbf{X}|^{-1}\alpha_r K + \beta mc^2$$

where $p_r = |\mathbf{X}|^{-1}(\mathbf{X}.\mathbf{P} - i\hbar)$, and $\alpha_r = |\mathbf{X}|^{-1}(\boldsymbol{\alpha}.\mathbf{X})$.

17.5° The matrices $\sigma_\pm(p)$ are defined in terms of the four-vector $p = (p_0, \mathbf{p})$ and the Pauli spin matrices σ_1, σ_2, and σ_3 by

$$\sigma_\pm(p) = p_0 1 \pm \boldsymbol{\sigma}.\mathbf{p}.$$

Show that $\sigma_+(p)$ is a self-adjoint matrix and that every self-adjoint 2×2 matrix is of this form for some four-vector p. Show also that

$$\det(\sigma_+(p)) = g(p,p) = \sigma_+(p)\sigma_-(p).$$

Hence or otherwise show that if A is a 2×2 matrix with unit determinant then

$$A\sigma_+(p)A^* = \sigma_+(\Lambda p)$$

for some Lorentz transformation Λ, and that

$$A^*\sigma_-(\Lambda p)A = \sigma_-(p).$$

The 4×4 matrices $\gamma(p)$ and $S(A)$ are defined by

$$\gamma(p) = \begin{pmatrix} 0 & \sigma_+(p) \\ \sigma_-(p) & 0 \end{pmatrix}, \qquad S(A) = \begin{pmatrix} A & 0 \\ 0 & A^{*-1} \end{pmatrix}.$$

Show that if $\phi(p)$ satisfies the Fourier-transformed Dirac equation

$$(D\phi)(p) = \gamma(p)\phi(p) = mc\phi(p)$$

then so does $S(A)\phi(\Lambda^{-1}p)$.

Show that

$$\Phi(p) = \begin{pmatrix} 0 & 1 \\ -1 & 0 \end{pmatrix} \phi(-p_0, \mathbf{p})$$

also satisfies the Dirac equation and deduce that the equation has both positive and negative energy solutions.

17.6° An electron of mass m satisfies the Dirac equation

$$i\hbar\frac{\partial\psi}{\partial t} = (c\boldsymbol{\alpha}.\mathbf{p} + \beta mc^2)\psi$$

where

$$\boldsymbol{\alpha} = \begin{pmatrix} 0 & \boldsymbol{\sigma} \\ \boldsymbol{\sigma} & 0 \end{pmatrix}, \qquad \beta = \begin{pmatrix} 1 & 0 \\ 0 & -1 \end{pmatrix}.$$

Setting $\tilde{\mathbf{k}} = \mathbf{k} + \boldsymbol{\omega} \times \mathbf{k}$ show that to first order in $\boldsymbol{\omega}$,

$$\tilde{\mathbf{k}}.\tilde{\mathbf{x}} = \mathbf{k}.\mathbf{x},$$

and also that

$$\left(1 - \tfrac{1}{2}i\boldsymbol{\omega}.\boldsymbol{\Sigma}\right)\boldsymbol{\alpha}.\mathbf{k} = \boldsymbol{\alpha}.\tilde{\mathbf{k}}\left(1 - \tfrac{1}{2}i\boldsymbol{\omega}.\Sigma\right) \quad \text{where} \quad \boldsymbol{\Sigma} = \begin{pmatrix} \boldsymbol{\sigma} & 0 \\ 0 & \boldsymbol{\sigma} \end{pmatrix}.$$

Deduce that if $\tilde{\psi}(\tilde{x}) = (1 - \frac{1}{2}i\boldsymbol{\omega}.\boldsymbol{\Sigma})\psi(x)$ then, to first order in ω, $\tilde{\psi}$ also satisfies the Dirac equation.

17.7 Show, using Theorem 17.7.2, that the acceleration, d^2X^k/dt^2, of a Dirac particle in the Heisenberg picture satisfies the equation

$$\frac{d^2X^k}{dt^2} + \frac{2i}{\hbar}\frac{dX^k}{dt}H_D = \frac{2ic^2}{\hbar}P^k.$$

Show that P^k and H_D are constants of the motion, and, using an integrating factor to integrate the equation of motion, find a relationship between the velocity and momentum. By integrating the equation a second time deduce that the position operator, X^k, can be written in the form

$$X_0^k + c^2P^kH_D{}^{-1}t + \frac{i\hbar}{2}\left(V^k - c^2P^kH_D{}^{-1}\right)H_D{}^{-1}\left(e^{-2iH_Dt/\hbar} - 1\right),$$

where X_0^k and V^k are the initial values of the position and velocity, respectively.

18 Dirac particles in electromagnetic fields

I . . . saw Dirac. He has now got a completely new system of equations for the electron which does the spin right in all cases and seems to be "the thing". His equations are first order, not second, differential equations! He told me something about them but I have not even succeeded in verifying that they are right for the hydrogen atom.

CHARLES GALTON DARWIN, letter to Bohr, 26 December 1927

18.1. Interacting Dirac particles

So far we have considered only free particles, moving relativistically without any external forces. However, since there are reasons to expect that electrons can be described by the Dirac equation, we wish to know how they are affected by an electromagnetic field. For this we use the fact, known from classical Hamiltonian theory (Exercise 18.1), that the equations of motion of a particle with charge e in an electromagnetic potential given by potentials ϕ and $\mathbf{A}$ can be obtained by replacing the energy E by $E - e\phi$ and the momentum $\mathbf{p}$ by $\mathbf{p} - e\mathbf{A}$. (Some books use $-e$ giving different signs throughout.) Relativistically this amounts to replacing the four-momentum p by $p - (e/c)\Phi$, where $\Phi = (\phi, c\mathbf{A})$ is the electromagnetic four-potential of Definition 17.1.5. Applying the same substitution to the Dirac theory we are led to the following equation:

Definition 18.1.1. The *Dirac equation* for a particle with charge e in an electromagnetic field with four-potential Φ is

$$\gamma\left(P - \frac{e}{c}\Phi\right)\psi = mc\psi.$$

As before this can be recast in Schrödinger form:

$$i\hbar\frac{\partial\psi}{\partial t} = \gamma^0\left[mc^2 - c\gamma_j\left(P^j - eA^j\right)\psi + e\phi\right]\psi. \tag{18.1}$$

Remark 18.1.1. We noted earlier that the electromagnetic potential is not unique, and the four potential $\Phi = (\phi, c\mathbf{A})$ could be replaced by

$$\Phi' = \left(\phi - \frac{\partial\chi}{\partial t}, c(\mathbf{A} + \nabla\chi)\right) = \Phi + \frac{ic}{\hbar}(P\chi), \tag{18.2}$$

where P has components defined by Definition 17.4.2.

Proposition 18.1.1. Let ψ be a solution of the Dirac equation $\gamma(P - e\Phi/c)\psi = mc\psi$. Then $\psi_\chi = \exp(ie\chi/\hbar)\psi$ satisfies

$$\gamma\Big(P - \frac{e}{c}\Phi'\Big)\psi_\chi = mc\psi_\chi,$$

where $\Phi' = \Phi + icP\chi/\hbar$.

Proof. This follows from a direct calculation, since

$$\begin{aligned}\gamma\left(P - \frac{e}{c}\Phi'\right)\left(e^{ie\chi/\hbar}\psi\right) &= e^{ie\chi/\hbar}\gamma\left(P + \frac{ie}{\hbar}P\chi - \frac{e}{c}\Phi'\right)\psi \\ &= e^{ie\chi/\hbar}\gamma\left(P - \frac{e}{c}\Phi\right)\psi = mc\left(e^{ie\chi/\hbar}\psi\right) \quad (18.3)\end{aligned}$$

□

The effect of a gauge transformation in the electromagnetic potential can therefore be absorbed by changing the argument of the wave function. This has no effect on the most important physical quantities, which depend only on moduli of components of ψ, and nowadays one tends to regard this gauge freedom to change the phase of the wave function as fundamental. The electromagnetic field is then a consequence of this freedom rather than vice versa.

One should not, however, assume that the potential is detectable only through the associated field. In 1959 Yakir Aharonov and David Bohm pointed out that the potential can introduce an observable phase factor even when the field vanishes, and this has subsequently been verified experimentally. Normally, the condition that curl $\mathbf{A}$ vanishes is sufficient to guarantee that $\mathbf{A} = \operatorname{grad}\chi$ for some χ, but if some portions of space are excluded this may fail. (Failures can occur whenever there is a closed curve that is not the boundary of a surface in the region.) Consider, for example, an electron moving in the region where $|\mathbf{C} \times \mathbf{r}| > a$ with a potential $\mathbf{A} = \mathbf{C} \times \mathbf{r}/|\mathbf{C} \times \mathbf{r}|^2$, where $\mathbf{C}$ is a constant. (Physically this could correspond to the potential due to a current in the direction $\mathbf{C}$, in a wire which is shielded to keep the electron away. The wire and shielding prevents one from putting a spanning surface across any closed curve that circumnavigates the wire.) One may check that curl $\mathbf{A}$ vanishes and that, formally,

$\mathbf{A} = \operatorname{grad}\theta/C$ where θ is a polar angle in the plane perpendicular to $\mathbf{C}$. However, although the gradient is well defined, $\chi = \theta/C$ and the corresponding phase shift in wave functions, $\exp(i\theta/C)$, are not, because changing θ by 2π introduces a phase factor $\exp(2\pi i/C)$. This Bohm–Aharonov phase is observable through interference effects. (This phase can also be regarded as a special case of the Berry phase of Definition 12.5.2.)

Following the same procedure of writing

$$\psi = \begin{pmatrix} \psi_+ \\ \psi_- \end{pmatrix}, \tag{18.4}$$

and using the explicit block form of the γ matrices given in Theorem 17.4.3, the Dirac equation reduces to the coupled equations

$$\begin{aligned} i\hbar\frac{\partial\psi_+}{\partial t} - e\phi\psi_+ &= mc^2\psi_+ + c\boldsymbol{\sigma}.(\mathbf{P} - e\mathbf{A})\,\psi_- \\ i\hbar\frac{\partial\psi_-}{\partial t} - e\phi\psi_- &= -mc^2\psi_- + c\boldsymbol{\sigma}.(\mathbf{P} - e\mathbf{A})\,\psi_+. \end{aligned} \tag{18.5}$$

We can subtract the rest energy from the component ψ_+ by a gauge transformation with $e\chi = mc^2 t$, replacing ψ by $\tilde{\psi} = \exp(imc^2 t/\hbar)\psi$. (We could achieve the same effect for ψ_- using $-\chi$.) The equations are now replaced by

$$\begin{aligned} \left(i\hbar\frac{\partial}{\partial t} - e\phi\right)\tilde{\psi}_+ &= c\boldsymbol{\sigma}.(\mathbf{P} - e\mathbf{A})\,\tilde{\psi}_-, \\ \left(i\hbar\frac{\partial}{\partial t} + 2mc^2 - e\phi\right)\tilde{\psi}_- &= c\boldsymbol{\sigma}.(\mathbf{P} - e\mathbf{A})\,\tilde{\psi}_+. \end{aligned} \tag{18.6}$$

In most everyday situations the kinetic energy and electrostatic potential are small compared with the rest energy, mc^2, so the second equation can be approximated by

$$2mc^2\tilde{\psi}_- = c\boldsymbol{\sigma}.(\mathbf{P} - e\mathbf{A})\,\tilde{\psi}_+. \tag{18.7}$$

Solving for $\tilde{\psi}_-$ and substituting into the other equation then we obtain

$$\left(i\hbar\frac{\partial}{\partial t} - e\phi\right)\tilde{\psi}_+ = \frac{1}{2m}\left(\boldsymbol{\sigma}.(\mathbf{P} - e\mathbf{A})\right)^2\tilde{\psi}_+. \tag{18.8}$$

In order to evaluate this we first note that the proof of Theorem 8.8.1 depends only on the properties of spin matrices and nowhere assumes that the components commute. It must therefore be true that

$$(\boldsymbol{\sigma}.(\mathbf{P} - e\mathbf{A}))^2 = (\mathbf{P} - e\mathbf{A}).(\mathbf{P} - e\mathbf{A}) + i\boldsymbol{\sigma}.(\mathbf{P} - e\mathbf{A}) \times (\mathbf{P} - e\mathbf{A}). \tag{18.9}$$

The big difference is that the vector product of two identical vectors with non-commuting components need not vanish. One has, for example,

$$(P_2-eA_2)(P_3-eA_3)-(P_3-eA_3)(P_2-eA_2) = [P_2-eA_2, P_3-eA_3]. \quad (18.10)$$

The components of **P** commute amongst themselves, as do those of **A**, but we still have

$$-e\left([P_2, A_3] + [A_2, P_3]\right) = -ie\hbar\left(\partial_2 A_3 - \partial_3 A_2\right), \quad (18.11)$$

which is the first component of $-ie\hbar \operatorname{curl} \mathbf{A} = -ie\hbar \mathbf{B}$, where $\mathbf{B} = \operatorname{curl} \mathbf{A}$ is the magnetic field. We have therefore proved the following result:

Corollary 18.1.2. For **A** and **P** as above and $\mathbf{B} = \operatorname{curl} \mathbf{A}$, we have

$$\boldsymbol{\sigma}.(\mathbf{P} - e\mathbf{A})^2 = |\mathbf{P} - e\mathbf{A}|^2 + e\hbar\boldsymbol{\sigma}.\mathbf{B}.$$

On substituting this result into the earlier equation for $\tilde{\psi}_+$, and transferring $e\phi$ to the right-hand side, we obtain the following result:

Corollary 18.1.3. When the kinetic and electromagnetic energies are small compared with mc^2, the two-component wave function $\tilde{\psi}_+$ approximately satisfies the equation

$$i\hbar\frac{\partial\tilde{\psi}_+}{\partial t} = \left(\frac{|\mathbf{P} - e\mathbf{A}|^2}{2m} + \frac{e\hbar}{2m}\boldsymbol{\sigma}.\mathbf{B} + e\phi\right)\tilde{\psi}_+.$$

This low energy approximation to the Dirac equation correctly gives the coupling $(e\hbar/2m)\boldsymbol{\sigma}.\mathbf{B}$ between the magnetic field, **B**, and the spin, $\frac{1}{2}\hbar\boldsymbol{\sigma}$, of the electron. Earlier attempts to deal with the problem non-relativistically had given only half the correct expression. This correct prediction of the so-called *gyromagnetic ratio* was another triumph of the Dirac theory.

18.2. Conserved currents

We want to define an analogue of the probability density and probability current which led to a conservation law in the non-relativistic case. Relativistically we might expect these to combine into a single probability current four-vector. To find probabilities, densities must be integrated over three-dimensional spatial slices of M, and we shall specify that we are interested in those of the form $\{x \in M : g(U, x) = k\}$ for constant k.

Definition 18.2.1. The *probability four-current density*, s, for the Dirac equation in the frame defined by the four-velocity U is the unique four-vector that satisfies

$$g(s, p) = c^{-1}\psi^\dagger \gamma(U)\gamma(p)\psi$$

for all $p \in M$, where $\psi^\dagger$ denotes the conjugate transpose of ψ.

If we make the usual choice of $U = (c, \mathbf{0})$, then this definition ensures that

$$s^0 = g(s, e_0) = \psi^\dagger {\gamma^0}^2 \psi = \psi^\dagger \psi, \tag{18.12}$$

which is the sum of the probability densities for the individual components of ψ.

Theorem 18.2.1. The probability current satisfies the conservation law

$$\frac{\partial s^\mu}{\partial x^\mu} = 0.$$

Proof.

$$i\hbar \frac{\partial s^\mu}{\partial x^\mu} = i\hbar \left[(\frac{\partial}{\partial x^0}\left(\psi^\dagger \psi\right) + \frac{\partial}{\partial x^j}\left(\psi^\dagger \gamma^0 \gamma^j \psi\right)\right]. \tag{18.13}$$

Each term contributes two pieces from differentiating ψ and $\psi^\dagger$. Those coming from ψ can be written as

$$i\hbar\psi^\dagger\gamma^0 \left(\gamma^0 \frac{\partial}{\partial x^0}\psi + \gamma^j \frac{\partial}{\partial x^j}\right)\psi = \psi^\dagger\gamma^0\gamma(P)\psi = \psi^\dagger\gamma^0\left(mc - \frac{e}{c}\gamma(\Phi)\right)\psi. \tag{18.14}$$

Those for $\psi^\dagger$ can be evaluated using the following identity, which follows immediately from the explicit matrix formulae $\left(\gamma^0\gamma^j\right)^\dagger = \gamma^0\gamma^j$. Remembering that the dagger involves a conjugation, we have

$$-\left[i\hbar\gamma^0\left(\gamma^0\frac{\partial}{\partial x^0} - \gamma^j\frac{\partial}{\partial x^j}\right)\psi\right]^\dagger \psi = -\left(\gamma^0\gamma(P)\psi\right)^\dagger \psi$$
$$= -\left[\gamma^0\left(mc - \frac{e}{c}\gamma(\Phi)\right)\psi\right]^\dagger \psi. \quad (18.15)$$

The ψ and $\psi^\dagger$ contributions therefore cancel giving the result. □

18.3. Charged antiparticles

In Section 17.5 we discussed the negative energy solutions of the Dirac equation. There is an alternative way of dealing with antiparticles that simplifies the discussion of their electromagnetic properties. It relies on the following result:

Lemma 18.3.1. For any complex four-vector R

$$\gamma^2\gamma(\overline{R})\gamma^2 = \overline{\gamma(R)}.$$

Proof. All the γ^j matrices except γ^2 are real, whilst that is purely imaginary so that

$$\overline{\gamma^j} = \begin{cases} \gamma^j & \text{for } j \neq 2 \\ -\gamma^j & \text{for } j = 2. \end{cases} \quad (18.16)$$

Now from the anticommutation relations on the γ^j we also have

$$\gamma^2\gamma^j\gamma^2 = \begin{cases} \gamma^j & \text{for } j \neq 2 \\ -\gamma^j & \text{for } j = 2, \end{cases} \quad (18.17)$$

so that $\overline{\gamma^j} = \gamma^2\gamma^j\gamma^2$, from which the result follows. □

Definition 18.3.1. The operator K that sends a Dirac wave function ψ to $\psi_c = \gamma^2\overline{\psi}$ is called the *charge conjugation operator* and ψ_c is called the *charge-conjugated wave function.*

The effect of conjugation is to reverse the sign of the energy (since, as for time reversal, Section 9.11, $\exp(-iEt)$ is changed to $\exp(iEt)$). The operator γ^2 interchanges the two pairs of components of ψ.

Theorem 18.3.2. The charge conjugation operator is conjugate linear, that is $K(\alpha\phi + \beta\psi) = \overline{\alpha}\phi + \overline{\beta}\psi$, and satisfies

$$\phi_c^\dagger \gamma^0 \psi_c = \psi^\dagger \gamma^0 \phi.$$

Proof. The conjugate linearity is easily checked, and

$$\begin{aligned} \phi_c^\dagger \psi_c &= \left(\gamma^2 \overline{\phi}\right)^\dagger \gamma^0 \gamma^2 \overline{\psi} \\ &= \overline{\phi^\dagger \gamma^{2\dagger} \gamma^0 \gamma^2 \psi} \\ &= \overline{\phi^\dagger \gamma^0 \psi} = \psi^\dagger \gamma^0 \phi. \end{aligned} \tag{18.18}$$

□

This means that probability densities are unchanged.

Theorem 18.3.3. Let ψ satisfy the Dirac equation $\gamma(P - e\Phi/c)\psi = mc\psi$ for mass m and charge e. Then ψ_c satisfies the Dirac equation

$$\gamma(P + e\Phi/c)\psi_c = mc\psi_c$$

for mass m and charge $-e$.

Proof. Conjugating the Dirac equation for ψ with the help of Lemma 18.3.1, we have

$$\gamma^2 \gamma \overline{(P - e\Phi/c)} \gamma^2 \overline{\psi} = mc\overline{\psi}. \tag{18.19}$$

Now $e\Phi$ is real but, thanks to its factor of $i\hbar$, P is imaginary, so multiplying by γ^2 we obtain

$$\gamma\Big(P + \frac{e}{c}\Phi\Big)\psi_c = mc\gamma^2\overline{\psi} = mc\psi_c.$$ □

We may interpret ψ_c as the wave function appropriate for the antiparticle; it describes a particle with the same mass but opposite charge.

18.4. The Dirac equation for central electrostatic forces

The following result shows that, as one might expect, the spin and orbital angular momentum are either added or subtracted depending on whether they are aligned or opposed.

Lemma 18.4.1. If the orbital angular momentum is $j\hbar$ then the eigenvalues of $|\mathbf{L} + \frac{1}{2}\hbar\boldsymbol{\sigma}|^2$ are $(j \pm \frac{1}{2})\hbar$ and $\boldsymbol{\sigma}.\mathbf{L}$ takes the values $j\hbar$ or $-(j+1)\hbar$.

Proof. According to the Clebsch–Gordan formula, Example 16.7.2, we know that when the angular momenta j and $\frac{1}{2}$ are combined we get either $j + \frac{1}{2}$ or $j - \frac{1}{2}$, so that

$$|\mathbf{L} + \tfrac{1}{2}\hbar\boldsymbol{\sigma}|^2 = \left(j \pm \tfrac{1}{2}\right)\left(j + 1 \pm \tfrac{1}{2}\right). \tag{18.20}$$

We therefore have

$$\begin{aligned} \hbar\boldsymbol{\sigma}.\mathbf{L} &= |\mathbf{L} + \tfrac{1}{2}\hbar\boldsymbol{\sigma}|^2 - |\mathbf{L}|^2 - \tfrac{1}{4}\hbar^2|\boldsymbol{\sigma}|^2 \\ &= \left[\left(j \pm \tfrac{1}{2}\right)\left(j + 1 \pm \tfrac{1}{2}\right) - j(j+1) - \tfrac{3}{4}\right]\hbar^2 \\ &= \left[\pm\left(j + \tfrac{1}{2}\right) - \tfrac{1}{2}\right]\hbar^2, \end{aligned} \tag{18.21}$$

as asserted. □

The operator $\boldsymbol{\sigma}.\mathbf{L}$ that appears in this result is useful because it is related closely to the expressions appearing in the Dirac equation and the following result will be of crucial importance when we consider central force problems.

Lemma 18.4.2. The position, momentum, and angular momentum operators satisfy the identities

$$\boldsymbol{\sigma}.\mathbf{P} = |\mathbf{X}|^{-2}(\boldsymbol{\sigma}.\mathbf{X})\,(\mathbf{X}.\mathbf{P} + i\boldsymbol{\sigma}.\mathbf{L}),$$

$$(\boldsymbol{\sigma}.\mathbf{X})\,(\boldsymbol{\sigma}.\mathbf{L} + \hbar) + (\boldsymbol{\sigma}.\mathbf{L} + \hbar)\,(\boldsymbol{\sigma}.\mathbf{X}) = 0.$$

Proof. The formula for products of spin matrices gives us

$$(\boldsymbol{\sigma}.\mathbf{X})(\boldsymbol{\sigma}.\mathbf{P}) = \mathbf{X}.\mathbf{P} + i(\boldsymbol{\sigma}.\mathbf{X}) \times (\boldsymbol{\sigma}.\mathbf{P}) = \mathbf{X}.\mathbf{P} + i\boldsymbol{\sigma}.\mathbf{L}. \tag{18.22}$$

The first identity follows on multiplying this equation on the left by $\boldsymbol{\sigma}.\mathbf{X}$. The second identity follows on multiplying the commutation relation of Proposition 8.1.1(i) by $\sigma_j\sigma_k = 2\delta_{jk} - \sigma_k\sigma_j$, and simplifying. □

We shall now show how to solve central force problems when there is no magnetic field. We shall therefore set $\mathbf{A} = 0$ and assume that the state is stationary so that $i\hbar\partial\psi/\partial t = E\psi$. Then the Dirac equation can be written in the coupled form

$$\begin{aligned}(E - mc^2 - e\phi)\psi_+ &= c\boldsymbol{\sigma}.\mathbf{P}\psi_- \\ (E + mc^2 - e\phi)\psi_- &= c\boldsymbol{\sigma}.\mathbf{P}\psi_+.\end{aligned} \tag{18.23}$$

Using Lemma 18.4.2 we can express this in terms of the angular momentum. Thus we may rewrite the equations as

$$\begin{aligned}(E - mc^2 - e\phi)\psi_+ &= c|\mathbf{X}|^{-2}(\boldsymbol{\sigma}.\mathbf{X})\,(\mathbf{X}.\mathbf{P} + i\boldsymbol{\sigma}.\mathbf{L})\,\psi_- \\ &= c|\mathbf{X}|^{-2}\,(\mathbf{X}.\mathbf{P} - i\,(\boldsymbol{\sigma}.\mathbf{L} + 2\hbar))\,(\boldsymbol{\sigma}.\mathbf{X})\psi_- \\ (E + mc^2 - e\phi)\psi_- &= c|\mathbf{X}|^{-2}(\boldsymbol{\sigma}.\mathbf{X})\,(\mathbf{X}.\mathbf{P} + i\boldsymbol{\sigma}.\mathbf{L})\,\psi_+.\end{aligned} \tag{18.24}$$

The second equation can be rewritten as

$$(E + mc^2 - e\phi)(\boldsymbol{\sigma}.\mathbf{X})\psi_- = c\,(\mathbf{X}.\mathbf{P} + i\boldsymbol{\sigma}.\mathbf{L})\,\psi_+. \tag{18.25}$$

This suggests the substitution $\Psi_- = \boldsymbol{\sigma}.\mathbf{X}/|\mathbf{X}|\psi_-$, whilst leaving $\Psi_+ = \psi_+$. Using radial coordinates, where, as usual, $\mathbf{X}.\mathbf{P} = -i\hbar\, r d/dr$, the two equations become

$$\begin{aligned}(E - mc^2 - e\phi)\psi_+ &= c\left(-i\hbar\, d/dr - ir^{-1}\,(\boldsymbol{\sigma}.\mathbf{L} + 2\hbar)\right)\Psi_- \\ (E + mc^2 - e\phi)\Psi_- &= c\left(-i\hbar\, d/dr + ir^{-1}\boldsymbol{\sigma}.\mathbf{L}\right)\Psi_+.\end{aligned} \tag{18.26}$$

We also know from Lemma 18.4.1 that $\boldsymbol{\sigma}.\mathbf{L}$ takes the values $j\hbar$ or $-(j+1)\hbar$, and we may as well concentrate on the former. By using the Pauli spin matrices, the equations for $\Psi_\pm$ can be combined into a single equation for the two-dimensional vector, ξ, formed from the $j\hbar$ eigenvector components of Ψ_+ and Ψ_-:

$$\begin{aligned}(E - e\phi - mc^2\sigma_3)\xi &= -ic\hbar\left[(d/dr + r^{-1}) + r^{-1}(j+1)\sigma_3\right]\sigma_1\xi \\ &= c\hbar\left[r^{-1}(j+1)\sigma_2 - i(d/dr + r^{-1})\,\sigma_1\right]\xi.\end{aligned} \tag{18.27}$$

Theorem 18.4.3. With the above conventions, the Dirac equation for motion in a central electrostatic force field can be written in the form

$$(E - e\phi - mc^2\sigma_3)\xi = -ic\hbar\left[\left(\frac{d}{dr} + \frac{1}{r}\right)\sigma_1 + i\frac{(j+1)}{r}\sigma_2\right]\xi.$$

This equation can be solved by the usual techniques. One can first look for asymptotic solutions, or, guided by the non-relativistic case, immediately guess the appropriate substitution, $\xi = \exp(-\kappa r)\eta$. The equation for η reduces to

$$(E - e\phi - mc^2\sigma_3)\eta = -ic\hbar\left[\left(\frac{d}{dr} + \frac{1}{r} - \kappa\right)\sigma_1 + i\frac{(j+1)}{r}\sigma_2\right]\eta, \quad (18.28)$$

or, equivalently, to

$$(E - mc^2\sigma_3 - ic\hbar\kappa\sigma_1)\eta = -ic\hbar\left[\left(\frac{d}{dr} + \frac{1}{r}\right)\sigma_1 + i\frac{(j+1)}{r}\sigma_2 + \frac{ie\phi}{c\hbar}\right]\eta. \quad (18.29)$$

It will be useful to introduce the notation

$$R = i(c\hbar)^{-1}(E - mc^2\sigma_3 - ic\hbar\kappa\sigma_1) \quad (18.30)$$

for this matrix, which appears often in the calculations. At this point we shall also specialize to the case $e\phi = c\hbar\alpha/r$. (The fine structure constant α is around $1/137$ for hydrogen, and smaller than 1 for all stable atomic nuclei.) With these conventions, our equation becomes

$$R\eta = \left[\left(\frac{d}{dr} + \frac{1}{r}\right)\sigma_1 + i\frac{(j+1)\sigma_2 + \alpha}{r}\right]\eta. \quad (18.31)$$

To obtain asymptotic solutions with η almost constant for large r we need $R\eta = 0$, and this can have non-trivial solutions only if $\det(R)$ vanishes, that is

$$\det\big(E - mc^2\sigma_3 - ic\hbar\kappa\sigma_2\big) = 0, \quad (18.32)$$

which forces κ to satisfy the equation

$$c^2\hbar^2\kappa^2 = m^2c^4 - E^2. \quad (18.33)$$

We must pick the positive root to get a normalizable solution.

To find η for smaller values of r we try the series solution

$$\eta = \sum r^{k+\delta} a_k, \tag{18.34}$$

where the coefficients a_k are two-dimensional column vectors. When this is substituted into (18.31), we obtain the equation

$$\sum r^{k+\delta} R a_k = \sum \left[(k+\delta+1)\sigma_1 + i(j+1)\sigma_2 + i\alpha\right] r^{k+\delta-1} a_k. \tag{18.35}$$

The indicial equation is

$$\left[(\delta+1)\sigma_1 + i(j+1)\sigma_2 + i\alpha\right] a_0, \tag{18.36}$$

which means that, for a non-trivial solution,

$$\det\left[(\delta+1)\sigma_1 + i(j+1)\sigma_2 + i\alpha\right] = 0, \tag{18.37}$$

or

$$(\delta+1)^2 = (j+1)^2 - \alpha^2. \tag{18.38}$$

For normalizability we must choose the positive root for $\delta + 1$. (Recalling that α is much smaller than 1, we see that there is no problem in taking the square root, and in fact $\delta \sim j$.) We can then study the recurrence relations for the coefficients, which as usual lead to the requirement that the series should terminate. If a_n is the last non-zero coefficient then, for consistency, we require that

$$R a_n = 0, \tag{18.39}$$

and that

$$R a_{n-1} = \left[(n+\delta+1)\sigma_1 + i(j+1)\sigma_2 + i\alpha\right] a_n. \tag{18.40}$$

Now, since R has trace $2iE/c\hbar$, and vanishing determinant, its characteristic equation is

$$R(R - 2iE/c\hbar) = 0. \tag{18.41}$$

Multiplying (18.40) by $R - 2iE/c\hbar$, the left-hand side vanishes to give

$$0 = (R - 2iE/c\hbar)\left[(n+\delta+1)\sigma_1 + i(j+1)\sigma_2 + i\alpha\right] a_n. \tag{18.42}$$

Now, exploiting the properties of the Pauli spin matrices, we have

$$(R - 2iE/c\hbar)\sigma_2 = -\sigma_2 R, \tag{18.43}$$

and

$$(R - 2iE/c\hbar)\sigma_1 = -\sigma_1 R + 2\kappa. \tag{18.44}$$

On substituting these relations and using (18.40) we arrive at

$$0 = [2\kappa(n + \delta + 1) + 2\alpha E/c\hbar]a_n. \tag{18.45}$$

This gives us the quantization condition that

$$\alpha E = -c\hbar\kappa(n + \delta + 1), \tag{18.46}$$

which with the earlier equations for κ and δ determines the energy levels. On substituting (18.46) into (18.33) we arrive at the following result:

Theorem 18.4.4. The energy levels of the relativistic hydrogen-like atom are given by

$$E = \frac{mc^2}{\sqrt{1 + [\alpha^2/(n + \delta + 1)^2]}},$$

where δ is determined by (18.38).

Since $\alpha < 1$ we may approximate this answer by the first terms in the binomial expansion to get

$$E \sim mc^2 - \frac{mc^2\alpha^2}{2(n + \delta + 1)^2}. \tag{18.47}$$

For hydrogen-like atoms one has $c\hbar\alpha = Ze^2/4\pi\epsilon_0$, which gives the approximation

$$E \sim mc^2 - \frac{Z^2e^2}{4\pi\epsilon_0 a}\frac{1}{2(n + \delta + 1)^2}, \tag{18.48}$$

where a is the usual Bohr radius.

Unlike its non-relativistic analogue, the relativistic energy does depend, through δ, on the angular momentum. This was another triumph of the Dirac theory, since it was known experimentally that the spectral lines were split into groups according to the angular momentum, and the relativistic formula correctly gives this fine structure.

18.5. The successes of the Dirac theory

- It correctly incorporates spin into the relativistic wave equation.
- It gives the correct coupling to the magnetic field.
- It correctly predicted the existence of antiparticles.
- It gives the fine structure in the spectrum of hydrogen-like atoms.

Exercises

18.1° The *classical* Hamiltonian for a charged particle in an electromagnetic field is

$$h = \frac{1}{2m}|\mathbf{p} - e\mathbf{A}|^2 + e\phi.$$

Show, using Hamilton's classical equations of motion, that

$$m\ddot{\mathbf{x}} = e(\mathbf{E} + \dot{\mathbf{x}} \times \mathbf{B}),$$

where $\mathbf{B} = \operatorname{curl}\mathbf{A}$ and $\mathbf{E} = -(\partial\mathbf{A}/\partial t + \operatorname{grad}\phi)$.
[*Hint*: $\operatorname{grad}(\frac{1}{2}|\mathbf{U}|^2) = \mathbf{U} \times (\operatorname{curl}\mathbf{U}) + \mathbf{U}\cdot\nabla\mathbf{U}$.]

18.2° The Dirac equation for a particle of rest mass m and charge e moving in an electromagnetic field with four-potential $a = (\phi, -c\mathbf{A})$ is

$$\left[\gamma^\mu\left(i\hbar\partial_\mu - \frac{e}{c}a_\mu\right) - mc\right]\psi = 0,$$

where

$$\gamma^0 = \begin{pmatrix} 1 & 0 \\ 0 & -1 \end{pmatrix}, \qquad \gamma^j = \begin{pmatrix} 0 & \sigma^j \\ -\sigma^j & 0 \end{pmatrix}, \qquad j = 1, 2, 3,$$

and σ_1, σ_2, σ_3 are the usual Pauli spin matrices and $\partial_\mu = \partial/\partial x^\mu$, $\mu = 0, 1, 2, 3$. Show that the equation can be rewritten in the form

$$i\hbar\frac{\partial\psi}{\partial t} = H_D\psi,$$

where H_D has the form

$$H_D = c\boldsymbol{\alpha}.(\mathbf{P} - e\mathbf{A}) + \beta mc^2 + e\phi.$$

Show that, when a vanishes, H_D commutes with $\mathbf{L} + \frac{1}{2}\hbar\boldsymbol{\sigma}'$, where $\mathbf{L}$ is the orbital angular momentum and

$$\boldsymbol{\sigma}' = \begin{pmatrix} \boldsymbol{\sigma} & 0 \\ 0 & \boldsymbol{\sigma} \end{pmatrix}.$$

Give a brief physical interpretation of this result.

Using the Heisenberg picture calculate the velocity dX_1/dt and show that

$$\frac{dX_1}{dt}^2 = c^2.$$

18.3° The Dirac equation for a particle mass m and charge e moving in the electromagnetic field with four-potential Φ is

$$[\gamma.(p - e\Phi) - mc]\psi = 0,$$

where p is the four-momentum operator and

$$\gamma^0 = \begin{pmatrix} 1 & 0 \\ 0 & -1 \end{pmatrix}, \qquad \boldsymbol{\gamma} = \begin{pmatrix} 0 & \boldsymbol{\sigma} \\ -\boldsymbol{\sigma} & 0 \end{pmatrix}.$$

The particles are constrained to move in the x_3 direction only, in the four-potential

$$\Phi_0 = \begin{cases} -V/e & |x_3| < a, \\ 0 & |x_3| > a, \end{cases} \qquad \mathbf{A} = 0,$$

where V is a positive constant and $a > 0$. Write down the differential equations satisfied by ψ in the regions $|x_3| < a$ and $|x_3| > a$. What are the connection conditions at $x_3 = \pm a$? When $mc > P_0 > mc - V > 0$ let ψ be given by

$$e^{-iP_0x_0/\hbar}\left[e^{iPx_3/\hbar}\begin{pmatrix} v \\ \frac{P}{P_0+V+mc}\sigma_3 v \end{pmatrix} + e^{-iPx_3/\hbar}\begin{pmatrix} v' \\ \frac{-P}{P_0+V+mc}\sigma_3 v' \end{pmatrix}\right]$$

when $|x_3| < a$, and by

$$e^{-iP_0x_0/\hbar}e^{-Kx_3/\hbar}\begin{pmatrix} w \\ \frac{iK}{P_0+mc}\sigma_3 w \end{pmatrix}$$

when $x_3 > a$, and v, v', and w are constant 2×1 matrices. Show that ψ defines a solution in $x_3 > -a$ if

$$(P_0 + V)^2 - P^2 = m^2c^2, \qquad {P_0}^2 + K^2 = m^2c^2$$

and v, v', and w satisfy the equations

$$e^{iPa/\hbar}v + e^{-iPa/\hbar}v' = e^{-Ka/\hbar}w,$$

$$e^{iPa/\hbar}v - e^{-iPa/\hbar}v' = i\frac{K(P_0 + V + mc)}{P(P_0 + mc)}e^{-Ka/\hbar}w.$$

Write down the solution in $x_3 < -a$ and deduce that the bound state energies are given by

$$\tan\left(\frac{2Pa}{\hbar}\right) = \frac{2\alpha}{1 - \alpha^2},$$

where $\alpha = |K|(P_0 + V + mc)/P(P_0 + mc)$ and $E = cP_0$.

19* Symmetries of elementary particles

> If, as I have reason to believe, I have disintegrated the nucleus of the atom, this is of greater significance than the war.
>
> ERNEST RUTHERFORD, apologizing for absence from a meeting of the International Anti-submarine Warfare Committee, June 1919.

19.1. The structure of matter

It will have become apparent over the last few chapters that the electrons, protons, and neutrons, which are the most prominent constituents of matter, are only some of the many particles that are now known. There are also neutrinos, which appear in some nuclear reactions, Yukawa's mesons, which help bind thc nucleus together, and then to every particle a corresponding antiparticle. With improved techniques for detecting particles in cosmic rays and the increasingly powerful particle accelerators built after the war more and more different kinds of subatomic particle were discovered. The various conservation laws found to hold during collisions between particles suggested that there must be some kind of symmetry principle linking different particles. For example, protons and neutrons belong to a class of particles called *baryons*, and during collisions the total number of baryons minus the total number of antibaryons is constant. A similar conservation law held for the class of *leptons* which included the lighter particles such as electrons and neutrinos. Scattering experiments at higher energies began to indicate that the proton has an internal structure suggesting that it might be composed of still smaller particles.

The theory that evolved during the early 1960s suggests that baryons and mesons are composed of particles called quarks and their antiparticles, the antiquarks. There are several different kinds of quark but all have spin $\frac{1}{2}$ and are fermions. Although they differ in such properties as their electric charge the quarks are remarkably similar. Already in the 1930s similarities between the proton and the neutron, whose masses (1836 and 1839 times the electron mass, respectively) are almost identical, and which seem to differ in little more than charge, had led Heisenberg to suggest that they might simply be different states of the same particle. The apparent differences between them would then be no more significant than those between electrons with different spins. In the current theories this idea is applied to quarks, which are all regarded as being states of a single particle. Rather than talking of Q different sorts of quark each with a two-component wave function (for the two different spin states), it is therefore more appropriate to talk of a $2Q$-component wave function describing the state of a quark.

Baryons are made of three fermionic quarks; the baryon states are described by $\wedge^3\mathbf{C}^{2Q}$-valued wave functions. (Experiments show that quarks nestle in close proximity within protons, indicating that fermionic antisymmetry affects the values of their wave functions rather than their spatial distribution.) Similarly mesons consist of a quark and an antiquark. The states of such a system lie in the space $\mathbf{C}^{2Q} \otimes \mathbf{C}^{2Q^*} \cong \mathcal{L}(\mathbf{C}^{2Q})$ (see Proposition 16.5.3(iv)), and, in fact, are always in the subspace of traceless operators.

For simplicity let us start by ignoring spin and concentrating attention on the space $\mathbf{C}^Q$ of quarks with a given spin. The unitary transformations of this space cannot be precise symmetries (that is, the Hamiltonian cannot intertwine them) or it would be quite impossible to distinguish the different sorts of quark. It is now believed that the qualities of quarks are of two kinds, known as *flavour* and *colour*. (As in the modern food industry colour and flavour are supposed to be independent of each other.) Colour symmetry is exact, and quarks that differ only in their colour cannot be distinguished. Flavour symmetry is only approximate, and different flavours of quark may have different masses and charges. In both cases, however, one is led to investigate the way in which the symmetry or approximate symmetry groups act on the states of the system, that is the representation theory of the groups of unitary operators.

19.2. Characters of unitary groups

Having explained the physical significance of the unitary groups we shall now investigate their properties mathematically.

Definition 19.2.1. The *unitary group* $\mathbf{U}(n)$ is the group of all unitary $n \times n$ matrices U, that is matrices satisfying $U^*U = 1 = UU^*$.

We shall start by recalling a well-known algebraic property of unitary matrices.

Theorem 19.2.1. Every matrix in $\mathbf{U}(n)$ is conjugate to a diagonal matrix. The entries on the diagonal are unique up to changes of order.

Proof. This is a reinterpretation of Theorem A1.3.2 in the first appendix, according to which each unitary operator U on $\mathbf{C}^n$ admits an orthonormal basis of eigenvectors, $v_1, v_2, \ldots, v_n$, such that $Uv_j = \alpha_j v_j$. Let V be the operator that maps the natural basis $\{e_j\}$ of $\mathbf{C}^n$ to the eigenvector basis. Then V must be unitary and

$$V^{-1}UVe_j = V^{-1}Uv_j = \alpha_j V^{-1}v_j = \alpha_j e_j, \tag{19.1}$$

so that $V^{-1}UV$, which is conjugate to U in $\mathbf{U}(n)$, has a diagonal matrix. Since the diagonal entries are the eigenvalues of U they are unique up to order. By reordering the basis elements any change of order is possible. □

We want to study finite-dimensional representations of a unitary group, G, that is homomorphisms of G into groups of unitary operators on inner product spaces. We saw in Section 9.7 that such representations can be described by their characters and that these are constant on conjugacy classes, which leads immediately to the following result.

Theorem 19.2.2. Each character χ of $\mathbf{U}(n)$ is uniquely determined by its restriction to the diagonal matrices. It can therefore be identified with a function on $\mathbf{T}^n$ which is invariant under permutations of its arguments.

Proof. The character χ is invariant under conjugation so that, in the notation of the previous proof,

$$\chi(U) = \chi(V^{-1}UV), \tag{19.2}$$

showing that χ is determined by its value on diagonal matrices. The diagonal entries $\alpha_1, \alpha_2, \ldots, \alpha_n$, being eigenvalues of a unitary matrix, have modulus 1, that is they lie in $\mathbf{T}$. Thus $\chi(V^{-1}UV)$ can be identified with a function of $(\alpha_1, \alpha_2, \ldots, \alpha_n) = \alpha \in \mathbf{T}^n$. Since the order of the α can be changed by conjugation χ must be invariant under permutations. □

Definition 19.2.2. We now introduce the *elementary symmetric polynomials* σ_j, $j = 1, 2, \ldots, n$, as the coefficients of the polynomial

$$\prod_{j=1}^{n}(1 + x\alpha_j) = 1 + \sum_{k=1}^{n} x^k \sigma_k.$$

Explicitly we have

$$\sigma_k = \sum_{j_1<j_2<\ldots<j_k} \alpha_{j_1}\alpha_{j_2}\ldots\alpha_{j_k}.$$

The first few symmetric polynomials are $\sigma_1 = \sum \alpha_j$, $\sigma_2 = \sum_{j<k} \alpha_j\alpha_k$, $\sigma_n = \alpha_1\alpha_2\ldots\alpha_n$.

Theorem 19.2.3. The characters of finite-dimensional representations of $\mathbf{U}(n)$ can be regarded as polynomials in $\sigma_1, \sigma_2, \ldots, \sigma_n$ and σ_n^{-1}.

Proof. The subgroup of diagonal matrices in $\mathbf{U}(n)$ is abelian and isomorphic to $\mathbf{T}^n$, and so the restriction of any finite-dimensional representation of $\mathbf{U}(n)$ will decompose into a direct sum of irreducibles. By equation (9.30) these are one dimensional and $(\alpha_1, \alpha_2, \ldots, \alpha_n)$ acts as $\alpha_1^{k_1}\alpha_2^{k_2}\ldots\alpha_n^{k_n}$ for some integers $k_1, k_2, \ldots, k_n$. The character $\chi(U)$, being a sum of such monomials, is therefore a polynomial in the α and their inverses. If $-k$ is the largest negative integer to appear as the exponent of any of the α then $\chi(U){\sigma_n}^k$ can contain only positive powers of the α and so is a polynomial. However, it is a classical theorem of algebra that any polynomial in the α that is invariant under permutations can be expressed as a polynomial in the σ. (This can be proved by induction on the number of variables and the degree of the polynomial.) Thus $\chi(U){\sigma_n}^k$ is a polynomial in $\sigma_1, \sigma_2, \ldots, \sigma_n$ as required. □

Remark 19.2.1.

$$\prod_{j=1}^{n}(1 + x\alpha_j) = \det(1 + xV^{-1}UV) = \det[V^{-1}(1 + xU)V] = \det(1 + xU), \tag{19.3}$$

so that the σ_j can easily be calculated directly from U; indeed they are closely related to the coefficients of the characteristic polynomial.

We have not so far used the fact that the α have modulus 1. This means that $\sigma_n^{-1} = \overline{\sigma_n}$, and more generally that $\sigma_{n-k} = \sigma_n \overline{\sigma_k}$.

Often we are only interested in subgroups of $\mathbf{U}(n)$, rather than the whole group.

Definition 19.2.3. The *special unitary group* is the subgroup of unitary matrices whose determinant is 1:

$$\mathbf{SU}(n) = \{U \in \mathbf{U}(n) : \det(U) = 1\}.$$

This is particularly important, since then

$$\sigma_n = \det(V^{-1}UV) = \det(U) = 1, \tag{19.4}$$

which leads to the following result:

Theorem 19.2.4. The characters of representations of $\mathbf{SU}(n)$ can be identified with elements of $\mathbf{C}[\sigma_1, \sigma_2, \ldots, \sigma_{n-1}]$.

Example 19.2.1. The characters of $\mathbf{U}(1)$ are described by elements of $\mathbf{C}[\sigma_1, \sigma_1^{-1}]$. Since $\mathbf{T} \cong \mathbf{U}(1)$, this accords with Example 9.4.2, where it is shown that the irreducible representations are just given by integer powers of $\alpha_1 = \sigma_1$.

Example 19.2.2. The characters of $\mathbf{SU}(2)$ are described by elements of $\mathbf{C}[\sigma_1]$, that is polynomials in $\sigma_1 = \alpha_1 + \alpha_2 = \alpha_1 + \alpha_1^{-1}$.

Example 19.2.3. The characters of $\mathbf{SU}(3)$ are described by elements of $\mathbf{C}[\sigma_1, \sigma_2] = \mathbf{C}[\sigma_1, \overline{\sigma_1}]$.

19.3. Representations of unitary groups

It is useful to preface our discussion of unitary groups with a general construction.

Definition 19.3.1. Let U be a representation of a group G on a space $\mathcal{H}$. The *contragredient representation* is defined on the dual vector space $\mathcal{H}^*$ by

$$(U^*(x)f)(\psi) = f(U(x^{-1})\psi)$$

for x in G, ψ in $\mathcal{H}$, and f in $\mathcal{H}^*$.

The importance of the contragredient representation lies in its application to antiparticles. If a symmetry group G for the particle has a representation U on the space of particle states, then the contragredient representation U^* acts on the states of the antiparticle. When U and U^* coincide then the particle is its own antiparticle.

When $\mathcal{H}$ is a finite-dimensional space so that we can write its elements as column vectors and the dual vectors as row vectors, then using $\top$ to denote the transpose we have

$$(U^*(x)f)^\top \psi = f^\top (U(x^{-1})\psi) = \left(U(x^{-1})^\top f\right)^\top \psi. \tag{19.5}$$

Then we can identify $U^*(x)$ with

$$U(x^{-1})^\top = U(x)^{*\top} = \overline{U(x)}, \tag{19.6}$$

since the adjoint is the complex conjugate of the transpose. A similar argument using dual bases shows that the character of the U^* is the complex conjugate of the character of U. If the character is real, as for the adjoint representation, then the contragredient representation U^* is equivalent to U.

We shall now describe some of the representations of $\mathbf{U}(n)$, which correspond to the characters already mentioned. We start with one obvious example.

Example 19.3.1. (The natural representation) One can also regard $\mathbf{U}(n)$ as the unitary operators on the space $\mathbf{C}^n$ equipped with the usual inner product, so there is a natural representation in which the matrix U

is mapped to itself regarded as an operator on $\mathbf{C}^n$. The character of this representation is

$$\operatorname{tr} U = \operatorname{tr}(V^{-1}UV) = \sum \alpha_j = \sigma_1. \tag{19.7}$$

This is the representation that describes the states of the quarks themselves.

The tensor powers of this representation can also be formed as in Definition 16.7.1. According to Proposition 16.7.1 the k-th tensor power has character ${\sigma_1}^k$. This means that all the representations of $\mathbf{SU}(2)$ can be constructed from tensor powers of the natural representation. The contragredient representation has character $\overline{\sigma}_1$. Example 19.2.3 then tells us that all the representations of $\mathbf{SU}(3)$ can be constructed from tensor products of the natural representation and its contragredient.

In practice the symmetric and exterior powers (Section 16.6) are more useful than general tensor powers.

Example 19.3.2. (The exterior power Λ^k) The exterior power is represented on $\Lambda^k\mathbf{C}^n$, which has basis $e_{j_1} \wedge e_{j_2} \wedge \ldots \wedge e_{j_k}$, with $j_1 < j_2 < \ldots < j_k$. This is an eigenvector for the diagonal matrix $V^{-1}UV$ with eigenvalue $\alpha_{j_1}\alpha_{j_2}\ldots\alpha_{j_k}$, so that the character is given by

$$\sum_{j_1<j_2<\ldots<j_k} \alpha_{j_1}\alpha_{j_2}\ldots\alpha_{j_k} = \sigma_k. \tag{19.8}$$

The baryons, which are made up of three quarks, are described by the representation of $\mathbf{U}(2Q)$ on $\wedge^3\mathbf{C}^{2Q}$, with character σ_3, and the antibaryons are described by the contragredient representation whose character is $\overline{\sigma_3}$.

It is a corollary of Theorem 19.2.4 that all the representations of $\mathbf{SU}(n)$ can be constructed from the various exterior powers of the natural representation.

Example 19.3.3. (The symmetric power S^k) The characters S_k of the symmetric powers are not so easily expressed, but by taking bases it is not difficult to show that

$$S_k(\alpha) = \sum_{k_1+\ldots+k_n=k} \alpha_1^{k_1}\alpha_2^{k_2}\ldots\alpha_n^{k_n}. \tag{19.9}$$

Although the symmetric powers would seem irrelevant to a discussion of fermions, they do appear as subrepresentations when we take into account the fact that flavour symmetry is not precise, but we really only need the cases of $k = 2$ and 3. When $k = 2$ the formula reduces to

$$S_2(\alpha) = \sum_j \alpha_j^2 + \sum_{j<k} \alpha_j\alpha_k = \sigma_1^2 - \sigma_2. \tag{19.10}$$

This confirms the decomposition of $\bigotimes^2 \mathbf{C}^n$ into a symmetric and an antisymmetric part ($\sigma_1^2 = S_2 + \sigma_2$). Setting $k = 3$ we obtain

$$S_3(\alpha) = \sum_j \alpha_j^3 + \sum_{j \neq k} \alpha_j^2 \alpha_k + \alpha_1 \alpha_2 \alpha_3 = \sigma_1^3 - 2\sigma_1\sigma_2 + \sigma_3. \tag{19.11}$$

It follows from this identity that $\sigma_1^3 = S_3 + \sigma_3 + 2\Pi$, where we have introduced $\Pi = \sigma_1\sigma_2 - \sigma_3$. In particular, this shows that the third tensor power contains more than just its symmetric and antisymmetric part. The character Π is of interest in its own right. We shall not describe an explicit form of the corresponding representation in general, but in low dimensions it often coincides with one of those already discussed. For example, in the case of $\mathbf{SU}(2)$ we have $\sigma_3 = 0$ and $\sigma_2 = 1$, so that $\Pi = \sigma_1$. For $\mathbf{SU}(3)$, we have $\sigma_3 = 1$ and $\sigma_2 = \overline{\sigma_1}$, so that $\Pi = |\sigma_1|^2 - 1$. This is the character of the adjoint representation, which we shall describe in the next example.

Example 19.3.4. (The adjoint representation) Mesons may be considered as bound quark–antiquark states, and we noted that these could be represented on the space of traceless elements of $\mathcal{L}(\mathbf{C}^2)$. The corresponding representation of $\mathbf{U}(2Q)$ is realized on the space of $n \times n$ matrices A whose trace vanishes, that is

$$\mathcal{L}_0 = \{A \in \mathcal{L}(\mathbf{C}^n) : \operatorname{tr}(A) = 0\}. \tag{19.12}$$

The space has the inner product

$$\langle A|B \rangle = \operatorname{tr}(A^* B), \tag{19.13}$$

and the adjoint representation on this space is defined by

$$\operatorname{ad}(U)A = UAU^{-1} = UAU^*. \tag{19.14}$$

To see that this is well defined we check that

$$\operatorname{tr}(\operatorname{ad}(U)A) = \operatorname{tr}(UAU^{-1}) = \operatorname{tr}(AU^{-1}U) = \operatorname{tr}(A) = 0. \tag{19.15}$$

Similarly we may show that $\operatorname{ad}(U)$ is unitary:

$$\begin{aligned} \langle \operatorname{ad}(U)A|\operatorname{ad}(U)B \rangle &= \operatorname{tr}((UAU^*)^*(UBU^*)) = \operatorname{tr}(UA^*U^*UBU^*) \\ &= \operatorname{tr}(ABU^*U) \\ &= \operatorname{tr}(A^*B) = \langle A|B \rangle. \end{aligned} \tag{19.16}$$

When $j \neq k$ the elementary matrix E_{jk}, with a 1 in the j-th row and k-th column and 0 elsewhere, has vanishing trace, and it is an eigenvector for the diagonal matrix $V^{-1}UV$ with eigenvalue $\alpha_j\overline{\alpha_k}$. Together with the matrices $E_{jj} - E_{nn}$ which also have vanishing trace and are fixed by $\mathrm{ad}(V^{-1}UV)$ these form an orthonormal basis of $\mathcal{L}_0$. The character of this representation is therefore

$$\chi_{\mathrm{ad}}(U) = \sum_{j\neq k} \alpha_j\overline{\alpha_k} + (n-1) = |\sigma_1|^2 - 1. \tag{19.17}$$

Proposition 16.7.1 provides a much easier way to calculate the character, for it tells us that the conjugation action on all $n \times n$ matrices is equivalent to the tensor product of the natural representation with its dual, and so has character $\overline{\sigma_1}\sigma_1$. But this is also the direct sum of the adjoint representation with the trivial representation on multiples of the identity, from which it follows that the adjoint representation must have character $|\sigma_1|^2 - 1$. As the character of the adjoint representation is real, it is equivalent to its contragredient. This means that the antiparticles of mesons are also mesons.

Example 19.3.5. (The rotation group characters) In Corollary 9.9.3 we noted that the rotation group characters Δ^j could be lifted to the special unitary group $\mathbf{SU}(2)$, where they make sense for half-integral j as well. These characters are linked to those for general unitary groups by the following relations:

Lemma 19.3.1. The following relations hold between characters of $\mathbf{SU}(2)$ and those of the rotation group:

$$R^*\Delta^0 = 1, \qquad R^*\Delta^{\frac{1}{2}} = \Pi, \qquad R^*\Delta^1 = \chi_{\mathrm{ad}}, \qquad R^*\Delta^{\frac{3}{2}} = S_3.$$

Proof. D^0 is the trivial one-dimensional representation, from which the first statement follows. We have already noted that for $\mathbf{SU}(2)$, $\Pi = \sigma_1$, which has the character

$$\alpha_1 + \alpha_2 = 2\cos\left(\tfrac{1}{2}t\right) = \frac{\sin t}{\sin\left(\frac{1}{2}t\right)}, \tag{19.18}$$

the character of $R^*D^{\frac{1}{2}}$. Similarly, the character of the adjoint representation is

$$|\sigma_1|^2 - 1 = |\alpha_1 + \alpha_2|^2 - 1 = 4\cos^2\left(\tfrac{1}{2}t\right) - 1 = 2\cos t + 1, \tag{19.19}$$

which is the character of R^*D^1. Finally, we have for S_3

$$\alpha_1^3 + \alpha_1^2\alpha_2 + \alpha_1\alpha_2^2 + \alpha_2^3 = \frac{\alpha_1^4 - \alpha_2^4}{\alpha_1 - \alpha_2} = \frac{\sin(2t)}{\sin(\frac{1}{2}t)}, \tag{19.20}$$

which is the character of $R^*D^{\frac{3}{2}}$. □

We shall henceforth find it useful to drop the R^* and refer to representations of $\mathbf{SU}(2)$ as though they were representations of the rotation group.

Example 19.3.6. (Power sums) Although every character is described by a polynomial in the eigenvalues $\alpha_1, \alpha_2, \ldots$ which is invariant under permutations, the converse is not true. The power sums,

$$s_r = \alpha_1^r + \alpha_2^r + \ldots + \alpha_n^r, \tag{19.21}$$

are clearly invariant under permutations, but apart from $s_1 = \sigma_1$, they are not themselves characters of representations. (This can be seen by checking that the character $s_2 = \alpha_1^2 + \alpha_2^2$ of $\mathbf{SU}(2)$ is lifted from the rotation group but cannot be expressed as a sum of any of the known characters given in Section 9.8.) In fact, we can easily check that

$$s_2 = S_2 - \sigma_2, \tag{19.22}$$

and in general, the permutation-invariant polynomials are formed from sums and differences of characters.

19.4. Subrepresentations of unitary groups

We are now ready to undertake a more detailed examination of quark symmetries. Following the discussion in Section 19.1, the number of quarks, Q, is the product CF of the number of colours, C, and the number of flavours, F. This must be doubled to obtain the total number of quark states, since there are two possible spin states for each. The full symmetries of the theory are therefore given by $\mathbf{U}(2CF)$, whilst the colour, flavour, and rotational symmetries are described by $U(C)$, $\mathbf{U}(F)$, and $\mathbf{SU}(2)$, respectively.

We have already noted the distinction between the exact colour symmetries and approximate or broken flavour symmetries, which arises because the Hamiltonian is invariant under the colour symmetries, but not under the full group of flavour symmetries. This means that we need information about what happens when the relevant representations of $\mathbf{U}(2CF)$ are restricted to the colour subgroup, $SU(C)$, and it is useful to consider this problem in a broader context.

For any positive integers m and n there is a representation I of $\mathbf{U}(m) \times \mathbf{U}(n)$ on $\mathbf{C}^m \otimes \mathbf{C}^n \cong \mathbf{C}^{mn}$ given by

$$I(V,W)(\phi \otimes \psi) = (V\phi) \otimes (W\psi), \tag{19.23}$$

for $V \in \mathbf{U}(m)$, $W \in \mathbf{U}(n)$. Since $I(V,W)$ is in $\mathbf{U}(mn)$, any representation U of $\mathbf{U}(mn)$ can be restricted to a representation $I^*U = U \circ I$ of $\mathbf{U}(m) \times \mathbf{U}(n)$. It is natural to ask how standard representations, such as the three-fold exterior and symmetric powers $\wedge^3$ and S^3, and the adjoint representation restrict to $\mathbf{U}(m) \times \mathbf{U}(n)$.

With respect to the natural bases $e_1, e_2, \ldots, e_m$ of $\mathbf{C}^m$ and $f_1, f_2, \ldots, f_n$ of $\mathbf{C}^n$ the diagonal matrices $\alpha \in \mathbf{T}^m \subseteq \mathbf{U}(m)$ and $\beta \in \mathbf{T}^n \subseteq \mathbf{U}(n)$ act as

$$I(\alpha,\beta)(e_j \otimes f_k) = \alpha_j e_j \otimes \beta_k f_k = \alpha_j \beta_k (e_j \otimes f_k), \tag{19.24}$$

so that $I(\alpha,\beta)$ is in the diagonal subgroup $\mathbf{T}^{mn}$ of $\mathbf{U}(mn)$. This makes it easy to restrict characters. For example, the natural representation gives

$$\sigma_1(I(\alpha,\beta)) = \sum_{j,k} \alpha_j \beta_k = (\sum_j \alpha_j)(\sum_k \beta_k). \tag{19.25}$$

More generally we see that

$$\begin{aligned} s_r(I(\alpha,\beta)) &= \sum_{j,k} {\alpha_j}^r {\beta_k}^r \\ &= (\sum_j {\alpha_j}^r)(\sum_k {\beta_k}^r) \\ &= s_r(\alpha) s_r(\beta). \end{aligned} \tag{19.26}$$

It is therefore possible to study the restriction simply by working with the characters.

Lemma 19.4.1. Let σ_3, S_3, Π, and χ_{ad} denote the characters of representations of unitary groups. Then

$$\begin{aligned} \chi_{\mathrm{ad}}(I(\alpha,\beta)) &= \chi_{\mathrm{ad}}(\alpha)\chi_{\mathrm{ad}}(\beta) + \chi_{\mathrm{ad}}(\alpha) + \chi_{\mathrm{ad}}(\beta) \\ \sigma_3(I(\alpha,\beta)) &= \sigma_3(\alpha)S_3(\beta) + \Pi(\alpha)\Pi(\beta) + S_3(\alpha)\sigma_3(\beta) \\ S_3(I(\alpha,\beta)) &= S_3(\alpha)S_3(\beta) + \Pi(\alpha)\Pi(\beta) + \sigma_3(\alpha)\sigma_3(\beta). \end{aligned}$$

Proof. Since $\sigma_1(I(\alpha,\beta)) = \sigma_1(\alpha)\sigma_1(\beta)$, we obtain

$$\begin{aligned}\chi_{\mathrm{ad}}(I(\alpha,\beta)) + 1 &= |\sigma_1(I(\alpha,\beta))|^2\\ &= |\sigma_1(\alpha)|^2|\sigma_1(\beta)|^2\\ &= [\chi_{\mathrm{ad}}(\alpha)+1][\chi_{\mathrm{ad}}(\beta)+1],\end{aligned} \tag{19.27}$$

whence we obtain the formula for χ_{ad}.

We saw in Example 19.3.3 that

$$s_1(\alpha)^3 = S_3(\alpha) + \sigma_3(\alpha) + 2\Pi(\alpha),$$

and it is easy to check the following identities in α:

$$\begin{aligned}s_3(\alpha) &= \sum \alpha_j^3 = S_3(\alpha) + \sigma_3(\alpha) - \Pi(\alpha)\\ s_1(\alpha)s_2(\alpha) &= (\sum \alpha_j)(\sum \alpha_j^2) = S_3(\alpha) - \sigma_3(\alpha).\end{aligned} \tag{19.28}$$

Using these and the similar identities for $s_k(\beta)$ and $s_k(I(\alpha,\beta))$ together with the preceding observation that $s_r(I(\alpha,\beta)) = s_r(\alpha)s_r(\beta)$, we obtain

$$\begin{aligned}&S_3(I(\alpha,\beta)) + \sigma_3(I(\alpha,\beta)) + 2\Pi(I(\alpha,\beta))\\ &\qquad = [S_3(\alpha)+\sigma_3(\alpha)+2\Pi(\alpha)][S_3(\beta)+\sigma_3(\beta)+2\Pi(\beta)]\\ &S_3(I(\alpha,\beta)) + \sigma_3(I(\alpha,\beta)) - \Pi(I(\alpha,\beta))\\ &\qquad = [S_3(\alpha)+\sigma_3(\alpha)-\Pi(\alpha)][S_3(\beta)+\sigma_3(\beta)-\Pi(\beta)]\\ &S_3(I(\alpha,\beta)) - \sigma_3(I(\alpha,\beta)) = [S_3(\alpha)-\sigma_3(\alpha)][S_3(\beta)-\sigma_3(\beta)].\end{aligned}$$

Taking the first equation of our triad plus twice the second gives

$$S_3(I(\alpha,\beta)) + \sigma_3(I(\alpha,\beta)) = [S_3(\alpha)+\sigma_3(\alpha)][S_3(\beta)+\sigma_3(\beta)] + 2\Pi(\alpha)\Pi(\beta), \tag{19.29}$$

which with the third gives

$$\begin{aligned}\sigma_3(I(\alpha,\beta)) &= \sigma_3(\alpha)S_3(\beta) + \Pi(\alpha)\Pi(\beta) + S_3(\alpha)\sigma_3(\beta)\\ S_3(I(\alpha,\beta)) &= S_3(\alpha)S_3(\beta) + \Pi(\alpha)\Pi(\beta) + \sigma_3(\alpha)\sigma_3(\beta).\end{aligned}$$

□

19.5. Particle multiplets

The group $\mathbf{U}(2Q)$ contains the subgroups $\mathbf{SU}(2)$ describing rotations of the spin $\frac{1}{2}$ particles, the subgroup $\mathbf{U}(C)$ of colour symmetries, and the subgroup $\mathbf{U}(F)$ of flavour symmetries. Since we do not need to allow for multiples of the identity more than once, we shall now investigate how the representations of $\mathbf{U}(2FC)$ restrict to the product subgroup $\mathbf{SU}(C) \times \mathbf{SU}(2) \times \mathbf{U}(F)$.

Before doing this we need to incorporate two further pieces of physical information. The first is that experiments indicate that $C = 3$. The second is that the colour symmetry seems to be so precise that only particles transforming under the trivial representation of $\mathbf{SU}(C) = \mathbf{SU}(3)$ seem to be observed in isolation. (At least if other representations do occur it is only at energies beyond the range of present-day experiments.) This is often expressed by saying that only colourless combinations of quarks are observable in isolation. (This means, in particular, that quarks themselves, which transform under the natural representation of $\mathbf{SU}(C)$, are not observable in isolation.)

Assumption 19.5.1. The number of colours, C, is three, and only the part, $[U]_C$, of a representation, U, that transforms trivially under $\mathbf{SU}(C)$, is relevant for isolated baryons and mesons.

In the following discussion the superscripts R, C and F refer to characters of $\mathbf{SU}(2)$, $\mathbf{SU}(C)$ and $\mathbf{U}(F)$.

Theorem 19.5.1. The parts of the adjoint and exterior cube representations that transform trivially under the colour subgroup $\mathbf{SU}(C)$ decompose under rotations and flavour transformations in the following way:

$$[\sigma_3]_C = \Delta^{\frac{3}{2}} S_3^F + \Delta^{\frac{1}{2}} \Pi^F;$$
$$[\chi_{\mathrm{ad}}]_C = \Delta^1(\chi_{\mathrm{ad}}^F + 1^F) + \Delta^0 \chi_{\mathrm{ad}}^F,$$

where the rotational representations are given by Lemma 19.3.1.

Proof. Applying Theorem 19.4.1 to $\mathbf{SU}(C) \times \mathbf{U}(2F)$ we obtain

$$\begin{aligned} \sigma_3 &= \sigma_3^C S_3^{2F} + \Pi^C \Pi^{2F} + S_3^C \sigma_3^{2F} \\ \chi_{\mathrm{ad}} &= \chi_{\mathrm{ad}}^C \chi_{\mathrm{ad}}^{2F} + \chi_{\mathrm{ad}}^C + \chi_{\mathrm{ad}}^{2F}. \end{aligned} \tag{19.30}$$

Since $C = 3$ we have $\sigma_3^C = 1$, and so the parts that transform trivially under $\mathbf{SU}(C)$ are

$$[\sigma_3]_C = S_3^{2F}$$
$$[\chi_{\text{ad}}]_C = \chi_{\text{ad}}^{2F}. \tag{19.31}$$

Now we can apply Lemma 19.4.1 again to $\mathbf{SU}(2) \times \mathbf{U}(F)$ to obtain

$$S_3^{2F} = S_3^R S_3^F + \Pi^R \Pi^F + \sigma_3^R \sigma_3^F;$$
$$\chi_{\text{ad}}^{2F} = \chi_{\text{ad}}^R \chi_{\text{ad}}^F + \chi_{\text{ad}}^R + \chi_{\text{ad}}^F. \tag{19.32}$$

Using Lemma 19.3.1 to identify the characters of the rotational $\mathbf{SU}(2)$ we obtain the results given. □

Corollary 19.5.2. The mesons and baryons fit into the following families:
(i) $F(F+1)(F+2)/6$ spin $\frac{3}{2}$ baryons transforming under the symmetric cube representation of $\mathbf{U}(F)$ on $S^3\mathbf{C}^F$;
(ii) $F(F^2-1)/3$ spin $\frac{1}{2}$ baryons transforming under the Π representation of $\mathbf{U}(F)$;
(iii) F^2-1 spin 1 mesons transforming under the adjoint representation of $\mathbf{U}(F)$ and a single spin 1 meson transforming trivially under $\mathbf{U}(F)$;
(iv) F^2-1 spin 0 mesons transforming under the adjoint representation of $\mathbf{U}(F)$.

Proof. The dimensions of S_3^F, Π, and the adjoint representations are $F(F+1)(F+2)/6$, $F(F^2-1)/3$, and F^2-1, respectively, so that this result is just giving a physical interpretation of the previous theorem. □

Remark 19.5.1. When this theory was first propounded it was believed that there were just three flavours, called *up*, *down*, and *strangeness*, giving $3.4.5/6 = 10$ spin $\frac{3}{2}$ baryons and $3(3^2-1)/3 = 8$ spin $\frac{1}{2}$ baryons as well as $3^2-1 = 8$ mesons of spin 0 and $3^2 = 9$ mesons of spin 1. In 1974 a fourth flavour, called *charm*, was discovered, and then a fifth, called *beauty* (or *bottom*). It has been generally believed that flavours come in pairs, and in 1995 the discovery of quarks of the sixth flavour, called *truth* (or *top*), was announced and confirmed. With a mass around 188 times that of a proton, the top quark is some forty times more massive than its partner, the bottom quark, giving a clear indication of the extent to which flavour symmetry is broken. One now expects that $F = 6$, giving $6.7.8/6 = 56$ spin $\frac{3}{2}$ baryons and $6(6^2 - 1)/3 = 70$ spin $\frac{1}{2}$ baryons as well as $6^2 - 1 = 35$ mesons of spin 0 and $6^2 = 36$ mesons of spin 1.

19.6. Conserved quantities

The evidence suggests that, although the Hamiltonian does not commute with the whole representation of $\mathbf{U}(F)$, it does commute with its diagonal subgroup. Since this subgroup is abelian and isomorphic to $\mathbf{T}^F$, Example 9.4.4 tells us that its irreducible representations are of the form $\alpha_1^{k_1}\alpha_2^{k_2}\ldots\alpha_F^{k_F}$. Their action commutes with the Hamiltonian, so the numbers $k_1, k_2, \ldots, k_F$ must be conserved quantities. These can be identified with the physical conserved quantities such as charge and baryon number as follows:

Definition 19.6.1. An eigenvector for the diagonal subgroup $\mathbf{T}^F \subseteq \mathbf{U}(F)$ that corresponds to the representation $\alpha_1^{k_1}\alpha_2^{k_2}\ldots\alpha_F^{k_F}$ corresponds to a state having *baryon number* $B = \frac{1}{3}\sum_j k_j$, *charge* $Z = \sum_j k_{2j} - B$, *strangeness* $S = -k_3$, *charm* $C = k_4$, *beauty* $-k_5$, and *truth* k_6.

Remark 19.6.1. By definition $\frac{1}{3}\sum k_j$ takes the value 1 for S_3 and Π, and vanishes for the adjoint and trivial representations. Using the definition of B the charge can also be expressed as

$$Z = \frac{2}{3}\sum_j k_{2j} - \frac{1}{3}\sum_j k_{2j+1}. \tag{19.33}$$

The non-trivial characters of the flavour group, $\mathbf{U}(F)$, that appear are

$$\begin{aligned} S_3 &= \sum_j \alpha_j^3 + \sum_{j\neq k} \alpha_j^2\alpha_k + \sum_{j<k<l} \alpha_j\alpha_k\alpha_l \\ \Pi &= \sum_{j\neq k} \alpha_j^2\alpha_k + 2\sum_{j<k<l} \alpha_j\alpha_k\alpha_l \\ \chi_{\text{ad}} &= \sum_{j,k} \alpha_j\overline{\alpha_k} - 1 = \sum_{j,k} \alpha_j\alpha_k^{-1} - 1. \end{aligned} \tag{19.34}$$

The proton and neutron are spin $\frac{1}{2}$ baryons and, according to Corollary 19.5.2, are therefore described by the representation Π. With our conventions their states correspond to the eigenvectors with eigenvalues $\alpha_1\alpha_2^2$

	n	p		$\alpha_1^2\alpha_2$	$\alpha_1\alpha_2^2$
Σ^-	Σ^0, Λ	Σ^+	$\alpha_1^2\alpha_3$	$\alpha_1\alpha_2\alpha_3$	$\alpha_2^2\alpha_3$
Ξ^-	Ξ^0		$\alpha_1\alpha_3^2$	$\alpha_2\alpha_3^2$	

FIGURE 19.1. The spin $\frac{1}{2}$ baryons corresponding to the representation Π, whose eigenvalues are shown on the right. The strangeness S is constant on the rows and takes the values 0, -1, -2 from top to bottom. The charge Z is constant on each column and takes the values -1, 0, $+1$, from left to right of each pattern.

and $\alpha_1^2\alpha_2$ respectively. Their charges are therefore $(2 \times 2 - 1)/3 = 1$ and $(2 \times 1 - 2)/3 = 0$, in agreement with the experimental evidence, and their other quantum numbers for strangeness, charm, truth, and beauty, all vanish. Next consider baryons and mesons whose eigenvalues involve only α_1, α_2, and α_3. Since $k_1 + k_2 + k_3 = 3B$ is known, only the charge Z and strangeness S are needed to specify the state. The possible values can be plotted on a diagram in the (Z, S)-plane (Figures 19.1 and 19.2).

When this theory was first introduced all but the Ω^- particle were known. It was the discovery of this baryon early in 1964 that convinced most physicists of the plausibility of the theory.

Just as the adjoint representation of $\mathbf{U}(F)$ describes the mesons (Figure 19.3) that are associated with the strong force, so the adjoint representation of $\mathbf{SU}(C)$, that is of $\mathbf{SU}(3)$, is believed to describe particles called gluons, which mediate the forces holding quarks together. There is therefore a diagram similar to Figure 19.3 showing the eight possible gluon states. However, since gluons have non-trivial colour transformations one would not expect to see them in isolation.

19.7. Historical note

As mentioned, Heisenberg's isospin theory treated the proton and neutron as two states of a single particle called the nucleon, which amounted to looking at a theory with $F = 2$. Since $\Pi = \sigma_1$ for $\mathbf{SU}(2)$, the nucleons themselves played the role of quarks. In the early 1960s Murray Gell-Mann and Yuval Ne'eman independently proposed that a theory with $F = 3$ would be more appropriate, and the discovery of the Ω^- particle soon provided strong evidence for this. The characters of $\mathbf{SU}(3)$ all lie in $\mathbf{C}[\sigma_1, \overline{\sigma_1}]$, so that all the representations can be derived from the natural representation and

$$\begin{array}{cccc}
N^{*-} & N^{*0} & N^{*+} & N^{*++} \\
\Upsilon^{*-} & \Upsilon^{*0} & \Upsilon^{*+} & \\
\Xi^{*-} & \Xi^{*0} & & \\
\Omega^{-} & & &
\end{array}
\qquad
\begin{array}{cccc}
\alpha_1^3 & \alpha_1^2\alpha_2 & \alpha_1\alpha_2^2 & \alpha_2^3 \\
\alpha_1^2\alpha_3 & \alpha_1\alpha_2\alpha_3 & \alpha_2^2\alpha_3 & \\
\alpha_1\alpha_3^2 & \alpha_2\alpha_3^2 & & \\
\alpha_3^3 & & &
\end{array}$$

FIGURE 19.2. The spin $\frac{3}{2}$ baryons corresponding to the representation S_3, together with the corresponding eigenvalues on the right. The strangeness S is constant along the rows, taking the values 0, -1, -2, and -3 from top to bottom. The charge Z is constant on the columns, taking the values -1, 0, $+1$, and $+2$ from left to right.

$$\begin{array}{ccc}
 & \kappa^0 & \kappa^+ \\
\pi^- & \pi^0,\ \eta & \pi^+ \\
\kappa^- & \overline{\kappa^0} &
\end{array}
\qquad
\begin{array}{ccc}
 & \alpha_1\alpha_3^{-1} & \alpha_2\alpha_3^{-1} \\
\alpha_1\alpha_2^{-1} & 1 & \alpha_2\alpha_1^{-1} \\
\alpha_3\alpha_2^{-1} & \alpha_3\alpha_1^{-1} &
\end{array}$$

FIGURE 19.3. The spin 1 mesons corresponding to the adjoint representation of the flavour group, together with the corresponding eigenvalues on the right. The strangeness S is constant on the rows and takes the values $+1$, 0, and -1 from top to bottom. The charge Z is constant on each column and takes the values -1, 0, $+1$, from left to right.

its contragredient. That suggested that all particles might be composed of quarks transforming with the natural representation and antiquarks transforming with its contragredient. (The name quark was taken by Gell-Mann from a passage mentioning 'three quarks' in James Joyce's book *Finnegan's Wake.*) As already observed the discovery of charm in 1974 raised the value of F to 4, and subsequent developments have suggested a figure of $F = 6$. For these larger groups some characters, such as σ_2, cannot be given by functions of σ_1 and $\overline{\sigma_1}$ alone, but it is nonetheless true that all the particles so far discovered can be regarded as combinations of quarks and antiquarks.

Early attempts to link spin and flavour led to the realization that all the baryons fitted together in the symmetric cube S^3 of $\mathbf{U}(2F)$. However, since the spin $\frac{1}{2}$ quarks ought to be fermions it was necessary to change this to $\wedge^3$, and colour was introduced as a device for doing this, and so rescuing Pauli's exclusion principle. However, the assumption that only colourless combinations are stable is much stronger than the Pauli principle itself. Indeed one can use the assumption to explain why quarks are only seen in threes or in combination with antiquarks, since these are the basic colourless combinations.

This chapter has focused on the baryons, but there is strong evidence that these are closely linked to the leptons too. There are three known massive leptons, the electron, the muon, and the tau particle, each associated with a very light or massless neutrino, and to each of these particles corresponds an antiparticle. (It has been amply confirmed that, although apparently massless, chargeless, and lacking in distinctive qualities, the e, μ, and τ neutrinos are really different from each other, and from their antiparticles.) There are also three *generations* of quarks, each consisting of a pair of quarks (up–down, strange–charmed, bottom–top) one of which has charge $\frac{2}{3}$ and the other charge $-\frac{1}{3}$, and for each quark there is an antiquark. The three pairs of quarks seem to correspond in some, as yet imprecisely understood, way to the three pairs consisting of a massive lepton and its neutrino. (For example, the difference in charge between two quarks in a pair is $\frac{2}{3} - \frac{1}{3} = 1$, and so is that between a neutrino and the corresponding massive lepton.)

There have been various attempts to explain this, including some that interpreted both leptons and baryons as bound states of still more fundamental, but as yet undiscovered, particles. But all have disadvantages and none has yet found widespread support. Whilst this chapter was being written, some new experimental evidence emerged which suggests that quarks may themselves have internal structure. Whether that particular result is confirmed or not it is clear that our understanding of the structure of matter is still far from complete.

Exercises

19.1 Prove that the power sum s_2 cannot be expressed as a sum of characters lifted from the rotation group.

19.2 Show that for U in the unitary group

$$\det(1 - tU)^{-1} = \sum_{r=0}^{\infty} t^r S_r(U).$$

Hence or otherwise, show that

$$\sum(-t)^r \sigma_r \sum t^k S_k = 0.$$

Deduce that

$$\sum(-1)^r \sigma_r S_{n-r} = 0,$$

for $n > 0$.

19.3 By writing

$$\prod(1 + t\alpha_j) = \exp\left(\sum \ln(1 + t\alpha_j)\right)$$

and expanding the logarithm, or otherwise, show that, for any unitary matrix U,

$$\sum t^r \sigma_r(U) = \exp\left(-\sum(-t)^r s_r(U)/r\right).$$

Appendix A1 A review of linear algebra and groups

It will interest mathematical circles that the mathematical instruments created by the higher algebra play an essential part in the rational formulation of the new quantum mechanics.

NIELS BOHR, lecture on *Atomic theory and mechanics*, 30 August 1925

A1.1. Inner product spaces

The basic algebraic ideas and results which we have used can be found in any textbooks on linear algebra and groups, but, for convenience, this appendix contains some of those most relevant to quantum mechanics. Although vector spaces can be defined over other fields we shall use only the complex numbers.

Definition A1.1.1. A (complex) *vector space* V consists of a set of elements called *vectors* which forms an abelian group with respect to an addition operation written $(u, v) \mapsto u + v$, and which has a scalar multiplication $\mathbf{C} \times V \to V$, denoted by $(\lambda, v) \mapsto \lambda v$, satisfying

$$\begin{aligned}(\lambda + \mu)v &= \lambda v + \mu v \\ (\lambda\mu)v &= \lambda(\mu v) \\ \lambda(u + v) &= \lambda u + \lambda v \\ 1v &= v.\end{aligned}$$

Definition A1.1.2. An *inner product space*, $\mathcal{H}$, is a vector space which has a map $\mathcal{H} \times \mathcal{H} \to \mathbf{C}$, written as $(u, v) \mapsto \langle u|v\rangle$, such that, for all $u, v, w \in V$, and for all $\alpha, \beta \in \mathbf{C}$,
(i) $\langle u|v\rangle = \overline{\langle v|u\rangle}$ for all $u, v \in V$;
(ii) $\langle u|\alpha v + \beta w\rangle = \alpha\langle u|v\rangle + \beta\langle u|w\rangle$ for all $u, v, w \in V$, and for all $\alpha, \beta \in \mathbf{C}$;
(iii) $\langle u|u\rangle > 0$ for all non-zero vectors u in V.
(It is a consequence of (ii) that $\langle u|0\rangle = 0$ for any vector $u \in V$ and so in particular $\langle 0|0\rangle = 0$.)

As emphasized in Section 6.1, quantum mechanics uses a different convention from that in most algebra texts in that the inner products are linear in the *second* variable and conjugate linear in the *first*.

Definition A1.1.3. The *norm* $\|v\|$ of a vector v is defined by

$$\|v\|^2 = \langle v|v\rangle.$$

If $\|v\| = 1$ then v is said to be normalized. Two vectors u and v in $\mathcal{H}$ are said to be *orthogonal* if $\langle u|v\rangle = 0$.

If there are no non-zero vectors orthogonal to all elements of a set S then S is said to span the space. In finite dimensions this is equivalent to the usual definition that every vector is a linear combination of elements of S, and in infinite dimensions that every vector ψ is a limit of such linear combinations.

Definition A1.1.4. A collection of vectors $\{v_j\}$ in $\mathcal{H}$ which satisfies $\langle v_j|v_k\rangle = \delta_{jk}$ for all j and k is said to be an *orthonormal set.* An orthonormal set of vectors is necessarily linearly independent, and if it also spans $\mathcal{H}$ then it is called an *orthonormal basis.*

Every finite-dimensional inner product space has an orthonormal basis. Suppose, now, that one wishes to expand a vector ψ in terms of an infinite orthonormal spanning set of vectors $\{\psi_k\}$. Setting $\phi_N = \psi - \sum_{k=1}^{N}\langle\psi_k|\psi\rangle$, we easily calculate that $\|\phi_N\|^2 = \|\psi\|^2 - \sum_{k=1}^{N}|\langle\psi_k|\psi\rangle|^2$, from which the Bessel inequality follows:

$$\|\psi\|^2 \geq \sum_{k=1}^{N}|\langle\psi_k|\psi\rangle|^2.$$

The sequence of partial sums on the right is monotonic increasing and bounded above, and so converges. This means that

$$\|\phi_N - \phi_M\|^2 = \sum_{k=M+1}^{N} |\langle\psi_k|\psi\rangle|^2 \tag{A1.1}$$

can be made arbitrarily small for M and N large enough. In infinite dimensions one makes a completeness assumption (see Section 6.2) which forces the sequence ϕ_N to converge to a vector ϕ. By construction, ϕ_N is orthogonal to ψ_k for $k \leq N$, and, going to the limit, ϕ must be orthogonal to all members of the sequence, and so zero. Since $\psi = \phi_N + \sum_{k=1}^{N} \langle \psi_k | \psi \rangle \psi_k$, we may take the limit to obtain $\psi = \sum_{k=1}^{\infty} \langle \psi_k | \psi \rangle \psi_k$.

The Cauchy–Schwarz–Bunyakowski inequality A1.1.1. For any x and y in $\mathcal{H}$

$$|\langle x|y\rangle|^2 \leq \|x\|^2 \|y\|^2$$

with equality if and only if x and y are linearly dependent.

A1.2. Linear transformations

Definition A1.2.1. Let V and W be vector spaces. A map T from V to W which preserves linear combinations, that is such that $T(\alpha u + \beta v) = \alpha T u + \beta T v$, for all $\alpha, \beta \in \mathbf{C}$ and all $u, v \in V$, is called a *linear transformation.*

The set of all linear transformations will be denoted by $\mathcal{L}(V, W)$. When $V = W$ we shall abbreviate $\mathcal{L}(V, V)$ to $\mathcal{L}(V)$. When W is the one-dimensional space $\mathbf{C}$ we write V^* for $\mathcal{L}(V, \mathbf{C})$, and refer to it as the *dual* of V. The elements of V^* are called *linear functionals.* A linear transformation is determined by its action on a basis.

Definition A1.2.2. The *matrix* $T_{j\alpha}$ of a linear transformation $T \in \mathcal{L}(V, W)$ with respect to bases $\{v_j\}$ for V and $\{w_\alpha\}$ for W is defined by the expansion

$$T v_j = \sum_\alpha T_{\alpha j} w_\alpha.$$

When $V = W$ the *trace*, tr$T$, is defined by tr$T = \sum T_{jj}$, and may be shown to be independent of the basis used.

Definition A1.2.3. The *sum* of two linear transformations S and T from V to W is the map defined by $(S+T)v = Sv + Tv$. Similarly, the product with a complex number λ can be defined by $(\lambda T)v = \lambda(Tv)$. Both $S+T$ and λT can themselves be shown to be linear transformations, so that these operations give $\mathcal{L}(V, W)$ the structure of a vector space. Using matrices it can be shown that $\dim \mathcal{L}(V, W) = \dim V \dim W$, so that in particular $\dim V^* = \dim V$.

Definition A1.2.4. The *image* or *range*, $\mathrm{im}(T)$, of T is the set of vectors in W of the form Tv for some v in V. The image is a subspace of W and its dimension is called the *rank* of T.

The *kernel* of T is defined by

$$\ker T = \{v \in V : Tv = 0\}.$$

The kernel is a subspace of V and its dimension is called the *nullity* of T.

Proposition A1.2.1. The linear transformation $T \in \mathcal{L}(V, W)$ is one–one if and only if $\ker T = \{0\}$, and is onto if and only if $\mathrm{im} T = W$.

Definition A1.2.5. Suppose that $T \in \mathcal{L}(V)$ and $v \in V$ is a non-zero vector for which there exists some complex number λ such that

$$Tv = \lambda v;$$

then v is said to be an *eigenvector* of T with *eigenvalue* λ.

Definition A1.2.6. The *characteristic polynomial*, χ_T, is defined by

$$\chi_T(t) = \det(t1 - T),$$

where the determinant can be calculated by choosing any basis and expressing T in matrix form.

If V is finite dimensional then a necessary and sufficient condition for λ to be an eigenvalue of T is that it be a root of the characteristic polynomial, that is $\chi_T(\lambda) = 0$.

Definition A1.2.7. If $S \in \mathcal{L}(U, V)$ and $T \in \mathcal{L}(V, W)$ then the composition or *product*, TS, is a linear transformation in $\mathcal{L}(U, W)$. If S and T are both in $\mathcal{L}(V)$ then so is TS. In particular, powers of $T \in \mathcal{L}(V)$ may be defined inductively by

$$T^1 = T, \qquad T^n = T(T^{n-1}).$$

By convention $T^0 = 1$, the identity operator.

Given any polynomial

$$p(t) = \sum_{k=0}^{N} a_k t^k,$$

we may form the corresponding operator

$$p(T) = \sum_{k=0}^{N} a_k T^k \in \mathcal{L}(V).$$

The Cayley–Hamilton theorem A1.2.2. If $\dim(V)$ is finite then

$$\chi_T(T) = 0.$$

A1.3. The spectral theorem

Definition A1.3.1. Let $\mathcal{H}$ and $\mathcal{K}$ be inner product spaces, and $T \in \mathcal{L}(\mathcal{H}, \mathcal{K})$ a linear transformation. The *adjoint*, $T^* \in \mathcal{L}(\mathcal{K}, \mathcal{H})$, is defined by the identity

$$\langle T^* u | v \rangle = \langle u | T v \rangle.$$

In finite dimensions this identity defines T^* uniquely. In infinite dimensions this and the next definition need some modification. (See Section 6.2.)

Definition A1.3.2. If $\mathcal{H} = \mathcal{K}$, then T and T^* are both in $\mathcal{L}(\mathcal{H})$. We define T to be *self-adjoint* if $T = T^*$, that is if

$$\langle T u | v \rangle = \langle u | T v \rangle$$

for all u and v in $\mathcal{H}$.

The spectral theorem A1.3.1. If $\mathcal{H}$ is a finite-dimensional space and $T \in \mathcal{L}(\mathcal{H})$ is self-adjoint then there exists an orthonormal basis $\{v_j\}$ of eigenvectors, that is vectors satisfying

$$T v_j = \lambda_j v_j \qquad \text{and} \qquad \langle v_j | v_k \rangle = \delta_{jk}.$$

Proof. We work by induction on $n = \dim \mathcal{H}$. First note that over $\mathbf{C}$ the characteristic polynomial has a root, λ, and so there exists a non-zero vector v such that $Tv = \lambda v$. We normalize this and set $v_n = v/\|v\|$. If $n = 1$ then v_n already provides an orthonormal basis of eigenvectors. Otherwise, define W to be the subspace of vectors orthogonal to v. For $w \in W$ consider

$$\langle T w | v \rangle = \langle w | T v \rangle = \langle w | \lambda v \rangle = \lambda \langle w | v \rangle = 0.$$

This shows that $Tw \in W$, and so $T(W) \subseteq W$. Let us write T_W for the restriction of T to W. The self-adjointness condition continues to hold when the vectors are restricted to lie in W, and that subspace has dimension less than that of $\mathcal{H}$, so applying the inductive hypothesis we may assume that W has an orthonormal basis $v_1, v_2, \ldots, v_{n-1}$ of eigenvectors for T_W, that is $Tv_j = T_0 v_j = \lambda_j v_j$ for $j = 1, 2, \ldots, n-1$. Adjoining v_n to this set provides a suitable orthonormal basis of eigenvectors for $\mathcal{H}$. □

Definition A1.3.3. If $T \in \mathcal{H}$ is self-adjoint then $\langle Tu|u\rangle$ is real for all $u \in \mathcal{H}$. If, further, $\langle Tu|u\rangle \geq 0$ for all $u \in \mathcal{H}$ then T is said to be *positive.*

Definition A1.3.4. A self-adjoint linear transformation $P \in \mathcal{H}$ which satisfies

$$P^2 = P = P^*$$

is called a *projection.*

Definition A1.3.5. A linear transformation $U \in \mathcal{H}$ which satisfies $U^*U = 1 = UU^*$ is said to be *unitary.*

This definition means that $\langle Uu|Uv\rangle = \langle u|v\rangle$, for all vectors u and v, so unitary transformations preserve the inner product.

The spectral theorem for unitary transformations A1.3.2. If $\mathcal{H}$ is a finite-dimensional space and $U \in \mathcal{L}(\mathcal{H})$ is unitary then there exists an orthonormal basis $\{v_j\}$ of eigenvectors, that is vectors satisfying

$$Uv_j = \lambda_j v_j \qquad \text{and} \qquad \langle v_j|v_k\rangle = \delta_{jk}.$$

This is proved by the same technique as the theorem for self-adjoint operators. With respect to the basis $\{v_j\}$ the matrix of U is diagonal. Since the diagonal entries are the eigenvalues this diagonal form is unique up to the ordering of the eigenvalues. Moreover, each λ_j has modulus 1, as one sees by noting that $|\lambda_j|^2\|v_j\|^2 = \|Uv_j\|^2 = \|v_j\|^2$.

We conclude this review of linear algebra with a variant of a well-known finite-dimensional result which shows that when two operators commute one may find simultaneous eigenvectors of both.

Proposition A1.3.3. Let A and B be self-adjoint operators on the inner product space $\mathcal{H}$, let $\mathcal{H}_A$ and $\mathcal{H}_B$ denote the subspaces spanned by eigenvectors of A and B, respectively, and let $\mathcal{H}_{A,B}$ denote the span of the vectors which are simultaneously eigenvectors for both A and B. If $AB = BA$ then $\mathcal{H}_{A,B} = \mathcal{H}_A \cap \mathcal{H}_B$.

Proof. If ψ is in the eigenspace $\ker(A - \alpha 1)$ for A then

$$(A - \alpha 1)B\psi = B(A - \alpha 1)\psi = 0,$$

so that $B\psi \in \ker(A - \alpha 1)$. This shows that B preserves each A-eigenspace, and so also preserves their span, $\mathcal{H}_A$. For $\phi \in \mathcal{H}_A^\perp$ and $\psi \in \mathcal{H}_A$ we have

$$\langle B\phi|\psi\rangle = \langle\phi|B\psi\rangle = 0,$$

so that $B(\mathcal{H}_A^\perp) \subseteq \mathcal{H}_A^\perp$, too.

By considering the components of a B-eigenvector in $\mathcal{H}_A \oplus \mathcal{H}_A^\perp$ we then get

$$\ker(B - \beta 1) = \mathcal{H}_A \cap \ker(B - \beta 1) \oplus \mathcal{H}_A^\perp \cap \ker(B - \beta 1).$$

Summing over all eigenvalues β we obtain

$$\mathcal{H}_B = \bigoplus_\beta (\mathcal{H}_A \cap \ker(B - \beta 1)) \oplus \bigoplus_\beta \left(\mathcal{H}_A^\perp \cap \ker(B - \beta 1)\right),$$

and so

$$\mathcal{H}_A \cap \mathcal{H}_B = \bigoplus_\beta (\mathcal{H}_A \cap \ker(B - \beta 1)).$$

On the other hand comparing the $\ker(A - \alpha 1)$ components of $B\psi$ and $\beta\psi$ for $\psi \in \mathcal{H}_A \cap \ker(B - \beta 1)$ we see that the $\ker(A - \alpha 1)$ component of ψ is in $\ker(A - \alpha\mathbf{1}) \cap \ker(B - \beta 1)$, so

$$\mathcal{H}_A \cap \mathcal{H}_B = \bigoplus_{\alpha,\beta} (\ker(A - \alpha 1) \cap \ker(B - \beta 1)) \subseteq \mathcal{H}_{A,B}.$$

The reverse inclusion is obvious, so the result now follows. □

Corollary A1.3.4. If $AB = BA$ and $\mathcal{H}$ admits an orthonormal basis of eigenvectors for A, then $\mathcal{H}_{A,B} = \mathcal{H}_B$.

Proof. Since $\mathcal{H}$ admits an orthonormal basis of eigenvectors for A we have $\mathcal{H}_A = \mathcal{H}$, so that

$$\mathcal{H}_{A,B} = \mathcal{H} \cap \mathcal{H}_B = \mathcal{H}_B.$$

□

A1.4. Groups

Definition A1.4.1. A *group* is a set G with a multiplication map $G \times G \to G$, denoted by $(x, y) \mapsto xy$, and an identity element $1 \in G$ such that
(i) $(xy)z = x(yz)$, for all x, y, and z in G;
(ii) $1x = x = x1$, for all x in G;
(iii) for each x in G there exists an inverse $x^{-1} \in G$, such that $xx^{-1} = 1 = -x^{-1}x$.

Definition A1.4.2. A group G is said to be *abelian* if $xy = yx$ for all x and y in G.

Definition A1.4.3. A *subgroup* H of a group G is a subset such that for all x and y in H, $x^{-1}y$ is also in H. It is a *normal subgroup* if, for all g in G and all x in H, $g^{-1}xg$ is in H.

Conjugation by $g \in G$ is defined to send $x \in G$ to gxg^{-1}. In abelian groups conjugation has no effect, since $x = gxg^{-1}$, so every subgroup of an abelian group is normal.

Definition A1.4.4. If N is a normal subgroup of G then the cosets $xN = \{xn \in G : n \in N\}$ can be given a multiplication

$$(xN)(yN) = xyN$$

with respect to which they form a group, G/N, called the *quotient group*.

Definition A1.4.5. Let G and H be groups. A map ϕ from G to H such that $\phi(x)\phi(y) = \phi(xy)$ for all x and y in G is called a *homomorphism*.

Definition A1.4.6. An *isomorphism* is a homomorphism which is one–one and onto. If there exists an isomorphism from G to H then G and H are said to be *isomorphic*.

Definition A1.4.7. The *kernel* of a homomorphism ϕ is the set

$$\ker \phi = \{x \in G : \phi(x) = 1\}.$$

The *image* is the set $\text{im}\phi = \{\phi(x) \in H : x \in G\}$.

Theorem A1.4.1. A homomorphism ϕ is one–one if and only if $\ker(\phi) = \{1\}$, and is onto if and only if $\text{im}(\phi) = H$.

The first isomorphism theorem A1.4.2. The kernel of a homomorphism $\phi : G \to H$ is a normal subgroup of G and the quotient group $G/\ker(\phi)$ is isomorphic to $\text{im}(\phi)$.

Appendix A2 Open systems

If I have understood correctly your point of view then you would gladly sacrifice the simplicity [of quantum mechanics] to the principle of causality. Perhaps we could comfort ourselves that the dear Lord could go beyond [quantum mechanics] and maintain causality.

WERNER HEISENBERG, letter to Einstein, 10 June 1927

In this section we shall show how to combine a two-dimensional quantum system with an 'environment' whose time evolution drives the original system asymptotically towards a collapse of the kind that occurs during a measurement. More precisely we prove the following result, stated in Section 10.7:

Theorem A2.0.1. Let V be a two-dimensional inner product space and Ω a vector in V. Then there exists an inner product space $\mathcal{H}$, a family of unitary operators U_t, a homomorphism $\phi : \mathcal{L}(V) \to \mathcal{L}(\mathcal{H})$ which respects adjoints, and a linear transformation which takes a vector $\psi \in V$ to $\Psi \in \mathcal{H}$ such that for all $A \in \mathcal{L}(V)$ we have

$$\langle \Psi | \phi(A) \Psi \rangle = \langle \psi | A \psi \rangle$$
$$\lim_{t \to \infty} \langle U_t \Psi | \phi(A) U_t \Psi \rangle = \langle \Omega | A \Omega \rangle,$$

where the inner products on the left are in $\mathcal{H}$ and those on the right are in V.

Proof. We take the space $\mathcal{H}$ to be the direct sum of spaces $\mathcal{H}_n$, for $n = 0, 1, 2, \ldots$, where $\mathcal{H}_0 = \mathbf{C}$, and, for $n \geq 1$, $\mathcal{H}_n$ is the space of antisymmetric normalizable wave functions ψ_n on $\mathbf{R}^n$, that is $\psi_n(s_1, s_2, \ldots, s_n)$ changes sign whenever any two of its arguments are interchanged. In particular, $\mathcal{H}_1$ is just the space of normalizable wave functions on $\mathbf{R}$, and $\mathcal{H}_2$ is the space of normalizable wave functions on $\mathbf{R}^2$ such that $\psi_2(s_1, s_2) = -\psi(s_2, s_1)$. The time evolution operators U_t are defined on $\mathcal{H}_n$ by

$$(U_t \psi_n)(s_1, \ldots, s_n) = \exp[-i(s_1 + \ldots + s_n)t] \psi_n(s_1, \ldots, s_n). \qquad (A2.1)$$

Let f be a normalized function on $\mathbf{R}$ and define for each n an operator $e : \mathcal{H}_n \to \mathcal{H}_{n+1}$ by

$$(e\psi_n)(s_1, s_2, \ldots, s_{n+1}) = \frac{1}{\sqrt{n+1}} \sum_{k=1}^{n+1} (-1)^k f(s_k) \psi_n(s_1, \ldots, s_{k-1}, s_{k+1}, \ldots, s_{n+1}). \tag{A2.2}$$

(This is essentially just multiplication by f; the sum is there only to ensure that $e\psi_n$ is an antisymmetric function.)

We also define a map e^* that sends $\mathcal{H}_0$ to 0 and maps $\mathcal{H}_{n+1}$ to $\mathcal{H}_n$, for $n > 0$, by

$$(e^*\psi_{n+1})(s_1, \ldots, s_n) = \sqrt{n+1} \int_{\mathbf{R}} \overline{f(s)} \psi_{n+1}(s, s_1, \ldots, s_n) ds. \tag{A2.3}$$

Using the antisymmetry of ψ_{n+1}, this can also be written as

$$\frac{1}{\sqrt{n+1}} \sum (-1)^k \int_{\mathbf{R}} \overline{f(s)} \psi_{n+1}(s_1, \ldots, s_{k-1}, s, s_{k+1}, \ldots, s_n) ds, \tag{A2.4}$$

from which it is easy to check that

$$\langle \phi_n | e^* \psi_{n+1} \rangle = \langle e\phi_n | \psi_{n+1} \rangle, \tag{A2.5}$$

so that e^* is the adjoint of e. Using the antisymmetry of the function ψ_{n+1} we see that $e^{*2}\psi_{n+1}$ is given by

$$\sqrt{n(n+1)} \int_{\mathbf{R}^2} \overline{f(s)f(t)} \psi_{n+1}(s, t, u_1, \ldots, u_{n-1}) ds dt = 0, \tag{A2.6}$$

so that $e^{*2} = 0$. Taking adjoints we also have $e^2 = 0$. The remaining important identity relates e and e^*. On the right-hand side of the identity

$$(e^* e\psi_n)(s_1, \ldots, s_n) = \int_{\mathbf{R}} \overline{f(s)} (e\psi_n)(s, u_1, \ldots, u_n) ds. \tag{A2.7}$$

The expression $e\psi_n$ is the sum of $f(s)\psi_n(u_1, \ldots, u_n)$ and

$$-\sum (-1)^k f(u_k) \psi_n(s, u_1, \ldots, u_{k-1}, u_{k+1}, \ldots, u_n). \tag{A2.8}$$

When we integrate against $\overline{f(s)}$ this latter expression gives

$$-(ee^*\psi_n)(u_1, \ldots, u_n), \tag{A2.9}$$

whilst the former gives $\|f\|^2\psi_n$. We therefore obtain $e^*e\psi_n = \|f\|^2\psi_n - ee^*\psi_n$, or $e^*e + ee^* = \|f\|^2 = 1$. Summarizing we have

$$e^2 = 0 = e^{*2} \qquad \text{and} \qquad ee^* + e^*e = 1. \tag{A2.10}$$

To describe the homomorphism ϕ we choose an orthonormal basis v_1, v_2 for V with $v_2 = \Omega$ and write $A \in \mathcal{L}(\mathcal{H})$ as a matrix. Then we set

$$\phi(A) = A_{11}ee^* + A_{21}e^* + A_{12}e + A_{22}e^*e. \tag{A2.11}$$

This clearly depends linearly on A and, using the formulae for products of e and e^*, it is easy to check that $\phi(AB) = \phi(A)\phi(B)$, $\phi(1) = 1$, and also that $\phi(A)^* = \phi(A^*)$, so that ϕ preserves self-adjointness.

Now, for any $\Psi \in \mathcal{H}$, we have

$$\begin{aligned}\langle U_t\Psi|\phi(A)U_t\Psi\rangle &= A_{11}\langle U_t\Psi|ee^*U_t\Psi\rangle + A_{21}\langle U_t\Psi|e^*U_t\Psi\rangle \\ &\quad + A_{12}\langle U_t\Psi|eU_t\Psi\rangle + A_{22}\langle U_t\Psi|e^*eU_t\Psi\rangle.\end{aligned} \tag{A2.12}$$

For any $\Phi \in \mathcal{H}_n$ and $\Phi' \in \mathcal{H}_{n+1}$ consider

$$\langle U_t\Phi|e^*U_t\Phi'\rangle = \int_{\mathbf{R}^{n+1}} e^{-ist}\overline{\Phi(u_1,\ldots,u_n)f(s)}\Phi'(s,u_1,\ldots,u_n)dsd^nu. \tag{A2.13}$$

By the Riemann–Lebesgue lemma this integral tends to 0 as $t \to \infty$. Similarly, for $\Psi \in \mathcal{H}_{n+1}$,

$$\langle U_t\Psi|ee^*U_t\Psi\rangle = \|e^*U_t\Psi\|^2 \tag{A2.14}$$

can also be written as

$$\int_{\mathbf{R}^{n+2}} e^{i(r-s)t}f(r)\overline{\Psi(r,u_1,\ldots,u_n)}\,\overline{f(s)}\Psi(s,u_1,\ldots,u_n)drdsd^nu, \tag{A2.15}$$

which tends to 0 as $t \to \infty$. The first three terms in the expression for $\langle U_t\Psi|\phi(A)U_t\Psi\rangle$ therefore tend to 0 as $t \to \infty$. The remaining term can be calculated by observing that

$$\langle U_t\Psi|e^*eU_t\Psi\rangle = \langle U_t\Psi|(1 - ee^*)U_t\Psi\rangle. \tag{A2.16}$$

It follows from the previous discussion that for large t this approaches

$$\|U_t\Psi\|^2 - 0 = \|\Psi\|^2. \tag{A2.17}$$

Recalling that $A_{22} = \langle \Omega | A\Omega \rangle$, we may combine these results for normalized Ψ to obtain

$$\langle U_t\Psi | \phi(A) U_t \Psi \rangle \to A_{22} \|\Psi\|^2 = \langle \Omega | A\Omega \rangle. \tag{A2.18}$$

Finally, given $\psi = \psi_1 v_1 + \psi_2 v_2 \in V$, we set

$$\Psi = \psi_1 f + \psi_2 \in \mathcal{H}_1 \oplus \mathcal{H}_0. \tag{A2.19}$$

This can also be written as $(\psi_1 e + \psi_2)1$ where 1 is the constant function in $\mathcal{H}_0$. By direct calculation we have

$$\langle \Phi | \Psi \rangle = \overline{\phi_1}\psi_1 \|f\|^2 + \overline{\phi_2}\psi_2 = \langle \phi | \psi \rangle. \tag{A2.20}$$

Bearing in mind the formulae for products of e and e^*, and the fact that $e^*1 = 0$, we see that

$$\begin{aligned}\phi(A)\Psi &= (A_{11}ee^* + A_{21}e^* + A_{12}e + A_{22}e^*e)(\psi_1 e + \psi_2)1 \\ &= (A_{11}\psi_1 e + A_{21}\psi_1 e^* e + A_{12}\psi_2 e + A_{22}\psi_2 e^* e)1 \\ &= (A_{11}\psi_1 + A_{12}\psi_2)\, f + (A_{21}\psi_1 + A_{22}\psi_2),\end{aligned} \tag{A2.21}$$

which is the image of $A\psi$. Using equation (*A*2.20) this means that

$$\langle \Psi | \phi(A) \Psi \rangle = \langle \psi | A\psi \rangle, \tag{A2.22}$$

as required. □

This argument merely shows that it is possible to mimic the effects of a projection asymptotically, but it can be generalized to give more plausible models. These occur naturally whenever one has a dissipative system (one that loses energy to its environment). The space $\mathcal{H}_n$ can be regarded as describing n-particle states of the system and its environment, with $s_1, \ldots, s_n$ defining the frequencies of the associated waves. The space V describes the system and its observables can also be regarded as observables on $\mathcal{H}$ by using the homomorphism ϕ. The sort of wave function, $f \in \mathcal{H}_1$, that arises in these cases is

$$f = \sqrt{\frac{2\eta}{\pi}} \left(\frac{\omega}{s^2 - \omega^2 + 2i\eta s} \right). \tag{A2.23}$$

The parameter ω plays the role of a natural frequency of the system (or of the system together with the measuring apparatus) and η is the strength of the coupling between the system and its environment. It is also possible to be more precise about the rate of convergence, which is actually proportional to $\langle f | U_t f \rangle$. This is a multiple of $\exp(-\eta t)$. The stronger the coupling to the environment the faster the collapse.

Some physicists and mathematicians

This list includes just some of those who contributed to the development of quantum theory or its mathematical techniques. The abbreviation N stands for a Nobel Prize in Physics (except in the case of Rutherford where the award was in Chemistry). Heisenberg's 1932 prize was actually awarded a year later in 1933.

John Stewart BELL (28 Nov 1928–1 Oct 1990)

Niels Henrik David BOHR (7 Oct 1885–18 Nov 1962, N 1922)

Max BORN (11 Dec 1882–5 Jan 1970, N 1954)

Louis-Victor Pierre Raymond, Prince de BROGLIE (15 Aug 1892–19 Mar 1987, N 1929)

Paul Adrien Maurice DIRAC (8 Aug 1902–20 Oct 1984, N 1933)

Albert EINSTEIN (14 Mar 1879–18 Apr 1955, N 1921)

Enrico FERMI (29 Sep 1901–18 Apr 1954, N 1938)

Richard Phillips FEYNMAN (11 May 1918–14 Feb 1988, N 1965)

Werner Karl HEISENBERG (5 Dec 1901–1 Feb 1976, N 1932)

David HILBERT (23 Jan 1862–14 Feb 1943)

Wolfgang PAULI (25 Apr 1900–4 Oct 1958, N 1945)

Max Karl Ernst Ludwig PLANCK (23 Apr 1858–4 Oct 1947, N 1919)

John William Strutt, Third Baron RAYLEIGH (12 Nov 1842–30 Jun 1919, N 1904)

Ernest, First Baron RUTHERFORD of Nelson and Cambridge (30 Aug 1871–19 Oct 1937, N 1908)

Erwin SCHRÖDINGER (12 Aug 1887–4 Jan 1961, N 1933)

John Louis von NEUMANN (3 Dec 1903–8 Feb 1957)

Hermann WEYL (9 Nov 1885–8 Dec 1955)

Eugene Paul WIGNER (17 Nov 1902–1 Jan 1995, N 1963)

YUKAWA Hideki (23 Jan 1907–8 Sep 1981, N 1949)

Further background reading

There are so many possible applications of quantum theory and developments of its mathematical structure, that we mention only the totally different approach described in *Quantum mechanics and path integrals* by R. Feynman and A. Hibbs, McGraw-Hill, 1965. We reluctantly omitted this very appealing description of quantum mechanics from the main text because it is much harder to give a mathematical justification of its methods except in simple cases. Those wishing to discover more about the mathematical approach to quantum mechanics could consult *Methods of mathematical physics*, Volume I, by M. Reed and B. Simon, Academic Press, 1972.

One can obtain a good impression of the range of further possibilities from *The quantum universe* by T. Hey and P. Walters, Cambridge UP, 1987, and Chapters 5, 10, and 13–18 in *The new physics*, edited by P. Davies, Cambridge UP, 1989 (particularly Chapter 13 on *The conceptual foundations of quantum mechanics* by A. Shimony). Most of the classical papers on quantum measurement and the paradoxes it engenders have been collected into the volume *Quantum theory and measurement*, edited by J.A. Wheeler and W.H. Zurek (Princeton UP, 1983). A selection of Bell's papers on the subject have been collected into *Speakable and unspeakable in quantum mechanics* (Cambridge UP, 1987). The book *The philosophy of quantum mechanics* by M. Jammer (Wiley Interscience, 1974) is still a useful account of various interpretations of the theory.

The series of volumes on *The historical development of quantum theory* by J. Mehra and H. Rechenberg (Springer-Verlag, 1982–) contains a wealth of historical information. Many of the original papers on quantum theory are printed (in translation) in *Sources of quantum mechanics* by B.L. van der Waerden, Dover, 1967. Another useful survey of the history is given by A. Pais in *Inward bound*, Oxford UP, 1986, and *The Born–Einstein letters*, edited by H. Born, Walker, 1971, provide an insight into the problems and excitement as the new discoveries were made.

Autobiographical works by quantum physicists include the following.

M. Born, *My life*, Taylor and Francis, 1978.
R.P. Feynman, *"Surely you must be joking Mr Feynman!"*, Unwin, 1985. and *"What do you care what people think?"*, Unwin, 1988.
W. Heisenberg, *Physics and beyond*, Harper and Row, 1971.
H. Yukawa, *Tabibito*, translated by L. Brown and R. Yoshida, World Scientific, 1979.

There are also numerous biographies of individual scientists, and the following represent just a small selection.

A.P. French and P.J. Kennedy, *Niels Bohr: Centenary volume*, Harvard, 1985.

A. Pais, *Niels Bohr's times*, Oxford UP, 1991.

H. Kragh, *Dirac a scientific biography*, Cambridge UP, 1990.

A. Pais, *"Subtle is the Lord"*, Oxford UP, 1982, and *Einstein lived here*, Oxford UP, 1994.

J. Gleick, *Genius*, Little, Brown and Company, 1992,

J. Mehra, *The beat of a different drum*, Oxford UP, 1994.

D. Cassidy, *Uncertainty*, W.H. Freeman, 1992.

J.L. Heilbronn, *The dilemmas of an upright man*, University of California Press, 1986.

D. Wilson, *Rutherford*, Hodder and Stoughton, 1983.

W. Moore, *Schrödinger*, Cambridge UP, 1989, and its abridgement.

Epilogue

When Hitler became German Chancellor in 1933, most of the German and Austrian physicists involved in the development of quantum theory left the country, whether Jewish or not. Of the more prominent, only Heisenberg and the elderly Planck remained, hoping to keep physics alive despite the political difficulties. (Planck's one remaining son was executed in February 1945 for conspiring to assassinate Hitler.) Johannes Stark, who as a young man had been an early proponent of relativity and the quantum hypothesis, was by now a bitter opponent of what he termed 'Jewish physics', which included anything modern and abstract. When Heisenberg wrote an article in which relativity theory was mentioned favourably, Stark denounced him to the Gestapo.

After fleeing from Berlin to Oxford, Schrödinger made an ill-advised return to his native Austria in 1936, only to become a refugee again when Hitler annexed the country two years later. He eventually became the first Professor of Theoretical Physics at the Institute of Advanced Studies in Dublin. Born left Germany in 1933 and after a couple of years in Cambridge became Professor at Edinburgh. Bohr managed to escape from Copenhagen during the Nazi occupation, and was smuggled out of Sweden in the bomb bay of a Mosquito, almost dying of asphyxiation on the way. He joined many of the other mathematicians and physicists who had taken part in the development of quantum theory in the United States. Fermi supervised the building of the first nuclear reactor which went critical on 2 December 1942 in the university squash courts in the centre of Chicago. Many other physicists were involved directly or indirectly in the Manhattan project to develop an atomic bomb, a project which combined the independent programmes started in Britain and the United States. In each case the project had been driven by the realization of some of the refugee physicists that the 1939 discovery of chain reactions by Hahn and Strassmann had made such weapons possible, and that Germany possessed the greater potential to build them. In fact, having extricated himself from the Gestapo, Heisenberg had been put in charge of the German bomb project and, for whatever reason, had carried it forward only slowly.

After the war the leading German scientific society, the Kaiser Wilhelm Gesellschaft, was renamed the Max Planck Gesellschaft in Planck's honour.

Hints for the solution of selected exercises

Chapter 1

1.1 (i) The energy of each photon is

$$\hbar\omega = (1.0546 \times 10^{-34}) \times (2\pi \times 2 \times 10^5) = 1.352 \times 10^{-28}\ \mathrm{J};$$

(ii)

$$\hbar\omega = (1.0546 \times 10^{-34}) \times (2\pi \times 4.95 \times 10^{14}) = 3.2800 \times 10^{-19}\ \mathrm{J};$$

(iii)

$$\hbar\omega = (1.0546 \times 10^{-34}) \times (2\pi \times 10^{28}) = 6.6242 \times 10^{-6}\ \mathrm{J}.$$

1.2 We have $200\,\mathrm{kW} = 2 \times 10^5\ \mathrm{J\,s^{-1}}$, which, using the answer to 1.1(i), is sufficient to create

$$\frac{2 \times 10^5}{1.3252 \times 10^{-28}} = 1.5092 \times 10^{33}$$

photons per second, if used with 100% efficiency. The aerial occupies 1 square metre out of $4\pi \times \left(10^6\right)^2$, so that we would expect it to be hit by

$$\frac{1.5092 \times 10^{33}}{4\pi \times 10^{12}} = 1.2010 \times 10^{20}$$

photons each second.

Were it to be at a distance of 3000 million kilometres, that is 3 million times further, then it should be struck by

$$\frac{1.2010 \times 10^{20}}{9 \times 10^{12}} = 1.334 \times 10^7$$

photons per second.

Chapter 2

2.1 The wave number, k, of the photon is

$$\frac{2\pi}{5.89 \times 10^{-7}} = 1.0668 \times 10^7\ \mathrm{m^{-1}},$$

so, writing v for the recoil velocity of the atom, the de Broglie law gives

$$(3.82 \times 10^{-26})v = (1.0546 \times 10^{-34}).(1.0668 \times 10^{7}) = 1.1250 \times 10^{-27},$$

from which we deduce that $v = 2.945 \times 10^{-2}$ metres per second, or a little over an inch per second.

2.2 The Schrödinger equation can be written in the form

$$-\frac{\hbar^2}{2m}\psi'' = (E - V_0)\psi,$$

which makes it obvious that it is the same as the square well, treated in Section 2.2, except that E has been replaced by $E - V_0$. Arguing as there, we get

$$E - V_0 = \frac{n^2\pi^2\hbar^2}{2ma^2},$$

from which the result folows.

2.3 When we substitute $\psi = X(x)Y(y)$ into the Schrödinger equation and divide by XY, we obtain

$$-\frac{\hbar^2}{2m}\left(\frac{X''}{X} + \frac{Y''}{Y}\right) = E,$$

in which the variables are separated. We may therefore deduce that $-(\hbar^2/2m)X''/X$ is a constant, which we may write as E_1, to get the equation

$$-\frac{\hbar^2}{2m}X'' = E_1 X,$$

which is the one-particle Schrödinger equation. The boundary conditions still force X to vanish when $x = 0$ or a, so that $E_1 = j^2\pi^2\hbar^2/2ma^2$, for some positive integer j. Combining this with the result of a similar argument for Y then gives

$$E = \frac{j^2\pi^2\hbar^2}{2ma^2} + \frac{k^2\pi^2\hbar^2}{2mb^2},$$

where k is also a positive integer. The energy $5\pi^2\hbar^2/2ma^2$ is obtained by taking $j^2 + k^2 = 5$, which means that $j = 2$ and $k = 1$ or $j = 1$ and $k = 2$. The wave functions are, in general,

$$\psi = XY = \frac{2}{\sqrt{ab}} \sin\frac{j\pi x}{a} \sin\frac{k\pi y}{b},$$

which reduces to

$$\psi = \frac{2}{a} \sin \frac{j\pi x}{a} \sin \frac{k\pi y}{a},$$

when $a = b$. The required probability is the integral of $|\psi|^2$ over the region where $x \leq y$, which turns out to be $\frac{1}{2}$.

2.4 Schrödinger's equation gives

$$-\frac{\hbar^2}{2m} \frac{1}{r} \frac{d^2(r\psi)}{dr^2} = E\psi,$$

so that $r\psi$ satisfies the one-dimensional Schrödinger equation. Moreover, we are told to assume that $\psi(a)$ vanishes, and if ψ is finite at the origin then $r\psi$ vanishes there, so the boundary conditions on $r\psi$ are the same as in one dimension. The only thing that changes is the normalization condition, which, on using the formula for the volume element in spherical polar coordinates and integrating out the angles, becomes

$$1 = \int |\psi|^2 \, dV = 4\pi \int |r\psi|^2 \, dr.$$

The probability is $\frac{1}{2}$.

2.5 The mean z coordinate is given by

$$\int_{r<a} z|\psi|^2 \, dV,$$

and this vanishes since the integrand is an odd function in the sphere. Arguing similarly for x and y, the mean position is at the origin. The variance of the height is therefore given by

$$\int_{r<a} z^2|\psi|^2 \, dV.$$

By symmetry this must be the same as the variances of x and y, and so all three are given by the average

$$\frac{1}{3} \int_{r<a} (x^2 + y^2 + z^2)|\psi|^2 \, dV = \frac{1}{3} \int_{r<a} r^2|\psi|^2 \, dV,$$

in which form it is easy to calculate the answer

$$\frac{a^2}{3} \left(\frac{1}{3} - \frac{1}{2n^2\pi^2} \right).$$

2.6 Notice that

$$\begin{aligned}\frac{1}{\sqrt{a}}\sin\left(\frac{\pi x}{a}\right)\left[1+2\cos\left(\frac{\pi x}{a}\right)\right] &= \frac{1}{\sqrt{a}}\left[\sin\left(\frac{\pi x}{a}\right)+\sin\left(\frac{2\pi x}{a}\right)\right] \\ &= \frac{1}{\sqrt{2}}\left[\psi_1+\psi_2\right],\end{aligned}$$

which evolves to

$$\frac{1}{\sqrt{2}}\left[e^{-iE_1t/\hbar}\psi_1+e^{-iE_2t/\hbar}\psi_2\right],$$

where $E_n = n^2\pi^2\hbar^2/2ma^2$.

2.7 After the barrier has been removed we must change the box length to $2a$; this gives the formula for the energies and shows that the normalized wave functions are

$$\psi_n(x) = \frac{1}{\sqrt{a}}\sin\left(\frac{n\pi x}{2a}\right),$$

for $x \in [0, 2a]$. One must then expand the wave function

$$\psi(x) = \begin{cases}\sqrt{\frac{2}{a}}\sin\left(\frac{\pi x}{a}\right) & \text{if } x \in [0,a] \\ 0 & \text{otherwise}\end{cases}$$

as $\sum c_n\psi_n$. For unchanged energy one needs $n = 2$, and so the probability is $|c_2|^2$ which turns out to be $\frac{1}{2}$.

2.8 The condition for a vanishing probability current is

$$0 = \overline{\psi}\psi' - \overline{\psi}'\psi,$$

from which we deduce that $\psi/\overline{\psi} = (\psi/|\psi|)^2$ is independent of x. Set $\lambda = |\psi|/\psi$.

Chapter 3

3.1 The normalized wave functions are

$$\psi_1(x) = \sqrt{\frac{2m\omega}{\hbar}}x\psi_0, \qquad \psi_2(x) = \frac{1}{\sqrt{2}}\left(2\frac{m\omega x^2}{\hbar}-1\right)\psi_0.$$

3.2 The potential energy can be written as

$$V(x,y) = \frac{1}{2}m\omega^2\,(x \quad y)\begin{pmatrix}10 & 6\\ 6 & 10\end{pmatrix}\begin{pmatrix}x\\ y\end{pmatrix}.$$

The eigenvalues of the square matrix are 4 and 16, so that the normal frequencies are $\sqrt{4}\omega = 2\omega$ and $\sqrt{16}\omega = 4\omega$ and the quantum mechanical energies are given by

$$E = \left(n_1 + \tfrac{1}{2}\right) 2\hbar\omega + \left(n_2 + \tfrac{1}{2}\right) 4\hbar\omega = [(n_1 + 2n_2) + 3]2\hbar\omega.$$

The degeneracies are determined by the number of ways of expressing a given positive integer as $n_1 + 2n_2$.

3.3 Separate variables by taking $\psi(x, y, z) = X(x)Y(y)Z(z)$. The oscillator then splits into three one-dimensional oscillators and the possible energies are found to be of the form

$$E = \left(n_1 + n_2 + n_3 + \tfrac{3}{2}\right) \hbar\omega.$$

The degeneracy of the energy level $(N + \frac{1}{2})\hbar\omega$ is $\frac{1}{2}(N+1)(N+2)$.

3.5 On substituting $y = x + \epsilon/m\omega^2$ into the Schrödinger equation

$$-\frac{\hbar^2}{2m}\psi'' + \frac{1}{2}m\omega^2 x^2 + \epsilon x = E\psi$$

we see that

$$-\frac{\hbar^2}{2m}\frac{d^2\psi}{dy^2} + \frac{1}{2}m\omega^2 y^2 = \left(E + \frac{\epsilon^2}{2m\omega^2}\right)\psi,$$

from which the result is easily deduced.

3.6 On substituting $\psi = R(r)\Theta(\theta)$ into Schrödinger's equation and multiplying by $r^2/R\Theta$, we see that Θ''/Θ must be constant. The condition that $\Theta(\theta + 2\pi) = \Theta(\theta)$ forces that constant to be of the form $-l^2$, with l an integer. The radial equation then takes the form

$$-\frac{\hbar^2}{2m}\left(\frac{1}{r}(rR')' - \frac{l^2}{r^2}\right) + \frac{1}{2}m\omega^2 r^2 R = ER.$$

One then looks for the asymptotic form ϕ and tries $R(r) = f(r)\phi(r)$, with f given by a series, exactly as in the one-dimensional case. The lowest energy solutions are given by

$$\psi(r, \theta) = Ne^{il\theta}e^{-m\omega r^2/2\hbar},$$

where $l = 0$ gives the ground state and $l = \pm 1$ gives the doubly degenerate first excited state.

3.7 Work in the momentum representation, where

$$(\mathcal{F}\psi_0)(p) = \sqrt{\frac{2a}{\pi\hbar}}\frac{1}{1 + p^2a^2/\hbar^2},$$

and time evolution is given by multiplication by $\exp(-ip^2/2m\hbar)$.

Chapter 4

4.1 The wave functions are

$$\psi_2(r) = \left(\frac{Z^3}{8\pi a^3}\right)^{\frac{1}{2}} \left(1 - \frac{1}{2}Zr\right) e^{-Zr/2a},$$

$$\psi_3(r) = \left(\frac{Z^3}{27\pi a^3}\right)^{\frac{1}{2}} \left(1 - \frac{2Zr}{3a} + \frac{2Z^2r^2}{27a^2}\right) e^{-Zr/3a}.$$

4.2 Substitute $\psi = R(r)\Theta(\theta)$ into Schrödinger's equation, and multiply by $r^2/R\Theta$ to separate the variables. Bearing in mind the fact that $\Theta(\theta) = \Theta(\theta + 2\pi)$, one deduces that $\Theta'' = -l^2\Theta$ for positive integral l. Treat the equation for R by the usual procedure of looking for a series solution times an asymptotic solution. The degeneracy is $2N + 1$ (recall that $\pm l$ give the same equation for R).

4.3,4 Treat the radial equation by the usual combination of asymptotic and series solution.

4.5 The energy eigenvalues are

$$E = \frac{\hbar^2}{2m}\left[4\kappa n + \kappa(2c_l + 1 - 2\kappa a^2)\right]$$

where

$$c_l = \tfrac{1}{2} + \sqrt{\left(l + \tfrac{1}{2}\right)^2 + \kappa^2 a^2}.$$

The factor of 4 rather than 2 results from the fact that the power series contains only even terms.

4.6 Set $\omega = 0$ and $\kappa = mw/\hbar$ in the previous question.

4.7 Substitute $\psi = UVW$ and multiply by uv/UVW to get an equation in which w separates from u and v. Since $w + 2\pi$ and w define the same point the resulting equation for w has solutions only if W is a multiple of $\exp(i\mu w)$ for some integer μ. When this is substituted, the equation for U can be written in the form

$$4(uU')' - \frac{\mu^2}{u}U + \left(\frac{2mE}{\hbar^2}\right)\mu U + AU = 0,$$

for some constant of separation A. A similar equation is satisfied by V but with A replaced by another constant B such that $A + B = 4Z/a$, with a the Bohr radius. The usual search for asymptotic and series solutions leads to the formula for E_n, and the degeneracy n^2.

Chapter 5

5.2 It is more efficient to assume that the potential takes the constant value V_2 inside the interval $[0, a]$ and V_1 outside it, so that part (a) is the special case of $V_1 = 0$ and $V_2 = V_0$, whilst in (b) it is V_2 which vanishes and $V_1 = V_0$.

5.4 The wave function ψ must vanish for $x \leq 0$, so the boundary conditions are that $\psi(0) = 0$ and that ψ and ψ' are continuous at $x = a$. Taking

$$\psi(x) = Ae^{-ikx} + Be^{ikx}$$

in the region $x > a$, the boundary conditions give

$$Be^{ika}\left(1 - \frac{ik}{k_0}\tan(k_0 a)\right) = Ae^{-ika}\left(1 + \frac{ik}{k_0}\tan(k_0 a)\right),$$

where $k_0^2 = k^2 + 2mV_0/\hbar^2$. Since k and k_0 are real this shows that $|B| = |A|$, and enables us to deduce that $B = A\exp(i\phi)$, where

$$\tan\left(\frac{1}{2}\phi + ka\right) = \frac{ik}{k_0}\tan(k_0 a).$$

5.5 Use the matching conditions at 0 and a and the condition that a bound state wave function must tend to 0 as $x \to \pm\infty$.

5.6 As in Exercise 2.4 the equation and boundary conditions for $r\psi(r)$ are similar to those in one dimension. For normalizable states $r\psi(r)$ must tend to 0 as $r \to \infty$. The limit of large V_0 can conveniently be studied by writing the condition in the form

$$\sin(ka) = \pm\sqrt{\frac{\hbar^2}{2mV_0}}k,$$

and considering where the graph of $\sin(ka)$ meets the line through the origin with slope $\pm\sqrt{\hbar^2/2mV_0}$.

5.7 The probability current is $\hbar\kappa(\overline{A}B - \overline{B}A)/im$.

5.8 Consider $\left(\overline{\psi} - \overline{a}\psi\right)/\overline{b}$.

Chapter 6

6.1 It is clear that $\mathsf{E}_\psi(A)$ is real if and only if $\langle\psi|A\psi\rangle$ is real, that is

$$\langle\psi|A\psi\rangle = \overline{\langle\psi|A\psi\rangle} = \langle A\psi|\psi\rangle.$$

We are told that this holds for all ψ, but what we really want is $\langle\phi|A\psi\rangle = \langle A\phi|\psi\rangle$ for all ψ and ϕ. To bridge the gap, first show that

$$\langle\phi|A\psi\rangle = \sum_{n=0}^{3} i^n\langle\psi + i^n\phi|A(\psi + i^n\phi)\rangle,$$

and then apply the known identity.

6.2 The expectation value of a self-adjoint operator is real, but in the state $\psi = \exp(-\kappa r^2/2)$ we find that

$$\mathsf{E}\left(-i\hbar\frac{\partial}{\partial r}\right) = i\kappa\hbar\int_0^\infty 4\pi r^3 e^{-\kappa r^2}\,dr,$$

which is imaginary. The inner product in $\mathbf{R}^3$ can be written as

$$\langle\phi|\psi\rangle = \int \overline{r\phi}\, r\psi \sin\theta\, dr d\theta d\phi,$$

and

$$\frac{i\hbar}{r}\frac{\partial}{\partial r}(r\phi) = i\hbar\left(\frac{\partial}{\partial r} + \frac{1}{r}\right)\phi,$$

from which the other part follows easily.

6.3 The eigenvectors of $\mathcal{P}$ with eigenvalue 1 are given by even wave functions and those with eigenvalue -1 are given by odd wave functions.

6.4 We have

$$(\mathcal{P}V\psi)(x) = (V\psi)(-x) = V(-x)\psi(-x) = V(-x)(\mathcal{P}\psi)(x),$$

which shows that even functions V commute with $\mathcal{P}$. By applying the chain rule or by Fourier transforming we can get $\mathcal{P}P = -P\mathcal{P}$, from which it easily follows that $\mathcal{P}$ commutes with the kinetic energy $P^2/2m$. If the system is in a non-degenerate eigenstate ψ, one first shows that this is also an eigenstate of $\mathcal{P}\psi$, and then uses

$$\langle\mathcal{P}\psi|P\mathcal{P}\psi\rangle = \langle\psi|\mathcal{P}P\mathcal{P}\psi\rangle = -\langle\psi|P\psi\rangle,$$

to deduce that the expectation value of P must vanish.

6.5 The normalized eigenvectors corresponding to the eigenvectors $3\hbar\omega$, 0, and $-3\hbar\omega$ can be taken to be

$$\phi_+ = \frac{1}{3}\begin{pmatrix}2\\2\\1\end{pmatrix}, \qquad \phi_0 = \frac{1}{3}\begin{pmatrix}2\\-1\\-2\end{pmatrix}, \qquad \phi_- = \frac{1}{3}\begin{pmatrix}1\\-2\\2\end{pmatrix},$$

respectively. The initial vector can be expanded as

$$\psi_0 = \langle\phi_+|\psi_0\rangle\phi_+ + \langle\phi_0|\psi_0\rangle\phi_0 + \langle\phi_-|\psi_0\rangle\phi_- = \tfrac{2}{3}\phi_+ + \tfrac{2}{3}\phi_0 + \tfrac{1}{3}\phi_-,$$

and therefore evolves to

$$\psi_t = \tfrac{2}{3}e^{-3i\omega t}\phi_+ + \tfrac{2}{3}\phi_0 + \tfrac{1}{3}e^{3i\omega t}\phi_-.$$

Writing ϵ_2 for the second standard basis vector, the probability, p_2, is given by

$$|\langle\epsilon_2|\psi_t\rangle|^2 = \left|\tfrac{2}{3}e^{-3i\omega t}\langle\epsilon_2|\phi_+\rangle + \tfrac{2}{3}\langle\epsilon_2|\phi_0\rangle + \tfrac{1}{3}e^{3i\omega t}\langle\epsilon_2|\phi_-\rangle\right|^2.$$

The result follows on calculating the inner products and simplifying the result.

6.6 This can also be done by expanding in eigenvectors, but it is easier to notice that, in terms of a standard basis ϵ_1 and ϵ_2, the initial vector $\psi_0 = \epsilon_1$ evolves to $\psi_t = \exp(-iHt/\hbar)\epsilon_1$ and that the required probability is

$$|\langle\epsilon_2|e^{-iHt/\hbar}\epsilon_1\rangle|^2.$$

It is easily checked that $H^2 = \hbar^2 e^2|\mathbf{B}|^2/4\mu^2$, from which one may calculate that

$$e^{-iHt/\hbar} = \cos\left(\frac{e|B|t}{2\mu}\right) - \frac{2i\mu H}{\hbar e|B|}\sin\left(\frac{e|B|t}{2\mu}\right),$$

and the lower left-hand matrix elements now give the result.

6.7 Use

$$i\hbar\frac{d}{dt}\langle\psi_t|A\psi_t\rangle = \langle -i\hbar\frac{d\psi}{dt}|\psi\rangle + \langle\psi|i\hbar\frac{d\psi}{dt}\rangle,$$

and substitute from Schrödinger's equation.

6.8 Suppose that orthonormal bound state eigenfunctions before the nuclear decay are $\psi_0, \psi_1, \ldots$, with ψ_0, the ground state, and afterwards are $\psi_0', \psi_1', \ldots$. The probability of finding the new ground state energy when one has the old ground state wave function is $|\langle\psi_0'|\psi_0\rangle|^2$. The normalized ground state wave functions are

$$\psi_0 = \sqrt{\frac{Z^3}{\pi a^3}}e^{-Zr/a}, \qquad \psi_0' = \sqrt{\frac{Z'^3}{\pi a^3}}e^{-Z'r/a},$$

so we calculate that

$$\langle\psi_0'|\psi_0\rangle = \frac{(Z'Z)^{\frac{3}{2}}}{\pi a^3}\int 4\pi r^2 e^{-2(Z'+Z)r/a}\,dr.$$

The answer follows on evaluating the integral.

Chapter 7

7.2 Use Proposition 7.1.2 repeatedly.

7.3 Use Lemma 7.2.1 with $A = P$ and $B = X$.

7.7 For part (iii) recall that $\mathsf{E}(T) + \mathsf{E}(V) = E$. For part (iv) note that $\mathsf{E}(T)$ must be positive.

7.10 Once one has shown that N is a projection, it follows that the expression $\psi = N\psi + (1 - N)\psi$ decomposes an arbitrary vector into eigenvectors of N with $N(N\psi) = N\psi$ and $N(1 - N)\psi = 0$. Using the identity

$$\|a\psi\|^2 = \langle\psi|N\psi\rangle$$

it is easy to see that the kernels of N and of a coincide. Let $\{\psi_1, \psi_2, \ldots\}$ be an orthonormal basis for the kernel of N. It can be shown from the commutation relations between N and a^* and a that $\{a^*\psi_1, a^*\psi_2, \ldots\}$ is an orthonormal basis for the image of N. The space is spanned by the image and kernel of N, so we may form a basis $\phi_{2j+1} = \psi_j$ and $\phi_{2j+2} = a^*\psi_j$, with respect to which we obtain a matrix representation of the operators.

7.11 Introduce the operators $P = P_1 + \frac{1}{2}eBX_2$ and $X = \frac{1}{2}X_1 - P_2/eB$, and show that they satisfy the canonical commutation relations. Since

$$H = \frac{P^2}{2m} + \frac{e^2B^2}{2m}X^2$$

the Hamiltonian has a harmonic-oscillator-type spectrum. (This example, due to Lev Landau, is of considerable importance in solid state physics.)

7.14 With generating functions we may calculate

$$\sum_{M,N=0}^{\infty} \frac{s^N}{N!}\frac{t^M}{M!}\langle\psi_N|X^k\psi_M\rangle$$
$$= \left(\frac{m\omega}{\pi\hbar}\right)^{\frac{1}{2}} e^{2m\omega st/\hbar} \int_{\mathbf{R}} [u - (s+t)]^k e^{-m\omega u^2/\hbar}\, du.$$

When $k = 1$ this gives

$$\sum_{M,N=0}^{\infty} \frac{s^N}{N!}\frac{t^M}{M!}\langle\psi_N|X\psi_M\rangle = -(s+t)e^{2m\omega st/\hbar}, \tag{19.24}$$

from which it follows, by comparing coefficients, that $\langle\psi_N|X\psi_M\rangle$ vanishes unless $|N - M| = 1$. Alternatively, we have the identity,

$$\langle\psi_N|P\psi_M\rangle \pm im\omega\langle\psi_N|X\psi_M\rangle = \langle\psi_N|a_\pm\psi_M\rangle. \tag{19.25}$$

This expression vanishes unless $|N - M| = 1$, because $a_{\pm}\psi_M$ is a multiple of $\psi_{M\pm 1}$, and therefore orthogonal to ψ_N unless $N - M = \pm 1$. Adding and subtracting, we see that $\langle \psi_N | X \psi_M \rangle$ and $\langle \psi_N | P \psi_M \rangle$ both vanish unless $|N - M| = 1$.

7.15 Write

$$\psi_0(x) = e^{-m\omega a^2/4\hbar} G\left(-\tfrac{1}{2}a, x\right)$$

and use the known evolution of G, to see that

$$\psi_t(x) = e^{-\frac{1}{2}\omega t} e^{-m\omega a^2/4\hbar} G\left(-\tfrac{1}{2}e^{-i\omega t}a, x\right).$$

7.16 This result, known as Floquet's theorem, and its three-dimensional analogue, Bloch's theorem, are basic to solid state physics. See also Section 12.6.

7.20 Multiply equation (7.52) by $\exp(-2m\omega a^2/\hbar)$ and integrate against a.

Chapter 8

8.4 Note that the mass in this case is M not m. There will not be bound states unless the potential is attractive, which means that the L_3 eigenvalue, m, must be negative. As usual we also need $l \geq |m|$.

8.5 It is simplest to find the eigenstates, ψ, which have L_3 eigenvalue $\pm\hbar$ as the solutions of $L_{\pm}\psi = 0$. The remaining state, with $L_3\psi = 0$, can be found as a multiple of $L_-\psi$ where $L_+\psi = 0$.

8.7 Show that

$$[L^2, X_j] = i\hbar\epsilon_{jkl}\left(L_k X_l + X_k L_l\right)$$

and recall that L^2 commutes with the components of $\mathbf{L}$.

8.8 When ψ is an eigenstate Exercise 8.3 gives

$$\langle \psi | L_1^2 \psi \rangle = \langle \psi | L_2^2 \psi \rangle = \tfrac{1}{2}\langle \psi | \left(L_1^2 + L_2^2\right)\psi \rangle,$$

which enables one to reduce $\langle \psi | H \psi \rangle$ to an expression which involves only expectation values of L^2, L_3^2, and L_3.

8.10 This can also be done by the method of Exercise 7.11.

8.11 Use the identity

$$\begin{aligned} e^{i\mathbf{a}.\boldsymbol{\sigma}} &= \sum \frac{(i\mathbf{a}.\boldsymbol{\sigma})^{2k}}{(2k)!} + \sum \frac{(i\mathbf{a}.\boldsymbol{\sigma})^{2k+1}}{(2k+1)!} \\ &= \sum \frac{(i|\mathbf{a}|)^{2k}}{(2k)!} + \frac{\mathbf{a}.\boldsymbol{\sigma}}{|\mathbf{a}|} \sum \frac{(i|\mathbf{a}|)^{2k+1}}{(2k+1)!}. \end{aligned}$$

8.12 You can define ϕ_+ as a normalized solution of $A^*\phi_+ = 0$, and $\phi_- = \hbar^{-1}A\phi_+$. Then you need only check that these are S_3-eigenvectors and that $A^*\phi_- = \hbar\phi_+$.

8.13 The case $\alpha = 0$ corresponds to the canonical commutation relations and $\beta = 0$ to the angular momentum operators.

8.14 The commutation relations given in (ii) provide a simple example of a generalization of a group, called a *quantum group*. These have recently been the object of intense study by mathematicians and physicists.

Chapter 9

9.2 The point of the first part is to show that if $g_1\mathbf{k} = g_2\mathbf{k}$ then

$$\langle U(g_1)\Omega|\psi\rangle = \langle U(g_2)\Omega|\psi\rangle,$$

so that T is well defined.

Chapter 10

10.1 First note that

$$\begin{aligned} PU_{t/n}PU_{t/n}\cdots PU_{t/n}P &= \left(PU_{t/n}P\right)^n P \\ &= \left[1 + \frac{t}{n}PHP + O\left(\frac{t^2}{n^2}\right)\right]^n P. \end{aligned}$$

10.5 For the last part note that

$$(n+1)\sin^2\frac{\pi}{2(n+1)} - 1 = \frac{\pi}{2}\frac{\sin^2\phi}{\phi} - 1,$$

where $\phi = \pi/2(n+1)$, and consider the behaviour for small ϕ.

Chapter 11

11.3 By equation (11.19) we see that

$$(P + im\omega X)_{t+i\pi/2\omega} = e^{i\omega(t+\pi/2\omega)}(P + im\omega X)_0 = i(P + im\omega X)_t,$$

from which it follows that $m\omega X_{t+\pi/2\omega} = P_t$.

11.4 Show that

$$[X(t) - \mathsf{E}(X(t))]/t = [P(0) - \mathsf{E}(P(0))]/m + [X(0) - \mathsf{E}(X(0))]/t.$$

11.5 It is easy to show from the commutation relations that

$$e^{itL_3/\hbar}L_+e^{-itL_3/\hbar} = e^{it}L_+,$$

and using this in the KMS condition we get

$$\mathrm{tr}\left[e^{-\beta L_3}L_+L_-\right] = e^{-\beta\hbar}\mathrm{tr}\left[e^{-\beta L_3}L_-L_+\right].$$

The differential equation follows on writing L_+L_- and L_-L_+ in terms of L^2 and L_3^2.

11.6 In the last part note that

$$\frac{\mathsf{E}(P)^2}{2m} + F\mathsf{E}(X) = \mathsf{E}\left(\frac{P^2}{2m} + FX\right) - \left(\frac{\mathsf{E}(P^2) - \mathsf{E}(P)^2}{2m}\right).$$

11.7 Show that the equation of motion for S implies that T is constant. In the last part show that

$$\mathsf{E}(J_1^2) = \mathsf{E}(J_2^2) = \frac{1}{2}\mathsf{E}(J_1^2 + J_2^2).$$

Chapter 12

12.2 Note that the first-order correction to the energy level $(n+\frac{1}{2})\hbar\omega$ is

$$\lambda\mathsf{E}(X^4) = \lambda\frac{\|X^2\psi\|^2}{\|\psi\|^2},$$

where the wave function is related to the ground state wave function, ψ_0, by $\psi = a_+^n\psi_0$.

12.4 One can reduce the work needed by noticing that all the matrix elements which appear in the calculation of the first-order energy corrections are of the form

$$\epsilon\int xy\frac{4}{a^2}\sin\left(\frac{j\pi x}{a}\right)\sin\left(\frac{k\pi y}{a}\right)\sin\left(\frac{r\pi x}{a}\right)\sin\left(\frac{s\pi y}{a}\right)\,dxdy,$$

for various values of j, k, r, and s, and these can be written as products of integrals of the form

$$\int x\sin\left(\frac{j\pi x}{a}\right)\sin\left(\frac{r\pi x}{a}\right)\,dx.$$

12.6 For the exact answer note that

$$BL_3 + CL_1 = \sqrt{B^2 + C^2}\,\mathbf{n}.\mathbf{L}$$

where $\mathbf{n}$ is the unit vector

$$\mathbf{n} = \left(\frac{C}{\sqrt{B^2 + C^2}}, 0, \frac{B}{\sqrt{B^2 + C^2}} \right),$$

and note that, by rotating coordinates, we may take $\mathbf{n}$ to $(0, 0, 1)$.

12.8 The first excited state is doubly degenerate so one must first calculate the elements of a 2×2 matrix, and then find its eigenvalues. The integrals giving the matrix entries can be written in terms of integrals of the form

$$\int \overline{\psi_j(x)} x^2 \psi_k(x)\, dx \langle \psi_r | \psi_s \rangle,$$

where ψ_j, ψ_k, ψ_r, and ψ_s are one-dimensional oscillator wave functions, with $j + r = 1 = k + s$. Unless $r = s$, and so also $j = k$, this expression vanishes, so the matrix is already diagonal. One readily sees that the diagonal entries are equal and opposite.

12.9 See the comments about Exercises 12.4 and 12.8.

12.11 With H_0 of the given form we have

$$H' = \lambda X^4 + \tfrac{1}{2} m(\alpha^2 - \omega^2) X^2 - \kappa,$$

and we can choose ω and κ so that $H'\psi_0$ is a multiple of ψ_3.

Chapter 13

13.2 Recall that

$$\int_{\mathbf{R}^3} e^{-i\mathbf{p}.\mathbf{x}} e^{-\frac{1}{2}\alpha x^2} = (2\pi/\alpha)^{\frac{3}{2}} e^{-p^2/2\alpha}.$$

13.3 Differentiate the hint for 13.2 with respect to α.

13.5 The exact ground state energy is

$$\frac{1}{2}\sqrt{\omega^2 + \lambda} = \frac{1}{2}\hbar\omega \left(1 + \frac{1}{2}\frac{\lambda}{\omega^2} - \frac{1}{8}\frac{\lambda^2}{\omega^4} + \frac{\lambda^3}{\omega^6} + \ldots \right).$$

The second-order Rayleigh–Schrödinger correction gives the terms in λ and λ^2, whilst the second-order Wigner–Brillouin approximation also gives the third-order term. (The third-order correction could also be obtained in Rayleigh–Schrödinger theory using Corollary 12.4.2.)

Chapter 14

14.2 The trial functions are not normalized, so one must divide by $\|\psi\|^2$ when calculating the expectation values. The integrals which arise are all of the form

$$\int x^{2k} e^{-cx^2}\, dx = \left(-\frac{d}{dc}\right)^k \int e^{-cx^2}\, dx.$$

Assume for the last part that the same formulae are valid when n is not integral.

14.3 The stationary value occurs when $A = -F/(E_0 + \hbar^2/2ma^2)$ and gives the bound

$$E_0 - \frac{a^2F^2}{(E_0 + \hbar^2/2ma^2)}.$$

14.8 The wave functions ψ_α and $X\psi_\alpha$ are (unnormalized) eigenvectors of the oscillator Hamiltonian $P^2/2m + \frac{1}{2}m\alpha^2\omega^2X^2$, with energies $\frac{1}{2}\alpha\hbar\omega$ and $\frac{3}{2}\alpha\hbar\omega$, respectively. Using the virial theorem, one easily calculates the expectations of $P^2/2m$ in the state ψ_α, and of $\frac{1}{2}m\omega^2X^2$, in the states ψ_α and $X\psi_\alpha$. Since

$$\langle\psi_\alpha|X^4\psi_\alpha\rangle = \mathsf{E}_{X\psi_\alpha}(X^2)\|X\psi_\alpha\|^2$$

and $\|X\psi_\alpha\|^2 = \langle\psi_\alpha|X^2\psi_\alpha\rangle$ is also related to the expectation of an oscillator potential energy, the result is now easily obtained with a minimum of calculation.

Chapter 15

15.1 Writing the energy as $E = \frac{1}{2}m\omega^2b^2$ one has

$$W = m\omega \int \sqrt{(b^2 - x^2)}\, dx = \tfrac{1}{2}m\omega\left(b^2 \sin^{-1}(x/b) + x\sqrt{b^2 - x^2}\right).$$

Since the classical turning points are $\pm b$, the Bohr–Sommerfeld rule gives $W(b) - W(-b) = (n + \frac{1}{2})\pi\hbar$, from which it follows that $E = (n + \frac{1}{2})\hbar\omega$, so that the energy is exact in this case.

15.2 In the same notation as the previous hint a is proportional to $(b^2 - x^2)^{-\frac{1}{4}}$, from which a''/a is easily calculated.

Chapter 16

16.1 In terms of the rotated coordinates $y = (x_1 + x_2)/\sqrt{2}$ and $z = (x_1 - x_2)/\sqrt{2}$, the potential can be expressed as $\frac{1}{2}Ky^2 + \frac{1}{2}(K-2)z^2$, whilst the Laplace operator is invariant under rotation. Note that the even wave functions represent bosons, and that an harmonic oscillator eigenstate ψ_n with energy $(n+\frac{1}{2})\hbar\omega$ is even if and only if n is even. (This may be proved from the generating function or from the fact that the wave function is a multiple of $a_+^n\psi_0$, and a_+ is odd.)

16.2 The dimension of the bosonic space is $\binom{2l+n}{n}$. The possible energies for distinguishable particles are

$$\frac{\gamma B\hbar^2}{4}\left[j(j+1)-2l(l+1)\right]$$

for integral j between 0 and $2l$, whilst for bosons only even values of j occur. (In each case the degeneracy of the level is $(2j+1)$.)

16.3 For the energies see Exercise 8.12. For distinguishable particles the possible energies are the triply degenerate level, $E_j+E_k-\kappa$, and the non-degenerate level, $E_j+E_k+3\kappa$, for each choice of j and k. For fermions j and k must be distinct in the case of the levels $E_j+E_k-\kappa$. Compare the lowest levels of each type when $4\kappa > (E_1-E_0) > 0$.

Chapter 17

17.1 After separation of variables the radial equation becomes

$$(rR)'' = \left(\frac{m^2c^2}{\hbar^2}-\frac{E^2}{c^2\hbar^2}\right)rR,$$

and for normalizability the solutions should be negative exponentials.

17.3 Show that $\mathbf{P}$ commutes with the Hamiltonian as well as $\mathbf{J}$. Notice also that $(\mathbf{J}.\mathbf{P})^2 = |\mathbf{P}|^2$, so that the possible eigenvalues of $\mathbf{J}.\mathbf{P}$ are related to those of $\mathbf{P}$.

17.4 Show that

$$\alpha_r(\alpha.\mathbf{p}) = p_r + \frac{i}{r}K.$$

17.5 For the last part show that

$$\begin{pmatrix} 0 & 1 \\ -1 & 0 \end{pmatrix}\gamma(-p_0,\mathbf{p}) = \gamma(p_0,\mathbf{p})\begin{pmatrix} 0 & 1 \\ -1 & 0 \end{pmatrix}.$$

Chapter 18

18.3 For a first-order equation, such as the Dirac equation, the appropriate matching conditions are just that the wave function itself has all its components continuous at $\pm a$. It is easy to check that the given vector-valued wave functions do satisfy the Dirac equation in the regions $|x_3| < a$ and $x_3 > a$, under the conditions that

$$(P_0 + V)^2 - P^2 = m^2c^2, \qquad \text{and} \qquad {P_0}^2 + K^2 = m^2c^2,$$

respectively. The continuity of each of the two-component pieces of ψ at a gives the simultaneous equations relating v, v', and w. For bound states we try

$$e^{-iP_0x_0/\hbar}e^{Kx_3/\hbar}\begin{pmatrix} u \\ \frac{iK}{P_0+mc}\sigma_3 u \end{pmatrix}$$

when $x_3 < -a$. (For general states we should add a second term with K replaced by $-K$ and u by u'.) There are then two more matching conditions at $-a$. Then u and w may be eliminated to give two equations linking v and v'. The final equation is the condition for these equations to admit non-trivial solutions.

Index